Computer Simulation of Polymeric Materials

Japan Association for Chemical Innovation
Editor

Computer Simulation of Polymeric Materials

Applications of the OCTA System

EXTRAS ONLINE

Springer

Editor
Japan Association for Chemical Innovation
Tokyo, Japan

Additional material to this book can be downloaded from http://extras.springer.com.

Original Japanese language edition published by The Chemical Daily Co., Ltd.
Kobunshi Zairyo Shimyureshon

ISBN 978-981-10-0814-6 ISBN 978-981-10-0815-3 (eBook)
DOI 10.1007/978-981-10-0815-3

Library of Congress Control Number: 2016945720

Printed on acid-free paper

This Springer imprint is published by Springer Nature
The registered company is Springer Science+Business Media Singapore Pte Ltd.

Preface

OCTA is software for simulating the behavior of polymeric materials in liquid, rubbery, and solid states. It was developed by a project of the Japanese Ministry of Economy, Trade and Industry. The aim of the project was to develop software that would be useful for the research and development of polymeric materials.

Polymeric materials are complex. Their development requires understanding and control of polymers at various-length scales, ranging from the atomistic scale (e.g., the structure and sequential regularity of functional groups) to the macromolecular scale (e.g., the degree of polymerization, structures of branching, and cross-links), and the mesoscopic scale (e.g., chain orientation, crystalline, and other morphological structures). OCTA was developed to be a useful tool for handling such diverse problems and thus has two distinctive features.

One is that OCTA has specific simulation engines that can deal with problems characteristic of polymers, such as coarse-grained molecular dynamics, self-consistent field dynamics, and reptation dynamics. The other is that OCTA has a common interface, called a platform, for all simulation engines. The role of the interface is to provide simulation engines with a convenient tool for data creation and analysis, and to transform data from one simulation engine to another, and thus to facilitate the collaboration and extension of simulation engines.

OCTA stands for Open Computational Tool for Advanced material technology. The name reflects our desire for advancing computational material design. Computational material design is a highly challenging technology, and we believe that it will be realized only with the coordination of many software programs developed all over the world in an open and free environment.

The first version of OCTA was released in 2002. Since then, OCTA has been maintained and upgraded by a group of volunteers. At the same time, tutorial courses and workshops have been organized for beginners and advanced users, and through such activities, experiences of using OCTA in real places of research for the development of polymeric materials have increased. This book is a summary of what we have learned. Readers will find that OCTA has been used for many materials (e.g., plastics, rubbers, polymer blends, composites, films, and dispersants), many physical properties (e.g., mechanical, thermal, and optical properties), and many physical processes (e.g., structural formation, diffusion, and permeation). The book will be useful to those who want an overview of the OCTA system and an understanding of OCTA and how it is used.

I thank the members of the editorial committee for their devotion and efforts toward publishing this book, and I sincerely hope for the further advance of the OCTA community.

Beihang University, Beijing, China
Nagoya University, Nagoya, Japan
The University of Tokyo, Tokyo, Japan
2015/10/09

Masao Doi

Contents

Part I Introduction of Computer Simulation of Polymeric Materials

1 **Expected Target of Polymer Simulation** 3
Takeshi Aoyagi

2 **Coarse-Grained Simulation** 5
Takeshi Aoyagi

Part II OCTA: Mesoscale Polymer Simulation System

3 **Overview of OCTA** 15
Jun-ichi Takimoto and Takeshi Aoyagi

4 **COGNAC: Coarse-Grained Molecular Dynamics Simulator** 29
Takeshi Aoyagi

5 **SUSHI: Density Functional Theory Simulator** 67
Takashi Honda

6 **PASTA and NAPLES: Rheology Simulator** 101
Yuichi Masubuchi

7 **MUFFIN: Multiphase Simulator** 129
Taku Ozawa

8 **KAPSEL: Colloidal Dispersion Simulator** 149
Ryoichi Yamamoto and John Jairo Molina

Part III Examples of the Application of OCTA

9 **Melt Viscoelasticity** 171
Naoki Kobayashi

10 **Crystallization of Polymers** 179
Takashi Yamamoto

11 Polymer Blends: Bulk Property 189
Naoki Kobayashi

12 Polymer Blends: Interfacial Strength 201
Taku Ozawa

13 Composites: Morphology 211
Kenta Chaki and Taku Ozawa

14 Composites: Interfacial Strength 221
Taku Ozawa and Hiroya Nitta

15 Cross-Linked Rubber 229
Hiroshi Shima

16 Thermoplastic Elastomers 249
Takeshi Aoyagi

17 Filler-Filled Rubbers 269
Hiroshi Morita

18 Structures of the Surface and Interface 283
Shigeru Yao

19 Glass Transition at the Surface and Interface 291
Hiroshi Morita

20 Evaporation from Polymer Solution 297
Hiroshi Morita

21 Crystallization in Thin Films of N-Alkanes 305
Takashi Yamamoto

22 Improvement of Adhesive Properties Through the Segregation of Oligomers and an Investigation of the Mechanism Using SUSHI Simulation 315
Hiroshi Sasaki

23 Adsorption of Polyelectrolytes 327
Hiroki Kubo

24 Adsorbed Structures and Surface Forces 337
Hiroki Kubo

25 Analysis of Relaxation Mechanism of Thread-Like Micelle Solution 347
Satoru Yamamoto, Taku Ozawa, and Kosuke Ohata

26 Vesicle Formation 359
Satoru Yamamoto and Taku Ozawa

27 Electrolyte Membranes 369
Satoru Yamamoto and Taku Ozawa

28 Orientation Birefringence 379
Makoto Wakabayashi

29 Lithography 389
Hiroshi Morita

Index 397

Part I
Introduction of Computer Simulation of Polymeric Materials

Chapter 1
Expected Target of Polymer Simulation

Takeshi Aoyagi

In the field of materials science and engineering, computer simulation is expected to contribute to the development of new materials that have required properties. If physical properties can be predicted by computer simulation from information about the chemical structure and composition of materials as well as processing conditions, it may be possible to develop new materials with screening and optimization of various conditions by simulation.

However, predicting the physical properties of polymeric materials (e.g., fiber, plastics, and rubber) is difficult because the required physical properties of polymeric materials vary greatly. For example, mechanical and thermal properties, like strength, elastic modulus, and heat resistance, as well as long-term properties (i.e., durability) are required for construction materials such as automobile parts. Meanwhile, electronic and optical properties, e.g., dielectric constant, refractive index, and birefringence, are required for electronic/optical devices including conductive polymers, optical lenses, and films. Moreover, properties such as fractionation performance and biocompatibility are required for separation membranes for water purification or medical use, while moisture-retaining and moisture-absorbing properties as well as suitable texture, which is difficult to describe using physical properties, are required for fibers and textiles for clothing.

The varied physical properties of polymeric materials arise from their different structures. The basic structure of a polymer is determined by the chemical structure of its monomers. However, the molecular weight and molecular weight distribution of a polymer generally depend on polymerization conditions. Furthermore, a

T. Aoyagi (✉)
Research Center for Computational Design of Advanced Functional Materials, National Institute of Advanced Industrial Science and Technology, Ibaraki, Japan
e-mail: aoyagi.t@aist.go.jp

Japan Association for Chemical Innovation, *Computer Simulation of Polymeric Materials*, DOI 10.1007/978-981-10-0815-3_1

copolymer consisting of multiple types of monomers may have the structure of a random or block copolymer depending on the sequence of monomers. Moreover, polymers are not limited to a linear chain; some polymers have branched chains.

In addition to their basic molecular structure, condensed polymer systems can exhibit higher-order structure (e.g., crystal or phase-separated structure), which can also affect their physical properties. For example, even a so-called crystal structure can have numerous levels, e.g., atomic coordinates in a unit lattice, a lamellar structure formed by folded polymer chains, and a spherocrystal structure formed by stacking lamellae. In particular, there are few cases where a higher-order structure observed in real materials is the most stable equilibrium structure; the structure of polymeric materials is usually frozen in a nonequilibrium state during production.

Thus, in computer simulation of polymeric materials, various polymer structures that are closely related to their physical properties must be handled as realistically as possible. In short, a wide range of length and time scales must be considered, ranging from chemical structures on the order of angstroms (10^{-10} m) to higher-order structure scales of up to nearly a millimeter (10^{-3} m) and, in terms of time scales of dynamics, from the scale of molecular vibration on the order of femtoseconds (10^{-15} s) to time scales corresponding to actual measurement of viscoelasticity. Therefore, there are a huge number of phenomena that never become clear simply by performing quantum chemical and molecular dynamics calculations. In other words, it is important to reproduce polymer structures that include higher-order structures when simulating polymeric materials. To achieve this, various ideas are necessary to handle structures beyond the molecular level. In fact, many simulation software products based on various theories have been used globally, and various analyzing procedures have been developed.

Application examples are described in Part 3 of this book, from which it becomes evident that the structures and physical properties that are the objectives of studies, and simulation methods, are both highly diverse. Therefore, when performing computer simulation of polymeric materials, suitable calculation models and methods must be selected depending on the length and time scales that control the physical properties being considered. In other words, considerable knowledge is needed to prepare inputs for computer simulation. Computer simulation of polymeric materials is difficult for those with little experience because of the complexity of initial modeling of focused problems. Even today, computer simulation of polymeric materials is not a simple research tool for experimental polymer chemists.

Nevertheless, researchers have been working on computer simulation of polymeric materials since around 1980, and the need for such simulations is certain to increase in the future. The simulation tool OCTA, which is the main subject of this book, has been in use for more than a decade. To make computer simulation more popular as a tool for research and development of polymeric materials in the future, it is important to provide upcoming users with information about the systems and applications that have been studied by researchers in this field. This book is intended to provide such information.

Chapter 2
Coarse-Grained Simulation

Takeshi Aoyagi

2.1 Coarse-Graining of Polymers

As described in Chap. 1, in simulation of polymeric materials, various length and time scales need to be handled because of the hierarchical structure of polymers and diversity of properties of polymeric materials. This would be simple if full atomistic simulation could be used to handle huge length and time scales. However, there are limits to the number of atoms and time that can be simulated even if using a state-of-the-art supercomputer.

Molecular dynamics (MD) simulation is often used in existing simulation software products, which employs a calculation model known as the full atomistic model. In this model, atoms are usually handled individually to simulate their dynamics. Therefore, MD simulation enables analysis of realistic polymer dynamics in which chemical structure is considered. However, handling atoms individually means that the interactions between all atoms are calculated, so the calculation time increases considerably as the number of atoms increases. Of course, the performance of computer hardware and software improves on a daily basis, so the length and time scales that can be studied are expanding steadily. However, the practical upper limits of the length and time scales that can be handled at present in a full atomistic MD simulation are on the order of 10 nm in terms of the length scale and on the order of 100 ns in terms of the time scale for laboratory-scale computer hardware.

The scales of chemical bonds (order of 0.1 nm) and monomer units (order of 1 nm) fall within the range that can be handled by a full atomistic MD simulation. However, it is difficult for full atomistic MD simulation to handle the size of a radius

T. Aoyagi (✉)
Research Center for Computational Design of Advanced Functional Materials, National Institute of Advanced Industrial Science and Technology, Ibaraki, Japan
e-mail: aoyagi.t@aist.go.jp

Japan Association for Chemical Innovation, *Computer Simulation of Polymeric Materials*, DOI 10.1007/978-981-10-0815-3_2

of gyration of a single polymer molecule that has a molecular weight of tens of thousands to hundreds of thousands (10–100 nm), the size of a microphase-separated structure of block copolymer (10 nm) or the size of a long period of crystalline lamellae (10–100 nm). Furthermore, the scales of macrophase-separated structures of polymer blends and dispersion structures of polymer–inorganic composites are larger still and cannot be handled in a full atomistic simulation. Also, the time scale of viscoelastic properties, which are measured by a rheometer and tensile testing, is on the order of about 10^{-2} s at the fastest and therefore is not on a scale to which a full atomistic simulation can be applied.

Thus, when simulating polymeric materials, the coarse-graining method is often used to handle the required length and time scales. In general, a system has many hierarchies, from the smallest to the largest, which can be described in terms of accuracy or coarseness. Coarse-graining refers to transfer from one stage to another of coarser observation (description) in those hierarchies. Moreover, it is also important to conduct coarse-graining with respect to length and time concurrently. In a molecular model, "coarse-graining" refers to the transition from a model in which atoms that constitute a molecule are handled individually to a coarser description, that is, handling some atoms or monomer units as a single unit. Furthermore, when conducting such spatial coarse-graining, any motion with an extremely short period, such as the stretching of chemical bonds, is ignored, so the time scale is also transferred to a coarse description in parallel with the length scale.

2.2 Examples of Coarse-Grained Molecular Models

The following sections detail some specific examples of coarse-grained molecular models. Many of the coarse-grained molecular models given here, such as the rigid-body model and ideal chain model, were originally used for theoretical analysis. However, such models will be explained within the context of coarse-grained molecular models for computer simulation.

2.2.1 United Atom Model

In the united atom model, a couple of atoms are handled as a single unit, while the characteristics of the chemical structure of a polymer chain are maintained. The united atom model usually omits hydrogen atoms of methylene (-CH_2-) and methine (-CH-) units and considers them as a single unit, as shown in Fig. 2.1. However, the united atom model usually does not omit all hydrogen atoms. The hydrogen atoms involved in hydrogen bonds such as those in hydroxyl and amide groups are kept to maintain the characteristics of the chemical structure.

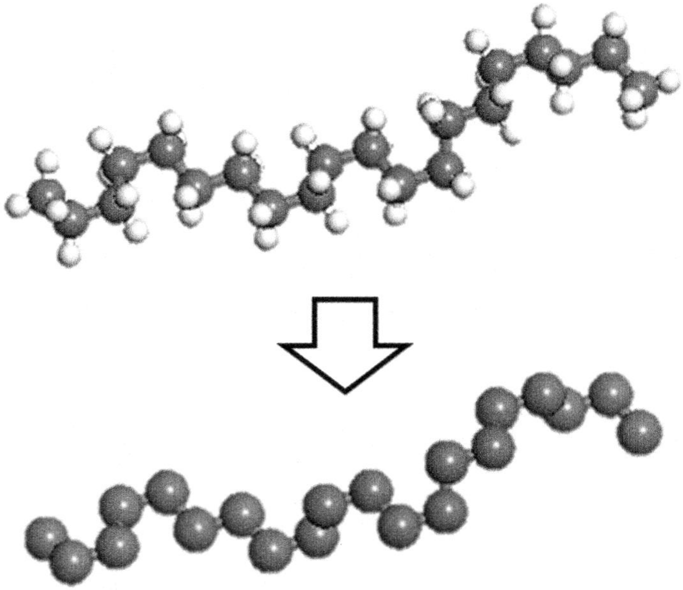

Fig. 2.1 Using the united atom model to represent a methylene chain

The united atom model is often used in MD simulation because it decreases the number of atoms being handled by 1/2–1/3 and lengthens time steps, thereby improving the efficiency of simulation. However, it is difficult to improve the computational efficiency to the order of magnitude compared with that of a full atomistic model.

2.2.2 *Rigid-Body Model*

The rigid-body model is a model in which a rigid molecule or partial structure of a molecule is represented by a rigid body such as a cylinder or ellipsoid. Using the rigid model, it is possible to conduct coarse-graining while maintaining a molecular shape by representing a rigid partial structure, such as a phenyl ring or diphenyl group, as an ellipsoid, as shown in Fig. 2.2.

Some improvements must be made to use the rigid-body model in MD simulation because estimating the force between rigid bodies is complicated. The rigid-body model is often used in Monte Carlo simulation, in which only interaction energy is considered. The Gay–Berne model [1] is an example of a rigid-body potential that can be used for MD simulation.

The rigid-body model is also used as a pure physical model to study topics like the phase transitions of liquid crystals, while ellipsoidal potential is used to represent chemical structures such as phenyl rings.

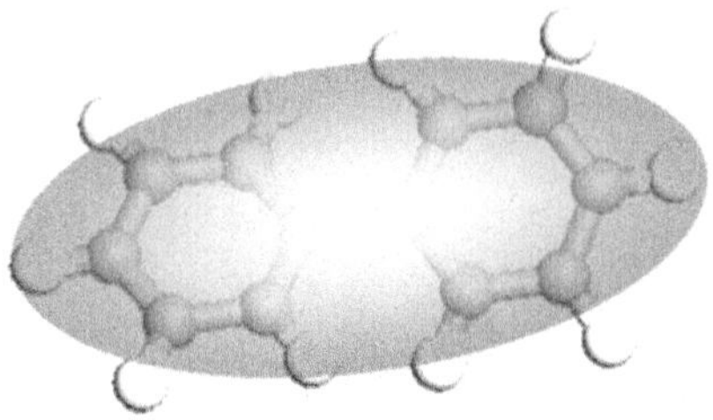

Fig. 2.2 Rigid-body model of an ellipsoid used to represent a biphenyl group

2.2.3 *Bead–Spring Model*

The bead–spring model is a simple molecular model in which beads with a specific volume are connected to a spring to make a chain-like structure. The bead–spring model is often used as a polymer model that does not consider chemical structure to study the universal behavior of polymer chains. The bead–spring model is a highly coarse-grained model, and several monomer units generally correspond to a single bead, as shown in Fig. 2.3.

In the bead–spring model, a molecular structure is represented by a small number of parameters, such as the size and mass of beads, equilibrium length of springs, spring constant, and interaction between beads. Therefore, it is generally difficult to use the bead–spring model to represent the characteristics of a chemical structure. However, it is possible to reflect a chemical structure by focusing on a property such as molecular size, entanglement, and miscibility and making that property correspond to the chemical structure. Therefore, the bead–spring model is useful for quantitative analysis when a coarse-grained molecular model is constructed with specific structures and properties.

Coarse-grained MD simulation mainly employs the bead–spring model. In some cases, the bead–spring model uses the potential functions of angle bending and dihedral angle to represent characteristics of polymer chains such as rigidity.

The Lennard-Jones potential is generally used to describe the interaction between beads. In the Lennard-Jones interaction, when two beads become too close, a large repulsive force acts on the beads to prevent their overlap. However, an extremely soft interaction is used in dissipative particle dynamics, so beads easily overlap and the bonds cross each other to reach an equilibrium structure faster.

The bead–rod model, in which the lengths of bonds are fixed and neither expand nor contract, is sometimes used as a molecular model similar to the bead–spring model. The bead–rod model is frequently used in Monte Carlo simulation.

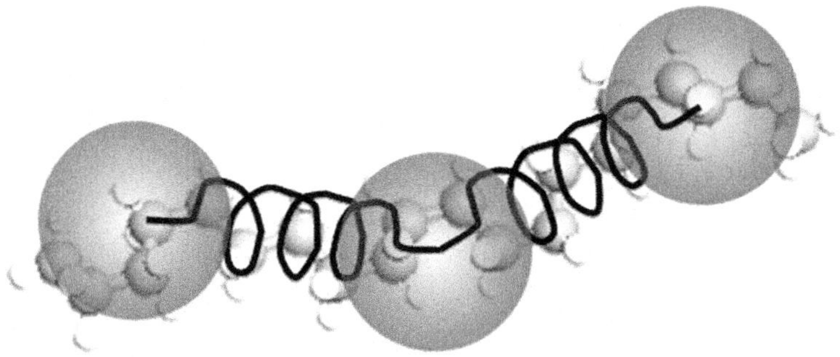

Fig. 2.3 Example of the bead–spring model used to represent a polymer chain

2.2.4 *Ideal Chain Model*

The ideal chain model is also a polymer chain model in which beads are connected by springs. However, the ideal chains do not have volume, whereas the beads in the bead–spring model do. In other words, if multiple ideal chains exist, they may completely overlap. Thus, this model seems to be unrealistic. However, when conducting coarse-graining of a flexible polymer chain to the utmost limit in a molten state, the overlap of atoms (excluded volume effect) can be ignored, thereby making it possible to analyze a polymer structure using the ideal chain model.

The interaction between molecules is not considered in actual simulations using the ideal chain model. Therefore, this model is not used in MD, in which simulation of a condensed state is conducted by considering a large number of molecules. However, the ideal chain model is used in the mean-field method in which the structure of a molecular chain of interest is focused on by considering the effect of surrounding molecular chains as a mean field. In addition, the distribution of distances between the ends of ideal chains follows a Gaussian distribution, and thus they are also called Gaussian chains.

2.2.5 *Slip-Link Model*

The aforementioned ideal chains do not have volume, and in such chains there is no interaction between molecules. Still, the ideal chain model is suitable to reproduce the equilibrium structure of a polymer in a molten state. However, entanglement of molecules is considered important in polymer melts. The slip-link model is a coarse-grained model that incorporates entanglement of polymer chains. As illustrated in Fig. 2.4, this model considers entanglement points as slip-links and partial chains between entanglement points as single springs [2]. Additionally, two chains are connected to form a pair by a slip-link, thus representing entanglement. The slip-links corresponding to entanglement points are formed or deleted as the molecular chains move.

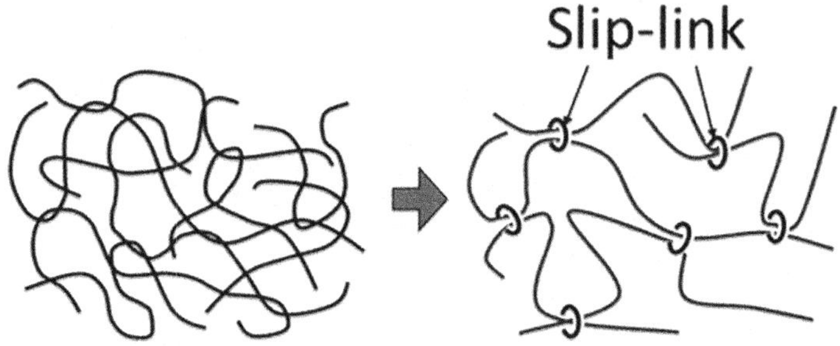

Fig. 2.4 Schematic illustration of the slip-link model

Computer simulation with the slip-link model is used to reproduce the viscoelasticity of a molten polymer and has greatly contributed to the molecular theoretical interpretation of viscoelasticity observed experimentally. More details about the slip-link model are introduced in Chap. 6.

2.3 Relation Between Coarse-Grained Models and Real Polymer Chains

When a simulation is conducted using one of the coarse-grained models, the translation of the model to real polymer chains is important. The relation between the united atom model and full atomistic structure is clear because the united atom model is derived from the original chemical structure. Moreover, the slip-link model defines a partial structure between entanglement points as a single unit, and therefore correspondence to the actual length scale is easy to estimate. Conversely, in the case of the bead–spring or ideal chain models, care must be taken to establish the relation between each coarse-grained model and real polymer chains. In particular, for the bead–spring model, the following two methods are mainly used to obtain correspondence with real molecular structures.

2.3.1 Coarse-Graining from Chemical Structures

This is a method used to derive the interaction between beads and the potential parameters of bond length and bending angle from a molecular structure. This method is used to determine parameters in coarse-grained MD simulation or interaction parameter χ in Flory–Huggins theory. For the purpose of coarse-graining for MD simulation, the interaction between coarse-grained units or a local structure of a polymer chain is usually determined with a full atomistic model or united atom model, and then the interaction parameters, bond length, and spring constant between coarse-grained units are determined to reproduce the results obtained by the atomistic simulation.

For example, the bond stretching potential $U^{\text{bond}}(L)$ between coarse-grained units is obtained by Eq. 2.1,

$$U^{\text{bond}}(L) = -k_B T \ln P^{\text{bond}}(L), \tag{2.1}$$

where $P^{\text{bond}}(L)$ represents the probability of bond length L, which is obtained by atomistic simulation, T represents the absolute temperature, and k_B represents the Boltzmann constant. Accordingly, coarse-grained potential can be derived by obtaining the distribution from full atomistic model simulation.

With respect to nonbonding potential, although some methods have been proposed [3], the nonbonding potential is also derived from the distribution or interaction energy between units obtained from full atomistic simulation in basically the same manner as bonding potential.

In this way, numerous studies to determine coarse-grained potential from a molecular structure have been carried out, and versatile enhanced methodologies [4] and a set of coarse-grained potentials [5] have also been proposed. However, even for the same molecules, the coarse-graining level and method are not unique. Thus, the force field parameters of a coarse-grained model must be determined individually based on the objective of the study. The preparation of versatile parameter sets remains an issue for future study.

2.3.2 Mapping Using the Scaling Concept

Real structures are mapped by comparing quantities of structures of interest and physical properties obtained by simulation of an arbitrarily determined coarse-grained model. For example, it is possible to map the length scale of a polymer by estimating the number of monomers corresponding to a single unit of a coarse-grained model using the degree of polymerization (molecular weight) between entanglement points. Kremer et al. [6] estimated that the degree of polymerization between entanglement points *Ne* in their bead–spring model was 35 from the diffusion of particles. Subsequent studies obtained somewhat larger values for *Ne*, and *Ne* is now commonly regarded to be about 50 [7]. By comparing the simulation results with the molecular weight between entanglement points of a real polymer, a single unit of the bead–spring model can be mapped to a certain molecular weight of the real molecule. Specifically, when experiments indicate that the molecular weight between entanglement points of polystyrene is 18,000, the molecular weight per single particle is about 515 (=18,000/35). Thus, in the case of polystyrene, about five monomer units correspond to a single unit of the bead–spring model.

Besides using the molecular weight between entanglement points to determine the number of monomer units per model unit, other examples, such as using the glass transition temperature to map the energy scale [8] and using the self-diffusion constant to map the time scale [6], have also been reported.

References

1. J.G. Gay, B.J. Berne, J. Chem. Phys. **74**, 3316 (1981)
2. J. Takimoto, H. Tasaki, M. Doi, Proc. XIIIth Int. Congr. Rheol. **2**, 97 (2000)
3. H. Fukunaga, J. Takimoto, M. Doi, J. Chem. Phys. **116**, 8183 (2002)
4. D. Reith, M. Pütz, F. Müller-Plathe, J. Comput. Chem. **24**, 1624 (2003)
5. S.J. Marrink, D.P. Tieleman, Chem. Soc. Rev. **42**, 6801 (2013)
6. K. Kremer, G.S. Grest, J. Chem. Phys. **92**, 5057 (1990)
7. R.S. Hoy, K. Foteinopoulou, M. Kroger, Phys. Rev. E. **80**, 031803 (2009)
8. H. Morita, K. Tanaka, T. Kajiyama, T. Nishi, M. Doi, Macromolecules **39**, 6233 (2006)

Part II
OCTA: Mesoscale Polymer Simulation System

Chapter 3
Overview of OCTA

Jun-ichi Takimoto and Takeshi Aoyagi

3.1 What is OCTA?

OCTA is a simulation system developed for the design of polymeric and other soft materials. It was first released in 2002 as an outcome of a Japanese national project, which started in 1998 and was financed by the Ministry of International Trade and Industry and the New Energy and Industrial Technology Development Organization.

Since its initial release, OCTA has been actively and continuously developed by many contributors, including the research group of Prof. Masao Doi (Nagoya University and The University of Tokyo) and the OCTA user group (including the original developers of simulation engines). A commercial version of OCTA was also developed by Japan Research Institute Ltd. and is now maintained and updated by JSOL Corp. Currently, OCTA is widely used by many researchers in academia and industry. The latest version of OCTA is available from http://octa.jp/.

Originally, OCTA consisted of four simulation engines (COGNAC, PASTA, SUSHI, and MUFFIN) and a graphical user interface (GUI) tool (GOURMET) as shown in Fig. 3.1. OCTA has been designed, however, such that new engines can be easily added; the only requirement of a new engine is that it uses user definable format (UDF) files (explained below) for input and output. Several new engines developed by other groups (including NAPLES and KAPSEL) have been added to OCTA using this method.

J.-i. Takimoto (✉)
Yamagata University, Yonezawa, Japan
e-mail: takimoto@yz.yamagata-u.ac.jp

T. Aoyagi
Research Center for Computational Design of Advanced Functional Materials,
National Institute of Advanced Industrial Science and Technology, Ibaraki, Japan
e-mail: aoyagi.t@aist.go.jp

Japan Association for Chemical Innovation, *Computer Simulation of Polymeric Materials*, DOI 10.1007/978-981-10-0815-3_3

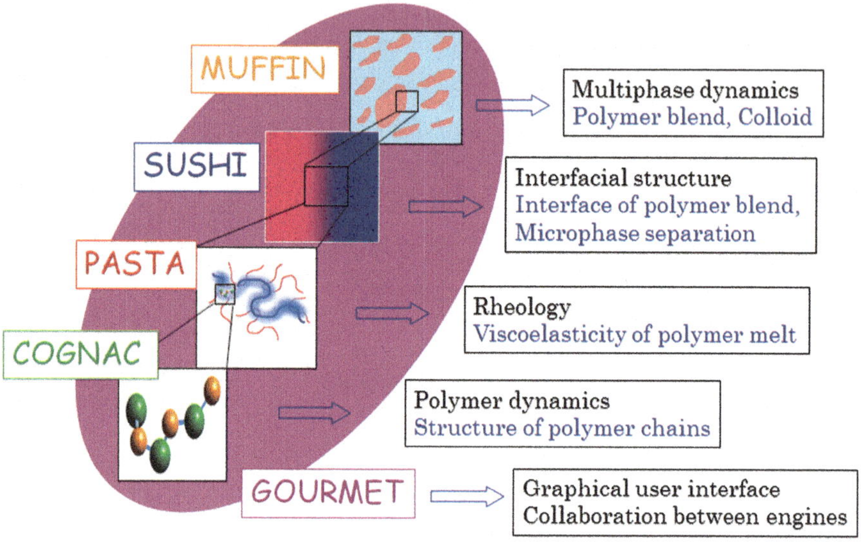

Fig. 3.1 Components of the original OCTA system

Details and the use cases of each simulation engine will be given in the following chapters. This chapter presents the overall structure of the OCTA system, followed by a brief introduction of the GUI tool GOURMET. A more detailed introduction and user manual of GOURMET can be found in the Portable Document Format documents included in the OCTA package (such as “GOURMET Primer”).

Polymeric (and other soft) materials, which are the main target of OCTA, have a hierarchy of structures and motions with very wide length and time scales. As a result, OCTA includes simulation engines from molecular dynamics to finite element/difference methods. However, many independent simulation engines are not sufficient for the analysis and design of polymeric materials. It is important that users can easily move between different time and length scales. The OCTA system was designed to help users achieve this “seamless zooming.”

To realize “seamless zooming,” it is necessary that the data can be easily shared among simulation engines. It is also necessary that users can analyze the data using their own methods, depending on the situation and their objectives; in other words, analysis tools should be more flexible and customizable than simple pre/post-processors. To meet this goal, a new file format, UDF, has been developed for OCTA.

A UDF file consists of both the definition of the data structure and the data themselves. All the simulation engines in OCTA use UDF files for input and output. Each datum in a UDF file has a name and can be accessed from the engines by the name. This enables engines to easily read the data written by other engines and also helps users understand the structure of input and output files. It is also possible to specify a unit for each datum.

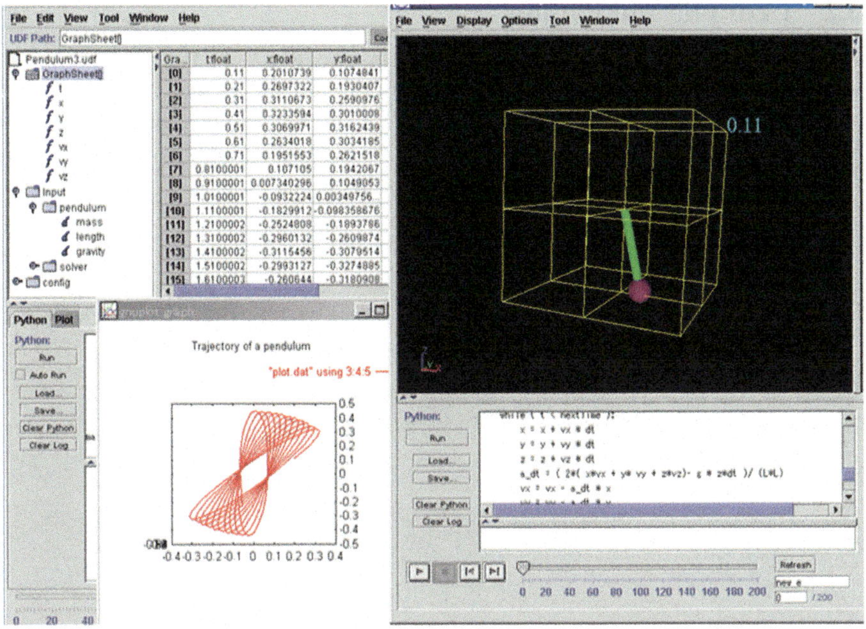

Fig. 3.2 An example session of GOURMET

GOURMET is a GUI tool for editing and analyzing UDF files. GOURMET can be used to create input files, run and control engines, and analyze output files. All the engines can be used with a single common interface. Figure 3.2 shows a typical session using GOURMET (an external program *gnuplot* is used to make plots). GOURMET is written in Java and runs on Windows and Linux.

GOURMET can also be used to convert physical quantities between systems having different units. Even if a simulation is run in dimensionless units, users can easily get all physical quantities in a physical unit by specifying a few basic units, such as those of length, mass, and energy.

The powerful scripting language Python can be used for the analysis of UDF files. Python in GOURMET is extended in such a way that it can easily access the data in UDF files. It has libraries for various analyses and for creating three-dimensional (3D) animations. It is also possible to automatically load and run Python scripts from a menu in GOURMET (the Action menu). UDF files designed for engines in OCTA already have many scripts and actions for analysis, and users can easily write their own scripts (or actions) for more flexible analysis. Python scripts can also be used to simplify (or automate) the preparation of input UDF files. Many examples of actions will be found later in the following chapters.

3.2 UDF File and Python

In this section, we briefly explain the structure of the UDF file and how to edit and analyze it using GOURMET and Python. As a simple example, let us consider a bank account in which we first deposit y_0 dollars (or billion dollars). If the interest rate per year is r, our balance y_n after n years satisfies

$$y_{n+1} = (1 + r)\, y_n.$$

Our task is to calculate y_n ($n = 1, 2, \ldots, N$) for a given N. The following UDF file (which is in "Ch3/2/account.udf" and can be opened by any text editor) describes this problem.

```
\begin{global_def}
input: {
    principal: double
    rate:      double
    years:     int
}
output: {
    y[]: double
}
\end{global_def}
\begin{data}
input: { 100.0, 0.04, 20 }
\end{data}
```

This UDF file consists of a *definition section* and a *data section*. The definition section is from `\begin{global_def}` to `\end{global_def}` and describes the structure of the data in our problem. The data section is from `\begin{data}` to `\end{data}` and contains the actual data (the definition can also be made using `def` instead of `global_def`; this will be explained in Sect. 3.6).

Let us first look into the definition section. In this section, two structures *input* and *output* (where a structure is a collection of data) are defined. The structure *input* has three member variables: *principal* (y_0), *rate* (r), and *years* (N). These three are sufficient to define the problem. The first two are real numbers with type double, while *years* is an integer. The structure *output* contains only one member *y[]*, where [] indicates that this variable is an array. (The number of elements in the array need not be specified.) When the calculation finishes, *y[n]* will contain y_n.

In the data section, the data for *input* are given as *principal* = 100.0, *rate* = 0.04 (4 % per year), and *years* = 20. Note that only the top-level structure name (*input*) need be specified. The structure *output* has no data yet.

If we open this UDF file using GOURMET, an *Editor window* will open as shown in Fig. 3.3. The upper half of this window (the *Editing pane*) shows the contents of the UDF file. There are two folder-like icons corresponding to the two structures

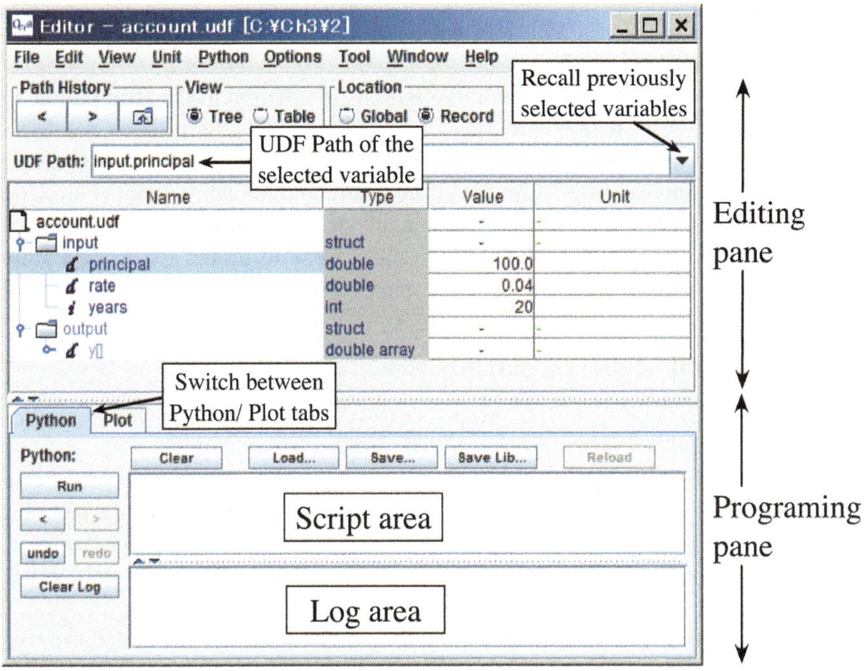

Fig. 3.3 An Editor window of GOURMET

input and *output*. These folders can be opened by either double-clicking on them or by clicking on a small bar-like icon to the left of the folder. Note that the output is displayed in a light-color font because there are no data for it yet. We can modify the data in *input* by simply typing new data in the column named **Value**.

The lower half of the Editor window (the *Programing pane*) is for data analysis; it has two tabs **Python** and **Plot**. Please select the **Python** tab if not selected. The **Python** tab has two text areas; the upper one (the *Script area*) is for entering a Python script, and the result of the script will be shown in the lower area (the *Log area*). As an example, we enter the following one-line script:

```
print $input.principal
```

into the Script area and click the **Run** button. An output 100.0 will be shown in the Log area. As this example indicates, we can obtain a value of a variable in the UDF file by prepending `$` to the *UDF path* of the variable. A UDF path of a member of a structure is given by concatenating the structure name and the member name by a dot ".". In the example above, the UDF path is *input.principal*. When a datum is selected in the Editing pane, the UDF path of the datum is shown in the text box named **UDF Path** near the top of the Editor window. If only the structure name is specified as in `$input`, then all the data in the structure are returned as a Python list [100.0, 0.04, 20]. We can also modify the data by assigning

```
$input.principal = 200.0
```

to the data.

The Programing pane has a few more buttons in addition to the **Run** button we have just used. The **Clear** button clears the contents of the Script area, while the **Clear Log** button clears the Log area. The "<" button below the **Run** button recalls the previous scripts into the Script area. The **Load** button loads a script from a file, while the **Save** button saves the current contents of the Script area into a file.

We now calculate the balance of our bank account using the following Python script:

```
$output.y[] = []
$output.y[0] = $input.principal
for n in range($input.years):
    $output.y[n+1] = (1 + $input.rate)*$output.y[n]
```

We click the **Clear** button to clear the old script, enter this script into the Script area, and click the **Run** button. The text of the *output* in the Editing pane has changed from a light to a dark color, which indicates that data are newly created for this structure. We open the structure *output* and then open the array *y[]* to reveal the data in *y[0]* ... *y[20]*.

Let us briefly explain each line of the script. The first line resets the array *y[]* (which may contain the result of the previous calculation) to an empty array. The second line assigns the *principal* to *y[0]*; this assignment automatically creates an array element *y[0]*. In the third line, the range() function gives a list [0, 1, .., 19] (if *input.years* is 20), and the fourth line is repeated with $n = 0, 1, 2, \ldots, 19$. This creates array elements *y[1]* ... *y[20]*.

We can now plot the growth of *y[n]* in the 20 years by taking the following steps (see Fig. 3.4). Near the top of the Editor window, there are radio buttons **Tree** and **Table** (under the title **View**). We select **Table** (instead of **Tree**, which should have been selected up to now), and select *y[]* in *output* to view all of its elements in a table. We now select the **Plot** tab (instead of the **Python** tab) in the Programing pane (lower part of the window), and click the **Make** button. This button automatically generates gnuplot commands for plotting the data currently shown in the Editing pane. Finally, we click the **Plot** button to run the gnuplot commands; a plot window will open, showing the growth of our deposit.

As this example shows, **Make** and **Plot** buttons in the **Plot** tab provide an easy way of creating a plot if the data are currently shown in Table View. If we want to create a plot with a more flexible choice of data, or if we want customize the appearance of the plot, then we can use one of the two Python modules (gnuplot and gnuplot2) that are installed with GOURMET. An example use of the gnuplot module can be found in the next section ("Ch3/3/calc-plot.act").

We can now try changing *principal*, *rate*, and/or *years* in *input* (we need *not* save the UDF file), run the script in the **Python** tab to recalculate *y[]*, and switch to the **Plot** tab and use the **Plot** button to see how the input parameters affect the growth of *y[]*.

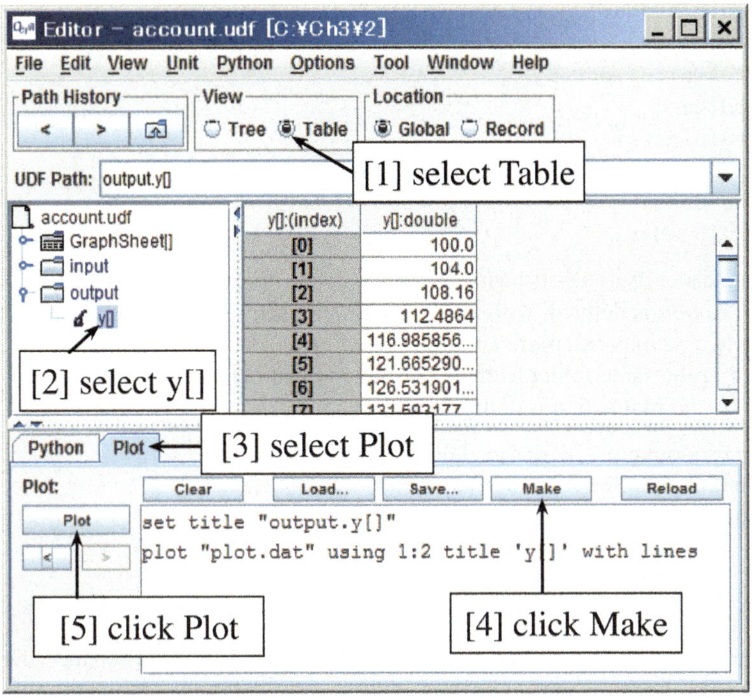

Fig. 3.4 Creating a plot by using GOURMET

3.3 Action Mechanism

Although the Python script used in the previous example was simple, it was still necessary to manually type the script into the Script area. Several steps were also necessary to make the plot of *y[]*. If we want to automate these steps, we can use the *Action* mechanism of UDF files. An example UDF file with two actions added can be found in "Ch3/3/account-action.udf." If we open this UDF file using GOURMET, we notice that *input* and *output* in the Editing pane are written in bold font. This indicates that these elements have *actions*, which can be invoked by right-clicking on them.

We first right-click on *input*; a menu that contains a single action **calculate** will pop up. We select it to calculate *y[]*. We then right-click on *output*, and select the **plot** action to create a plot of *y[]*. We see that all the steps we followed in the previous section are automated by the two actions.

How these actions work can be easily understood if we look into the UDF file. We open "account-action.udf" using a text editor, and notice that the following header is added at the beginning of the file.

```
\begin{header}
\begin{def}
    Action: string
\end{def}
\begin{data}
    Action: 'calc-plot.act'
\end{data}
\end{header}
```

The header itself has its own definition section and data section. A single variable named *Action* is defined, with a value "calc-plot.act." This is the name of a file in which two actions **calculate** and **plot** are defined. The action file "calc-plot.act" is located in the same folder as the UDF file and can be opened by a text editor. For example, the **plot** action is defined as follows:

```
action output: plot(): \begin
import gnuplot
gnuplot.plot(
        data=[range($input.years+1),$output.y[]],
        labels=['year', 'y'])
\end
```

The first line indicates that an action named **plot** is defined for the UDF path *output*. The body of the action is a Python script between `\begin` and `\end`. The script first imports the gnuplot module (which is installed with GOURMET). It then calls the plot() function in the module, with arguments *data* and *labels*; *data* specifies the data for the horizontal and vertical axes, while *labels* specifies the label on each axis. Note that the data for the horizontal axis are created using the range() function because they are not in the UDF file.

3.4 Unit Conversion

GOURMET has a unit conversion facility. As an example, let us consider a UDF file with the following definition section ("Ch3/4/unit.udf"):

```
\begin{global_def}
    mass:     double  [kg]
    volume:   double  [m3]
    density: double   [kg/m^3]
\end{global_def}
```

We open this UDF file using GOURMET and notice that units of *mass*, *volume*, and *density* are displayed in the column **Unit**. If we right-click on the unit of *mass* [kg], a dialog will open as shown in Fig. 3.5, and we can select a new unit from the drop-down menu. For example, if we select g (gram) and click the **OK** button, the **Unit** column of *mass* will change to [g] and the **Value** column will show 1000.0.

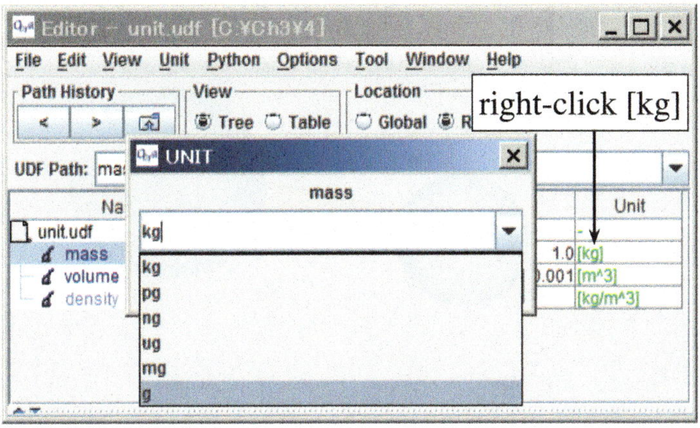

Fig. 3.5 Unit conversion by using GOURMET

It is important to note that although the displayed value changes according to the selected unit, the data stored in the UDF file do not change. If we run

```
print $mass
```

then it will print 1.0, which is the value measured in the *default* unit (the unit defined in the definition section of the UDF, i.e., [kg] for *mass*). If we run

```
$density = $mass/$volume
```

then the value of *density* in the UDF file will be in unit of [kg/m^3], irrespective of the units currently selected for displaying *mass*, *volume*, and *density*.

3.5 3D Graphics and the Select Mechanism

GOURMET has a set of functions for 3D display. As an example, we enter the following script into the Script area and click the **Run** button.

```
sphere([0, 0, 0], [0, 1, 1, 1, 0.8])
sphere([2, 2, 2], [0, 1, 1, 1, 0.8])
cylinder([0, 0, 0], [2, 2, 2], [1, 1, 0, 1, 0.2])
```

A new *Viewer window* will open, where a diatomic molecule is drawn as in Fig. 3.6. We can rotate the molecule by dragging with the left mouse button, zoom in/out by right-dragging, and translate by shift-dragging (left-dragging while the shift key is pressed). In the example above, [0, 0, 0] and [2, 2, 2] are the coordinates of the centers of the two spheres (or two ends of the cylinder), and [0, 1, 1, 1, 0.8] indicates the color (R, G, B = 0.0, 1.0, 1.0), the transparency (1.0 = opaque), and the radius 0.8 of the spheres. We can find many more drawing functions in

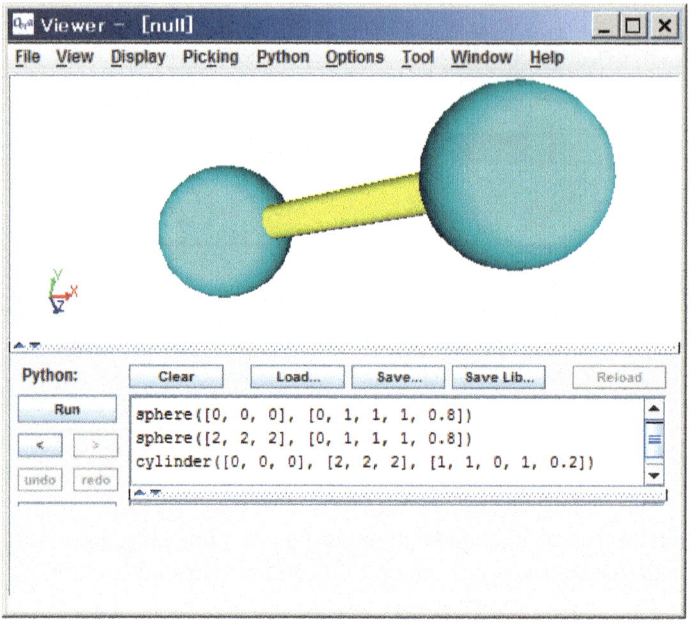

Fig. 3.6 A Viewer window of GOURMET showing a diatomic molecule

Sect. 3.3.3 of *GOURMET Python Script Reference Manual.* There are also utility functions for displaying molecules in the UDF file created by COGNAC; see Sect. 7.4.1 of *COGNAC User Manual.*

We can change the background color of the Viewer window using the **Background** submenu of the **Display** menu of the window. In the same menu, **Draw Size/Width/Divisions**... can be used, for example, to change the smoothness of the spheres. If we uncheck the **Wire Frame Movement** in the same menu, the molecules are drawn with solid models even while rotating using a mouse. Finally, we use **ImageCapture** in the **Options** menu to capture the image as a file (in either png or jpg format, which can be selected in the **Save Options**... submenu).

We now open "Ch3/5/darw3d.udf" in GOURMET, right-click on the file name *draw3d.udf* in the leftmost column **Name**, and select the action **show** (remembering that if the file name or data name is in a bold font then it has actions). The same diatomic molecule as in the previous example will be displayed. The only data in this UDF file are an array *Object[]* with three elements. In each element, a variable *kind* specifies the type of object, Sphere for *Object[0]* and *[1]* and Rod for *Object[2]. Object[0]* and *[1]* have a structure *Sphere* (which specifies the center and radius), while *Object[2]* has a structure *Rod* (which specifies two ends of the cylinder and the radius). Note that the **Type** of the variable *kind* is *select*. A variable of this type can only take a set of string values. If we click on the **Value** column of *Object[2].kind* (currently the value is Rod), a drop-down menu will appear and we can select from three values: Sphere, Line, and Rod. These three values are the only possible values of the variable *kind*. If we change the value from Rod to Line, the structure *Rod* in *Object[2]* is replaced by the structure *Line*, i.e., the appropriate data member is automatically selected depending on the current value of a variable of type select. We enter values of *start* and *end* in the structure *Line*, and rerun the action **show** to redraw the molecule with the new data.

Many of the UDF files used by simulation engines in OCTA use this select mechanism so that users need to specify only those parameters necessary for their specific purpose.

If we want to add an element to the array *Object[]*, we highlight the *last* element of the array (i.e., *Object[2]* in our example) in the **Name** column, and select **Add and Array Element** in the **Edit** menu. This will create a new element *Object[3]*. We can enter all the necessary data into this new element, but we may want to copy the data from another (already existing) element and modify only some of them. To do this, we first select **Deep Copy** in the **Edit** menu (otherwise only the top-level data of the structure will be copied). We then select the element from which we want to copy from, such as *Object[1]*, and select **Copy** in the **Edit** menu. Finally, we select *Object[3]* and select **Paste** in the **Edit** menu. This will copy all the contents of *Object[1]* into *Object[3]*, and we can modify any member in *Object[3]* as we like; for example, we may change *Object[3].center*. We then rerun the action **show**, which will draw four objects.

3.6 Record and Animation

In molecular dynamics and similar simulations, the configuration (e.g., the coordinates and velocities of the particles) of the system will change as the simulation proceeds, and we may want to output many configurations at some time interval. For this purpose, we can save each configuration in a separate record of the output UDF.

As an example, we open "Ch3/6/MonteCarlo.udf" using GOURMET. This example will generate many configurations of a single Gaussian chain (one end of the chain is fixed at the origin) employing a simple Monte Carlo (MC) method. The model is described by the structure *Model*. *N* is the number of monomers in the chain, and *k* is the force constant of the spring connecting two neighboring monomers. The simulation conditions are specified in the structure *Condition*; *delta* is the maximum distance a monomer can move in a single MC step, *T* is the temperature, *maxMCS* is the total number of MC steps to calculate, and *outInt* is the output interval.

To start the simulation, we right-click the file name *MonteCarlo.udf* at the top of the **Name** column (it is in bold font), and select the action **calculate** in the pop-up menu to start a simulation. When it completes, 101 records are created in the UDF file, and a slider will appear at the bottom of the Editor window as shown in Fig. 3.7. The slider is at the rightmost position, indicating the last record (record No. 100) is selected. We can select other records by moving the slider. In the structure *Output*,

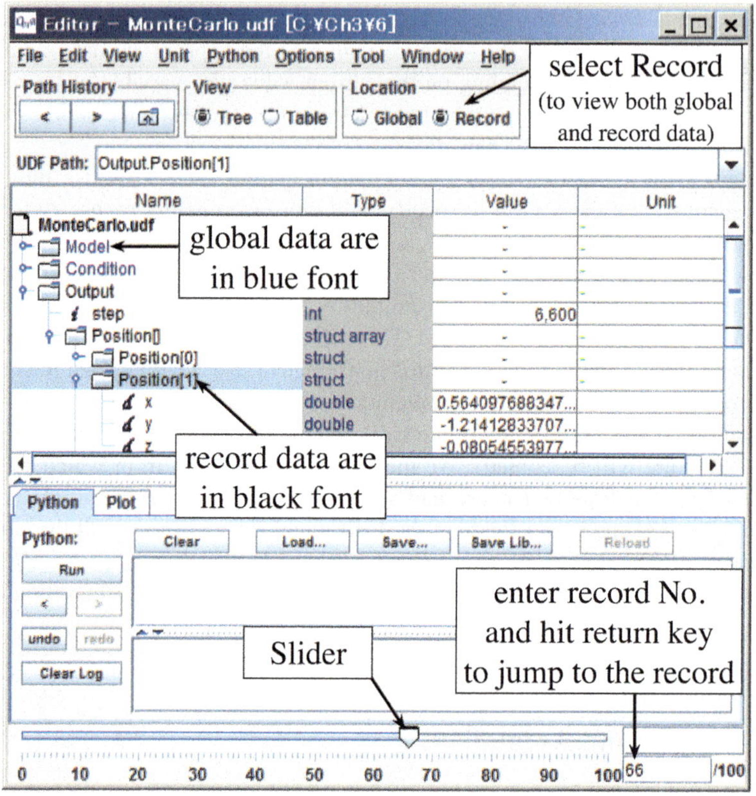

Fig. 3.7 Editing a UDF file with record data

there is a variable *step* (MC step) and an array *Position[]* (coordinates of monomers), whose values change as we move the slider. These data are called "*record data*" and are written in each record of the UDF file. Meanwhile, the data in the structures *Model* and *Condition* are independent of the record selected. These data are written outside of any records and are called "*global data*." Note that the names of global and record data are displayed in blue and black, respectively.

We next open "MonteCarlo.udf" using a text editor. We will see that the global data are defined using `global_def` as before, while the record data *Output* are defined as follows:

```
\begin{def}
Output: {
    step: int   'current MC step'
    Position[]: { x:double, y:double, z:double }
}
\end{def}
```

Note that a string “current MC step” can be appended at the end of the definition of the variable *step*. This string is displayed as a tooltip when we move the mouse cursor over the variable.

We now go back to GOURMET, right-click on the filename *MonteCarlo.udf*, and select the action **show** in the pop-up menu. A Viewer window will open, and the chain configuration in the currently selected record will be displayed. The Python script used for the display, which is just a few lines of code, is shown in the Script areas of both Editor and Viewer windows. The Viewer window also has a slider at the bottom, and if we move the slider, the display changes accordingly. There are also four animation buttons (start, stop, forward, and backward) to the left of the slider. We click the start button to start the animation. We can rotate the molecule by dragging the mouse even during the animation.

As this example shows, the simulation results can be easily viewed in animation just by writing a short Python script, whose task is to draw a single animation frame using data in a single record. Most of the simulation engines in OCTA have already implemented this as actions.

3.7 Closing Remarks

One of the advantages of OCTA is that it allows users to analyze simulation results with ease and flexibility using GOURMET and Python. Although various analysis actions are already prepared in the UDF files for each engine, users can write their own analysis scripts for more flexible, detailed, and extensive analyses. Python is an intuitive and easy-to-learn script language, with many modules available for numerical analysis. Writing analysis scripts requires only a basic knowledge of Python, which can be obtained from Chap. 15 of *GOURMET Primer*. More detailed information can be found in *Gourmet Python Script Reference* or in many books and websites on Python. We hope the reader will take full advantage of OCTA using GOURMET and Python scripts.

Chapter 4
COGNAC: Coarse-Grained Molecular Dynamics Simulator

Takeshi Aoyagi

4.1 What Is COGNAC?

COGNAC (COarse-Grained molecular dynamics program by NAgoya Cooperation) is a general-purpose molecular dynamics program that was developed to handle various coarse-grained and atomistic models. In addition to conventional molecular dynamics, COGNAC can also be used to study the higher-order structure and physical properties of polymeric materials. Figure 4.1 shows examples that COGNAC can handle including microphase-separated structure, polymer–inorganic compounds, and pseudo-chemical reactions. More examples of the applications of COGNAC will be discussed in Part 3.

This chapter introduces the features and functions of COGNAC and includes a brief example of an operation of COGNAC. Readers are recommended to refer to the COGNAC user manual in the OCTA package for more information.

4.2 Potential Functions

COGNAC provides various potential functions to handle a wide range of molecular models from full atomistic to highly coarse-grained bead–spring models. This section introduces some popular functions that are available in COGNAC.

T. Aoyagi (✉)
Research Center for Computational Design of Advanced Functional Materials,
National Institute of Advanced Industrial Science and Technology, Ibaraki, Japan
e-mail: aoyagi.t@aist.go.jp

Japan Association for Chemical Innovation, *Computer Simulation of Polymeric Materials*, DOI 10.1007/978-981-10-0815-3_4

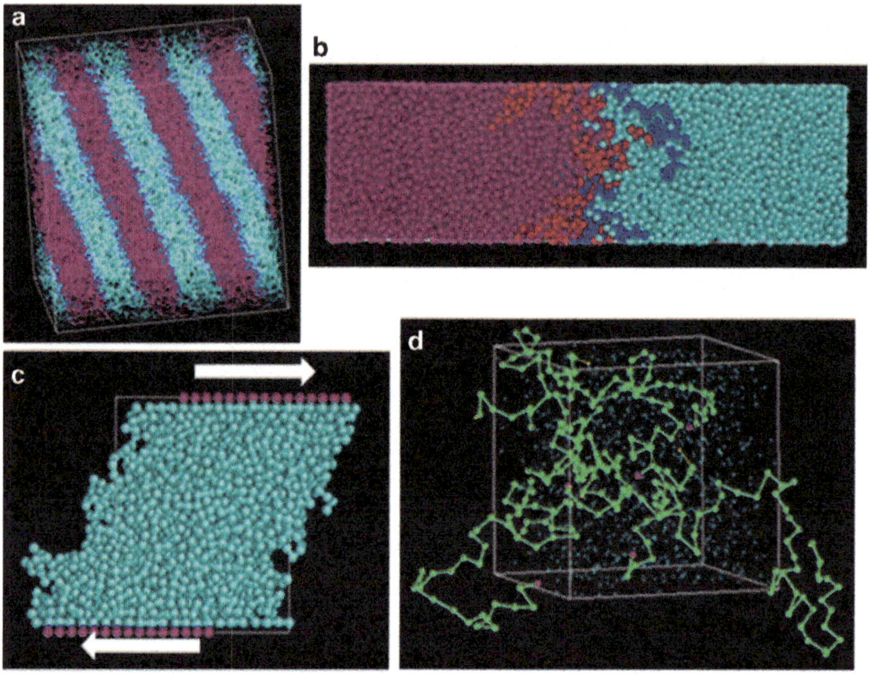

Fig. 4.1 Examples of the outputs of COGNAC including (**a**) microphase-separated structure of block polymers, (**b**) interface of a polymer blend, (**c**) shear flow of a polymer melt confined between solid walls, and (**d**) molecular dynamics considering polymerization

4.2.1 Bond-Stretching Potential Functions

Harmonic This is a simple harmonic spring widely used for atomistic and coarse-grained models.

FENE + LJ This function consists of a combination of finite extensible springs and repulsive cores with a Lennard–Jones potential and is usually used for Kremer–Grest-type bead–spring models [1].

Gaussian This function reproduces a Gaussian distribution of bond lengths.

DPD This function is usually used to simulate dissipative particle dynamics (DPD) [2] and is essentially the same as the Gaussian function with a different interface of input parameters.

Table The potential and force are calculated from a given numerical table with this function.

4.2.2 Angle Bending Potential Functions

Theta Harmonic Harmonic spring of bending angle θ. This function is widely used from full atomistic potentials to the bead–spring model.

Theta Harmonic 2 When the equilibrated angle θ_0 is set to zero with the theta harmonic potential, the force around the equilibrated angle becomes unstable. Theta harmonic 2 is used to avoid this problem.

Cosine Harmonic A harmonic spring of $\cos\theta$ is used instead of θ. Some force fields for atomistic models use this function.

4.2.3 Torsion Potential Functions

Cosine Polynomial Uses the polynomial function of $\cos\phi$, where ϕ is a torsional angle. This function is used for the united atom model to reproduce complex torsion potential profiles induced by omitting hydrogen atoms with the simple torsion angle of carbon–carbon bonds.

Amber AMBER-type potential function [3].

Dreiding Dreiding-type potential function [4].

4.2.4 Nonbonding Potential Functions

In addition to the bonding potential, COGNAC also has nonbonding potential functions. To calculate nonbonding potentials, COGNAC uses "interaction sites." Interaction sites are defined by atom(s), and an arbitrary function can be defined between the interaction sites using the coordinates of atoms. This capability enables the center of nonbonding interactions to be set at different positions from the position of a single atom.

Some popular nonbonding interactions that can be implemented in COGNAC are as follows:

Lennard–Jones The most popular nonbonding interaction. The current version of COGNAC can set arbitrary numbers for the coefficient of power of repulsive and attractive terms.

Lennard–Jones with excluded volume An extended Lennard–Jones potential in which the distance of the Lennard–Jones potential is shifted from the center of the interaction site [5]. This function is used to handle a mixture of large particles such as inorganic particles and small particles such as atoms.

Gay–Berne Ellipsoidal potential function [6]. This potential is often used to simulate liquid crystal mesogens.

DPD A soft potential that is used in DPD simulation. The coefficient of this potential can be derived from the interaction parameter χ of Flory–Huggins theory.

4.2.5 External Potential Functions

COGNAC can apply external potentials that act on a single interaction site such as a solid wall or gravity field.

The examples of external potential functions that can be implemented in COGNAC are as follows:

Lennard–Jones atomic-type potential A solid wall potential where Lennard–Jones particles are placed on a square lattice on the wall. This wall potential can apply shear deformation by moving the wall.

Lennard–Jones-type flat wall A wall potential where Lennard–Jones particles are placed an infinite distance from a wall with a given density [7]. This potential is derived from the analytical integration of the Lennard–Jones function, and the wall has a completely flat surface.

Density-biased potential This potential reads volume fraction data obtained from SUSHI (see Chap. 5) and MUFFIN (see Chap. 7) calculations and applies the potential and force on the interaction site from the volume fraction [8]. This potential is used to constrain the morphology of molecules to the results of SUSHI and MUFFIN calculations.

4.2.6 Electrostatic Interaction

COGNAC can calculate the electrostatic interaction between point charges and between dipoles.

Some electrostatic interactions that can be implemented in COGNAC are:

Cutoff Simple Coulomb interaction with distance cutoff

Ewald Conventional Ewald method [9]

PPPM Particle–particle and particle–mesh methods [10]

4.3 Equations of Motion

COGNAC has the following equations of motion in addition to the conventional Newton's equation of motion.

Langevin Dynamics Langevin dynamics are defined in Eq. 4.1 and consider a friction constant and random noise:

$$m\frac{d^2\boldsymbol{r}}{dt^2} = \boldsymbol{F} - m\Gamma\frac{d\boldsymbol{r}}{dt} + W(t), \tag{4.1}$$

where m and $\boldsymbol{r}$ are the mass and position of a particle, respectively, and $\boldsymbol{F}$ is a force acting on the particle. This formula is often used to represent a heat bath used for temperature control in coarse-grained dynamics.

DPD In DPD simulations, the force $\boldsymbol{f}_i$ acting on particle i is obtained from Eq. 4.2:

$$\boldsymbol{f}_i = \sum \left(\boldsymbol{F}_{ij}^C + \boldsymbol{F}_{ij}^D + \boldsymbol{F}_{ij}^R\right), \tag{4.2}$$

where $\boldsymbol{F}_{ij}{}^C$, $\boldsymbol{F}_{ij}{}^D$, and $\boldsymbol{F}_{ij}{}^R$ are conservative, dissipative, and random forces, respectively.

4.4 Ensembles

COGNAC has numerous methods to control temperature and pressure. Here, the methods and instructions for the usage of each method will be introduced. Readers are recommended to refer to the COGNAC manual and references for further details.

4.4.1 Temperature Control

Velocity Scaling The velocity of atoms is scaled at a specified interval of steps to maintain the temperature of a system. This is a robust method to keep the temperature constant. However, this method does not reproduce the canonical ensemble.

Loose Coupling [11] This method scales the velocity of atoms with a coupling constant. The coupling constant is described as a relaxation time, and as the relaxation time gets shorter, the coupling becomes stronger.

Nosé–Hoover [12] This is a modified algorithm of the original Nosé Hamiltonian method.

Langevin Dynamics This method is often used in coarse-grained molecular dynamics simulations with the bead–spring model.

Normally, the Nosé–Hoover method is suitable to control the temperature of an equilibrated system. However, it is often the case that it is more effective to use velocity scaling or the loose-coupling method for non-equilibrated systems such as for relaxation from initial structures.

4.4.2 Pressure/Stress Control

Loose Coupling Similar to temperature control, unit cell size will be scaled to control pressure/stress with a given coupling constant. COGNAC provides Berendsen-type isotropic scaling and Brown–Clarke-type anisotropic scaling [13].

Extended Hamiltonian Method In this method, unit cell size is changed within an extended Hamiltonian. COGNAC provides Andersen-type isotropic control [14] and Parrinello–Rahman-type anisotropic control [15].

Extended Hamiltonian-type pressure/stress control is suitable for equilibrated systems, while for non-equilibrated systems, the loose-coupling method is more robust.

4.5 Boundary Conditions

COGNAC can use the following boundary conditions:

Periodic boundary condition This is the most popular type of boundary condition used in molecular dynamics simulation. This boundary condition can be applied independently for each axis.

Lees–Edwards boundary condition [16] This boundary condition is used when studying shear flow.

Staggered reflective boundary condition This is a modified version of the reflective boundary condition, which was developed originally in the OCTA/COGNAC project [8]. In the case of simple reflective boundary conditions, the coordinate of real molecules and image molecules will overlap, as shown in Fig. 4.2a. Staggered reflective boundary conditions avoid such overlap by shifting the coordinate of image molecules by half of the cell size (L) in the direction of the boundary plane (Fig. 4.2b). This boundary condition can be applied only when the system is homogenous in the direction of the boundary plane.

It should be noted that COGNAC was developed to study condensed systems, in which some boundary conditions should be applied. Thus, if the boundary condition is set to NONE, a solid wall or other external field should be applied to prevent the molecules from leaving the unit cell.

Periodic Chain An infinite chain can be defined by applying periodic boundary conditions to bonding potential in addition to nonbonding interactions as illustrated in Fig. 4.3.

Periodic bonds are usually used to study infinite chains of crystals. Also in the case of bond formation, this method can be used to model network structure.

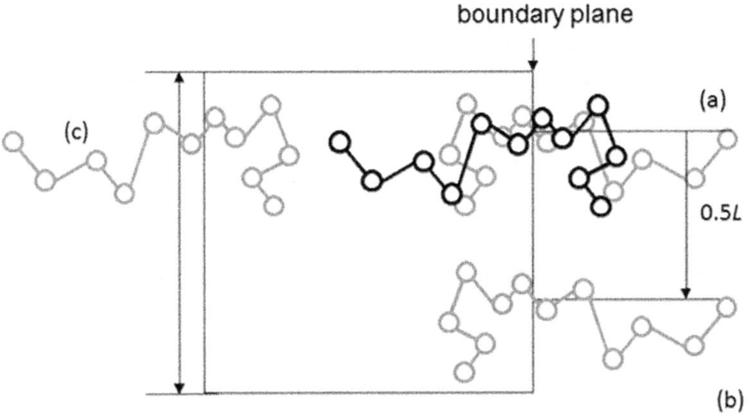

Fig. 4.2 Schematic illustration of staggered reflective boundary condition: black molecule shows real coordinates and gray molecules show the position of image molecules by each boundary conditions, (**a**) reflective boundary condition, (**b**) staggered reflective boundary condition, (**c**) periodic boundary condition

Fig. 4.3 Schematic illustration of periodic bonds

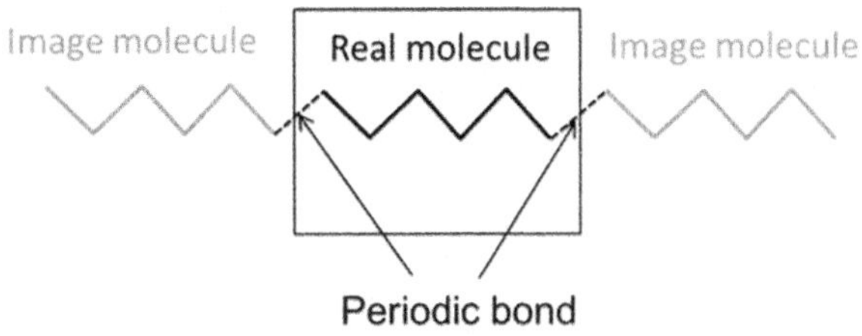

4.6 Generation of Initial Coordinates

To start a molecular dynamics simulation, the coordinates of all atoms must be given. While COGNAC can read the coordinates from another data file, COGNAC also has some simple functions to generate the coordinates:

Random This function basically generates the coordinates of atoms by random walk with a specified equilibrated bond length. In the case of random generation, bending and torsional angles can be restricted by specified equilibrated angles. Furthermore, the coordinate of one end of an arbitrary chain can be specified so that a grafted chain with an end fixed on a surface is modeled. This random walk method in COGNAC does not consider the excluded volume effect. Thus, a long relaxation run is usually needed to implement excluded volume.

Helix This function generates helical chain coordinates in a regular lattice. Simple atomic crystals and regularly placed linear molecules can also be generated by this function. The direction of alignment and stereoregularity such as *meso*- and *racemic*-type structures can also be specified.

Crystal This function generates arbitrary crystal structures using a lattice constant, symmetry operation, and fractional coordinates in a different universal disc format (UDF) file. The symmetry operation cannot be specified automatically by specifying space group; it must be given manually.

Some more extended functions will be discussed in Sect. 4.8.

4.7 SILK

OCTA/COGNAC does not have molecular modeling capability with a graphical user interface (GUI), as is usually provided by commercial software for molecular simulation. Instead of such a GUI, COGNAC provides the script-based modeling tool SILK (Set of molecules Interpreter of Light Kits) and an action tool called Action SILK, in which part of the function of SILK is implemented. The functions of Action SILK are limited. However, it has a user-friendly interface and can be used without much difficulty. In contrast, the original SILK script has various functions, but it is difficult for beginners to use. Section 4.12 will introduce the usage of SILK and Action SILK. For more information, please refer to Chap. 6 of the COGNAC manual.

4.8 Extended Functions to Study Polymeric Materials

In addition to basic functions for molecular dynamics simulations, COGNAC has various functions that can be used to study the structure and properties of polymeric materials. This section introduces some of these advanced functions.

4.8.1 Deformation

To study mechanical properties such as the Young's modulus and stress–strain curves of polymeric materials, COGNAC provides a couple of methods to apply strain with deformation:

Lees–Edwards Shear deformation can be studied by applying the Lee–Edwards boundary conditions. In addition to steady shear, sinusoidal shear deformation can also be applied.

Deformation rates In this approach, the unit cell is deformed by applying a strain rate tensor. An arbitrary type of flow such as elongation or shear can be specified by applying the strain rate tensor.

Simple elongation The unit cell is elongated or compressed with constant deformation speed. The strain rate is not constant because the cell size will change during the deformation, which is the same as in experimental tensile testing. Uniaxial and biaxial deformation can be applied. During the deformation, the cell size normal to the deformation can be changed by the specified Poisson's ratio or stress control with constant pressure and temperature (NPT) ensembles. In the case of selecting an NPT ensemble, the size of unit cell parallel to the deformation is restricted by the specified deformation speed, and the angles of the unit cell should be restricted by 90°.

Oscillation Similar to simple elongation, the unit cell is deformed uniaxially or biaxially. However, this function can apply sinusoidal deformation by specifying amplitude and frequency.

4.8.2 Bond Formation and Breakage

COGNAC can form and break bonds during simulation to mimic chemical reactions.

A+B→A–B-type Bond Formation This is formation of a simple bond between atoms A and B. Typical reactions are often studied with this model, e.g., termination of radicals, condensation, reaction of epoxy rings, and vulcanization to generate cross-links. The real reaction of an epoxy ring is shown in Fig. 4.4a. This reaction is modeled for molecular dynamics simulation as represented in Fig. 4.4b.

The criterion of reaction is simply the distance between atoms. When the distance becomes shorter than the threshold value, a new bond will be formed between atoms. If an atomistic force field is used, corresponding new angle and torsion potential will be added. Thus, it should be noted that the simulation will become unstable because of the large energy and force from the new angle and torsion potential, which are far from equilibrated values. Another factor to note is that the value of the point charge is not changed after the reaction.

I+M→P_1 or P_n+M→P_{n+1}-type Bond Formation This type of bond formation mimics radical or ion polymerization. The bond will be formed between an initiator or active end and monomer. After the new bond has formed, the monomer will be

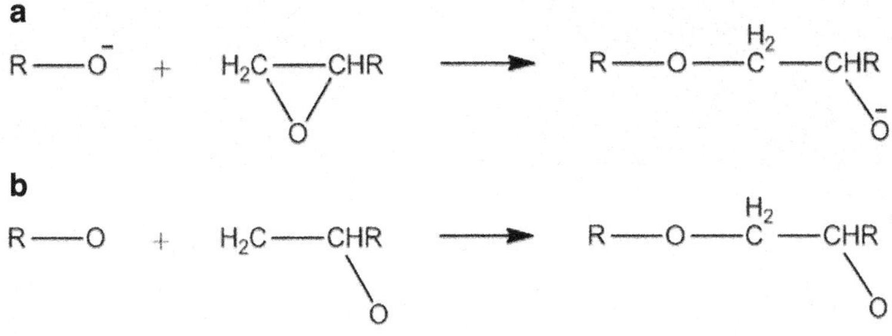

Fig. 4.4 Model reaction of epoxy ring opening, (**a**) real reaction, (**b**) modeled reaction used in COGNAC

the active end and undergo a new reaction with another monomer. The criterion of this type of bond formation is also the distance between atoms.

Bond Breakage An existing bond can be broken when the bond length becomes longer than a threshold value. This can simulate thermal degradation. When a bond is broken, a nonbonding interaction is applied between the atoms at the ends of the broken bond. It should be noted that a large force might be applied between the atoms if the stable distance of the nonbonding interaction is much larger than the bond length.

4.8.3 Zooming

The time scale that can be handled by molecular dynamics simulation is usually insufficient to study the formation of higher-order structures such as phase separation. Thus, it is necessary to prepare an initial structure that is close to an equilibrated structure.

OCTA was developed so that collaborative operations between engines can be performed easily; this process is called "zooming." Following this concept, COGNAC has the ability to use the equilibrated volume fraction of segments obtained by SUSHI and generate initial coordinates by random walk biased by volume fractions (density-biased Monte Carlo algorithm [8]). For details of the algorithm, refer to the COGNAC manual and reference [8].

Using this function, higher-order polymer structures like the microphase-separated structure of block copolymers (Fig. 4.5), distribution of atoms near solid walls, and distribution of grafted chains can be generated effectively.

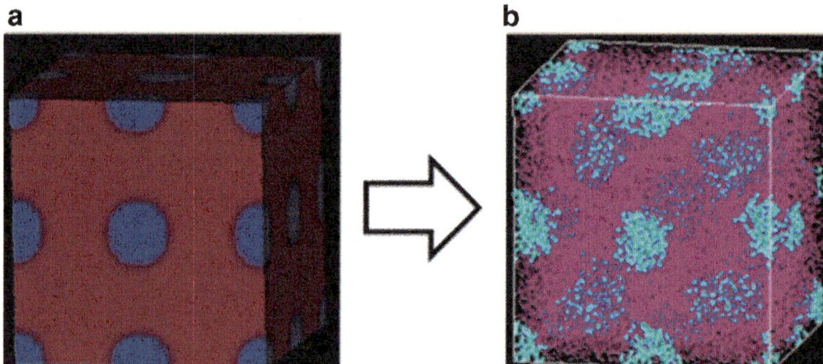

Fig. 4.5 Example of zooming from SUSHI to COGNAC. (**a**) The distribution of segment volume fraction obtained by SUSHI is used to generate (**b**) molecular coordinates that are used by COGNAC

4.8.4 Data Conversion

Input and output data of engines in OCTA are described in UDF and all UDF data can be processed by GOURMET.

In addition to UDF, GOURMET/COGNAC can read and write other data formats as follows:

Import from PDB and MOL format files GOURMET has the ability to import Protein Data Bank (PDB) and MDL mol files. PDB and mol files can be imported via the command *Tool → Molecular Builder* in the Editor Window of GOURMET.

However, it should be noted that when a PDB file is imported, bonds cannot be defined automatically. Thus, CONECT information is necessary to import PDB data with correct bond information.

Export to other formats Output coordinate data of COGNAC can be exported by the Action tool *EXPORT_data*. PBD, UDF (extracting one record from the trajectory data), and some other formats are supported.

Data transfer with LAMMPS COGNAC is parallelized with OpenMP Architecture and can run effectively on a small to medium number of core machines with shared memory. Instead of performing a massive parallel calculation by itself, COGNAC provides the capability to transfer data with large-scale atomic/molecular massively parallel simulator (LAMMPS) [17]. The procedure of transfer is outlined as follows:

Step 1:
Preparation of topological data of molecules (*Set_of_Molecules*) and initial coordinates using COGNAC and SILK.

Step 2:
Export topology and coordinate data in COGNAC in LAMMPS data format using the Action tool *EXPORT_data*.

Step 3:
Execution of LAMMPS. The default command *EXPORT_data* only exports bonding information and coordinates in *Set_of_Molecules* and *Structure* and does not export various simulation conditions. Thus, the user needs to prepare input data of simulation conditions independently.

There is an enhanced version of *EXPORT_data* to LAMMPS in the gifts folder of the COGNAC package. This enhanced function can export part of the input parameters. For further details, see the "Readme" file in the gifts folder.

Step 4:
Import trajectory data and dump file from LAMMPS to a COGNAC UDF file. The COGNAC UDF file must have the same *Set_of_Molelules* data as that in the UDF file used for exporting LAMMPS data.

4.9 Output Information

The results of simulation are output as UDF and other types of files:

UDF File An output UDF file contains a carbon copy of input data and the trajectory of various output data such as coordinates, energy, pressure, density, and other physical properties. This UDF file is used to visualize molecular trajectory and for further analysis using GOURMET.

dat File This file contains physical properties in comma-separated value (CSV) format, which is useful for post processing with other software. The contents of the output can be specified in *Simulation_Conditions.Output_Flags.Statistics*.

In addition to the file output, some physical properties such as energy, pressure, and density are output to standard output for monitoring the simulation.

4.10 Analysis of Results

The results of a simulation can be analyzed by Python script and Action commands on GOURMET. The contents of commands and their usage are described in Chap. 7 of the COGNAC manual. Here, some more details of some Action commands are introduced.

ANALYSIS_1D_profile This command plots profiles of number density (*density*) or velocity (*velocity*) of specified atoms or molecules in the direction of specified axis.

When molecules are specified, the position and velocity are those of the center of mass. When *normalize_flag* is set on, density is normalized by the density at homogeneous distribution. Thus, the number density of atom A at position r is given by

$$\phi_A(r) = \frac{N_A\left(r - 1/2\Delta r; r + 1/2\Delta r\right)/\Delta V}{N_A/V}, \tag{4.3}$$

where N_A is the total number of atom A in the system, V is the volume of the unit cell, and $N_A\,(r - 1/2\Delta r;\ r + 1/2\Delta r)$ is the number of atom A that exist between the positions $r - 1/2\Delta r$ and $r + 1/2\Delta r$. ΔV is the volume of the unit cell between $r - 1/2\Delta r$ and $r + 1/2\Delta r$, and Δr is the length of the unit cell divided by the number of bin (*num_of_bin*).

ANALYSIS_R2_Rg2 This command calculates the square of the end-to-end distance R^2 and radius of gyration Rg^2 of specified molecules. When current is selected in *use_record*, all R^2 and Rg^2 values of specified molecules are output to a

Python log window. When range is selected in *use_record*, the averaged values of R^2 and Rg^2 of specified molecules in each record are output.

ANALYSIS_autocorrelation This command calculates autocorrelation functions of various properties. When a physical property at time t is given by $A(t)$, the autocorrelation function $C(t)$ is defined as

$$C(t) = \frac{\langle A(t)A(0)\rangle}{\langle A(0)A(0)\rangle}. \tag{4.4}$$

$C(t)$ is a normalized function and is 1.0 at $t = 0$.

The Action command of COGNAC can calculate the autocorrelation function of the following physical properties:

Normal coordinates (*Normal_mode*): The normal coordinate defined by Eq. 4.5 corresponds to the normal vibration of molecules:

$$\boldsymbol{X}_p(t) = \frac{1}{N}\sum_{i=1}^{N}\boldsymbol{r}_i(t)\cos\frac{p\pi\,(i-1)}{N-1} - \frac{1}{2N}\left\{\boldsymbol{r}_1(t) + (-1)^p\boldsymbol{r}_N(t)\right\}. \tag{4.5}$$

A larger number of mode p corresponds to higher vibration. Because $p = 0$ indicates translational motion, $p = 1$ represents the slowest and largest motion of molecules itself, corresponding to vibration and deformation of whole molecules. The relaxation of polymer chains can be analyzed by this autocorrelation function of normal coordinates.

Vector: This is an autocorrelation function of the vector between an arbitrary pair of atoms. When it is the vector between both end atoms of a chain, the autocorrelation function becomes similar to that of normal coordinates with $p = 1$.

Atom Velocity/Molecular Velocity: This is the autocorrelation function of the velocity of atoms or center of mass of molecules. The self-diffusion coefficient (D) can be calculated from the function based on the Green–Kubo theory as shown in Eq. 4.6:

$$D = \frac{1}{3}\int_0^{\infty}\langle \boldsymbol{v}_i(t)\boldsymbol{v}_i(0)\rangle dt, \tag{4.6}$$

where $v_i(t)$ is the velocity of atom i or center of mass of molecule i at time t. However, using mean-square displacement (MSD) to obtain the diffusion coefficient is more popular than using atom velocity because the statistical error is small in usual cases.

Force/stress: This is the autocorrelation function of force acting on each atom or stress of the system. Viscosity can be calculated from the autocorrelation function of the stress. However, non-equilibrated dynamics with shear flow are typically used to calculate the shear viscosity of polymeric systems.

ANALYSIS_msd This command calculates the MSD of atoms or the center of mass of molecules. MSD is defined by

$$\mathrm{MSD}\,(\Delta t) = \left\langle (\boldsymbol{r}_i\,(t + \Delta t) - \boldsymbol{r}_i(t))^2 \right\rangle, \tag{4.7}$$

where $\boldsymbol{r}_i(t)$ is the position of atom i at time t. This commands use all pairs of times that have an interval Δt in trajectory. Thus, a longer Δt tends to increase statistical error because the number of samples decreases. It is recommended that the maximum Δt for analysis should be 1/3 to 1/2 of the total simulation time.

The self-diffusion coefficient D is calculated from Einstein's equation:

$$D = \lim_{t \to \infty} \frac{1}{6t} \mathrm{MSD}(t). \tag{4.8}$$

D is often used to study the diffusion of small molecules in polymer matrices.

ANALYSIS_order_parameter This command calculates the second-order orientational order parameter $P2$ defined by

$$P2 = \frac{3\left\langle \cos^2\theta \right\rangle - 1}{2}, \tag{4.9}$$

where θ is an angle between a specified vector and referential axis. $P2$ is 1.0, −0.5, or 0 when the two vectors are parallel, perpendicular, or disordered, respectively.

COGNAC can specify the following vectors to calculate the order parameter:

Principal axis of moments of molecules: This vector corresponds to the anisotropy of molecules. The order of whole molecules can be analyzed using this vector.

Vector between arbitrary atoms: This method is used to study local order in a molecule, such as the order of liquid crystal mesogens.

The referential vector can be selected from the following:

x/y/z axis: The referential axis is taken from the axis in real coordinates. This criterion is used when constant flow and deformation are applied to the system.

Average: The referential axis is defined as an average of specified vectors to calculate the order parameter. When there is no flow or deformation, the direction of order is undefined even though the total molecular system is ordered. In such a case, the order parameter between each vector and averaged vector indicates the order of the system.

ANALYSIS_pair_distribution This command calculates pair distribution functions $g_{ij}(r)$. $g_{ij}(r)$ is the probability of the existence of atom j within distance r from atom i. Numerically, r has a range from $r - 1/2\Delta r$ to $r + 1/2\Delta r$ as shown in Fig. 4.6 and Eq. 4.10:

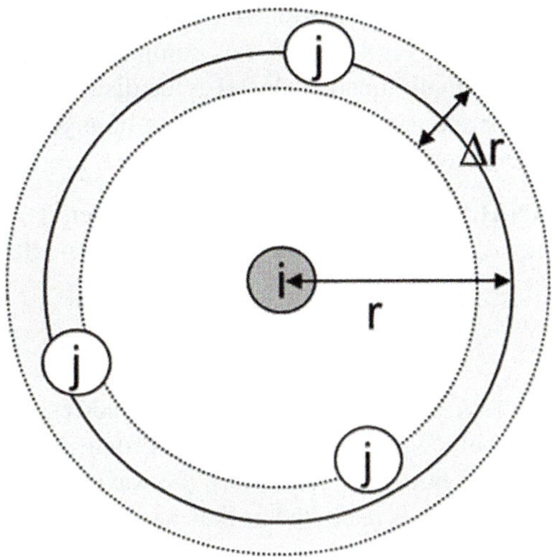

Fig. 4.6 Schematic illustration of a pair distribution function

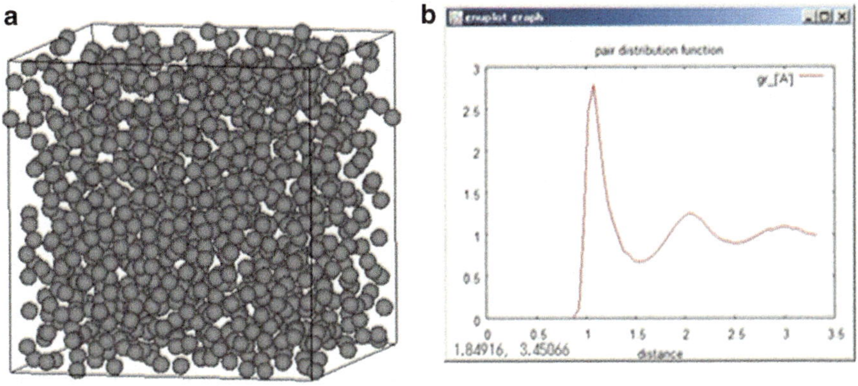

Fig. 4.7 (**a**) Snapshot of the structure of a Lennard–Jones liquid and (**b**) pair distribution function of the snapshot

$$g_{ij}(r) = \frac{V}{N_i N_j} \sum^{N_i} \frac{n_{ij}\left(r - \Delta r/2; r + \Delta r/2\right)}{4\pi r^2 \Delta r}, \tag{4.10}$$

where N_i and N_j are the number of atoms i and j in volume V, respectively, and $n_{ij}(r-\Delta r/2;r+\Delta r/2)$ is the number of atom j in the range of distance $r \pm 1/2\Delta r$ from atom i. Equation 4.10 normalizes the distribution by the homogenous distribution.

This pair distribution function corresponds to the order of atoms and is often used to evaluate the local order of liquids and amorphous materials. Figure 4.7 shows an example of the pair distribution of a Lennard–Jones liquid. Although the pair distribution becomes unity at long range, order over short ranges such as nearest neighbor and second nearest neighbor is observed.

This pair distribution function also indicates higher-order structure like phase separation. Furthermore, this command can calculate not only the averaged distribution in all directions but also the distribution of the coordinates for each axis and sector average of a specified range of angles. This capability is useful to study the orientation of molecular structures.

ANALYSIS_scattering_function When r_i is the position of atom i, the distribution of atoms in a Fourier space can be calculated by Eq. 4.11:

$$\rho(k) = \sum_{i=1}^{N} \exp\left(i\boldsymbol{k} \cdot \boldsymbol{r}_i\right), \tag{4.11}$$

where $\boldsymbol{k}$ is a wave vector and $\rho(k)$ corresponds to the fluctuation of the density of atoms. Normally, because molecular dynamics simulations are conducted with periodic boundary conditions, the wave number corresponding to the periodicity in a real space is restricted as indicated in Eq. 4.12:

$$\boldsymbol{k} = \left(\frac{2\pi}{L}\right)\left(k_x, k_y, k_z\right), \tag{4.12}$$

where k_x, k_y, and k_z can only take integer values. In particular, the lower limit of k is restricted by the cell size L. The structure factor $S(k)$ is calculated from $\rho(k)$:

$$S(k) = \frac{1}{N}\left\langle \rho(k)\rho\left(-k\right)\right\rangle. \tag{4.13}$$

This structure factor corresponds to the intensity obtained by X-ray scattering and is often used to study the regularity and periodicity of a structure. $S(k)$ is essentially the same as the Fourier transformation of a pair distribution function as introduced above.

Actual X-ray scattering is obtained from the distribution of electrons, which is different from that of atoms. This Action command can specify the relative electron density of each atom type so that the simulation results can be quantitatively compared with experimental values.

Because this command uses fast Fourier transformation of field data obtained from the distribution of atoms, the resolution is not fine enough to study atomistic configuration obtained from wide-angle X-ray scattering measurements. Instead, this command is effective to study higher-order structure, which is evaluated by small-angle X-ray scattering.

4.11 Total Flow of COGNAC Execution

Figure 4.8 outlines the process of COGNAC execution. The details of each step are described in this section.

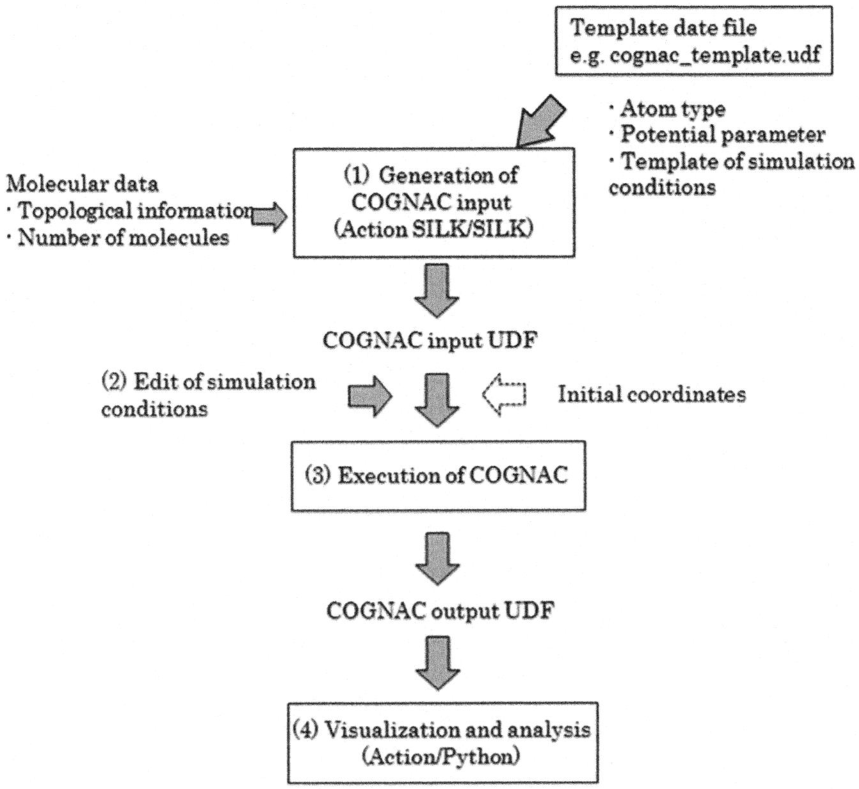

Fig. 4.8 Outline of COGNAC execution

4.11.1 Step 1: Generation of a COGNAC Input UDF File

A COGNAC input UDF file consists of at most the following nine categories:

Simulation_Conditions: General simulation conditions (e.g., temperature, pressure, time step, and ensemble).

Initial_Structure: Selection of the initial structure (e.g., random, helix, crystal, and restart from existing data).

Molecular_Attributes: List of atom types and bonding potentials that will be used in the simulation.

Interactions: List of nonbonding, electrostatic, and external potential interactions that will be used in the simulation.

Set_of_Molecules: Molecular structure (e.g., bonding information, type of potential of each bond, and point charge). This data can be read from different files.

Structure (*optional*): Molecular coordinates (position, velocity, and force acting on an atom) corresponding to the molecular definition in "set of molecules" and unit cell information. This data also can be read from different files.

When the generation of structure is selected in "initial structure," this data is not needed.

React_Conditions (*optional*): Information about bond formation and breakage.
Unit Parameter (*optional*): Information about unit conversion. COGNAC itself does not use this information.
Draw_Attributes (*optional*): Information for drawing the molecular structure using GOURMET. COGNAC itself does not use this information.

Because it is not trivial to create input data from scratch, it is recommended that users refer to the samples in the OCTA system and web pages as a template. Furthermore, COGNAC has a template file "potential_map.udf" in which some template inputs for various coarse-grained models are included.

The "potential_map.udf" is the same format of data as the input of COGNAC, and different models and parameters are included in each record. In addition, each record includes typical simulation conditions and potential parameters without topological and coordinate data for molecules. Molecular information can be added to the templates using SILK and Action SILK.

4.11.2 Step 2: Editing of Simulation Conditions

A COGNAC input file that is created by SILK or Action SILK from the template "potential_map.udf" can be executed using the default conditions. However, the input may be modified for the specific objective of a study.

It is often the case that simulation conditions such as time step, temperature and pressure, ensemble for temperature and pressure control, parameters for the Lennard–Jones potential, and the preparation of initial structure will be modified.

4.11.3 Step 3: Execution of COGNAC

After editing the input UDF file, COGNAC will be executed. There are two methods to execute COGNAC: (1) interactive run from GOURMET and (2) execution in command line.

Interactive execution can be used for demonstration and training. However, it takes communication time and lengthens the throughput time of the simulation. Thus, COGNAC should be executed in command line in the case of serious studies to maximize performance.

UDF files can be transferred between Windows and Linux. Thus, a UDF file that is prepared in GOURMET on Windows can be used in Linux environment.

4.11.4 Step 4: Visualization and Analysis

If COGNAC terminates normally, an output UDF file is created in addition to log and data files, which have been described in Sect. 4.9. Visualization of the trajectory and various analyses can be executed on GOURMET by reading the output UDF file.

The basic output of a molecular dynamics simulation is very simple: trajectory of position, velocity of atoms, force acting on atoms, and physical properties of the system such as temperature, pressure, and potential energy. COGNAC has various functions that allow further analysis of the results, which were partially introduced in Sect. 4.10. Users can also execute their own analysis using Python-based scripts on GOURMET.

4.12 Example: Study of the Molecular Shape and Size of a Polymer

This section describes an example of a study of a simple bead–spring polymer model. The method used to model the molecular topology using SILK and Action SILK, and various analyses such as $\langle R^2 \rangle$ and $\langle Rg^2 \rangle$, pair distribution function, and autocorrelation function are introduced.

4.12.1 Step 1: Generation of COGNAC Input UDF File Using Action SILK

1. *Loading of "potential_map.udf" and selection of record*
   A template file "potential_map.udf" was used to generate a COGNAC input UDF. The "potential_map.udf" is located in the folder C:\OCTA8\ENGINES\COGNAC90\python\silk\sample in the default installation in Windows. The template file "potential_map.udf" has a similar structure to the COGNAC input UDF file, as shown in Fig. 4.9.
   As described in Sect. 4.11, "potential_map.udf" contains different models and simulation conditions in each record, a brief description of which can be seen in the label at the right side of the bottom of the window in Fig. 4.9. Any records labeled as BS*** can be used for Kremer–Grest-type bead–spring models. Here, record 4 "BS_LAMELLA" was used.
2. *Execution of Action SILK*
   Action SILK was executed to create *Set_of_Molecule* data. Action SILK commands will appear as illustrated in Fig. 4.10 by clicking the *Set_of_Molecules* icon with the right button of the mouse. **SILK_CREATE_LinearPolymer_Bead_ Spring_1_Homo . . .** should be selected to define the linear chain of the bead–spring model.

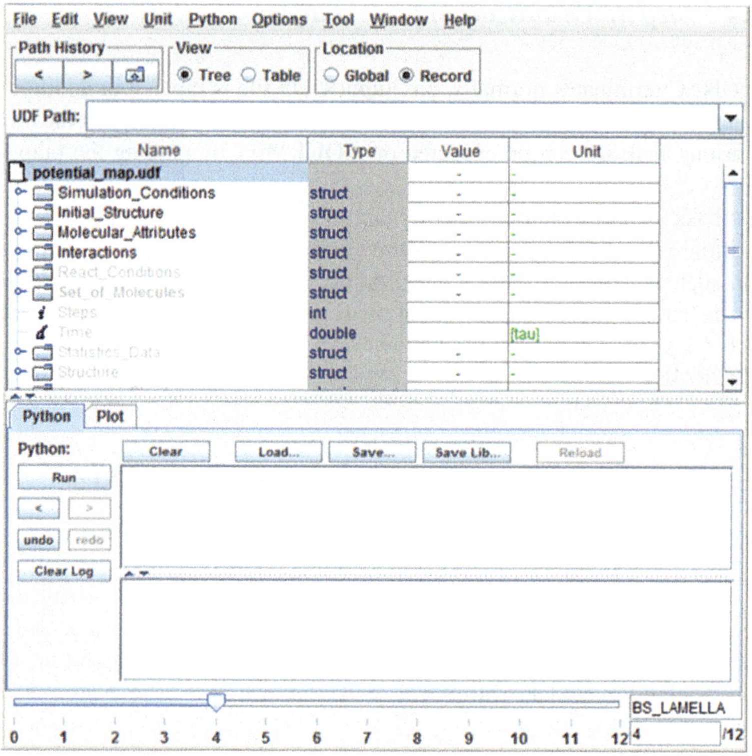

Fig. 4.9 Edit view of "potential_map.udf". Record is moved to No. 4

3. *Specification of the number of chains and number of beads in a chain*
   A new window like that shown in Fig. 4.11 will appear when the command **SILK_CREATE_LinearPolymer_Bead_Spring_1_Homo . . .** is selected. The number of chains is set in *numMol* and the number of beads in a chain (degree of polymerization) is set in *atom_num*. In this example, the polymer consisted of 50 chains with a degree of polymerization of 100. Arbitrary molecular and atom names can be set in *name* and *atom_name*, respectively. However, the types of atoms, bonds, and interaction sites, which are set in *atom_type*, *bond_type*, and *interactionSite_Type*, respectively, should be chosen from the list in *Molecular_Attributes*. Here, "atom1", "bond1", and "siteType1", which are listed in *Molecular_Attributes* in record 4, were set in *atom_type*, *bond_type*, and *interactionSite_Type*, respectively. This window was then closed by selecting **OK**.

   Different types of molecules can be added by repeating this procedure.
4. *Output to Set_of_Molecules*
   Molecular information, which was defined in Sect. 4.11, is not output as UDF data on GOURMET yet. Instead, a new window like that displayed in Fig. 4.12 will appear by again clicking *Set_of_Molecules* with the right button of the

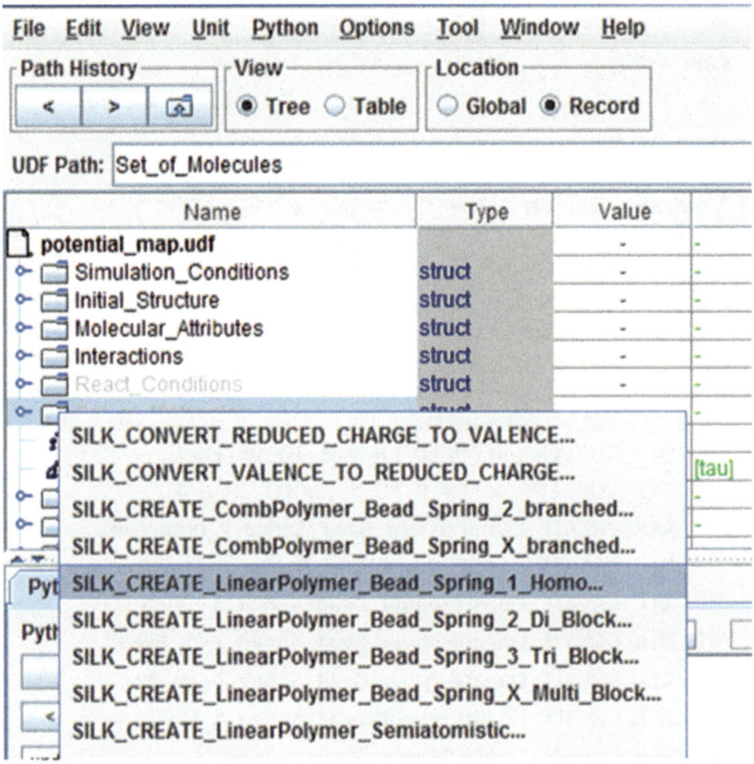

Fig. 4.10 Commands of Action SILK

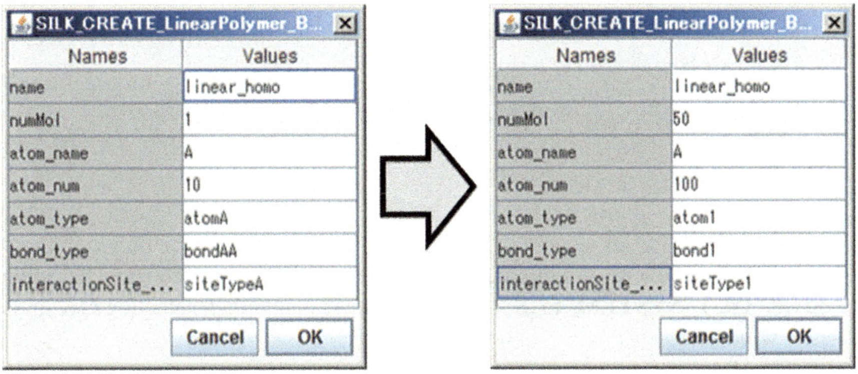

Fig. 4.11 Setup for the number of chains, degree of polymerization, and other inputs

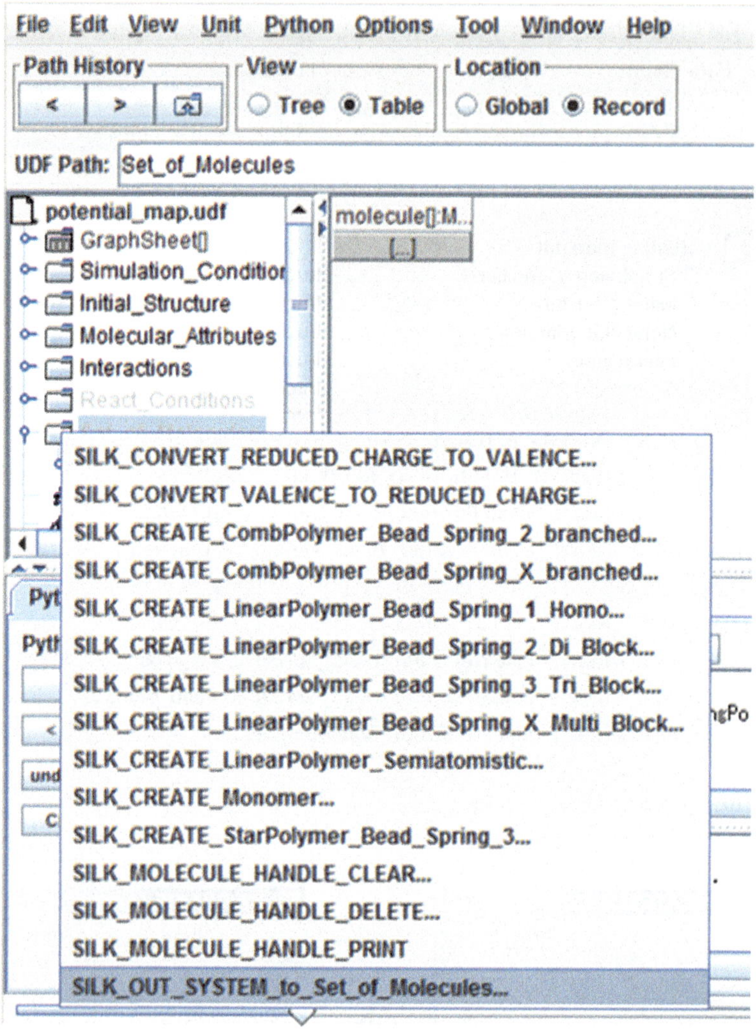

Fig. 4.12 Output to Set_of_Molecules

mouse. Molecular information is output to a UDF file by choosing the menu **SILK_OUT_SYSTEM_to_Set_of_Molecules . . .** and then selecting **OK** in the new window.

The characters in *Set_of_Molecules* turn black when the process is completed.

5. *Output to UDF file*
The defined topological information was output to the record in "potential_map.udf" on GOURMET. This record should be output to an independent UDF file for use as an input of COGNAC.

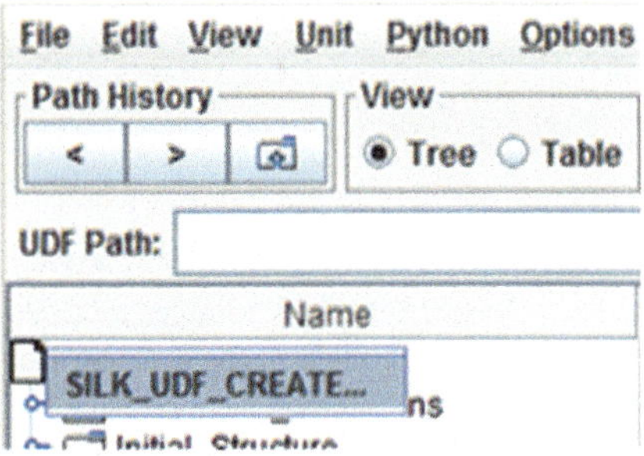

Fig. 4.13 Output to UDF file

When the icon "potential_map.udf" is clicked by the right button of the mouse and **SILK_UDF_Create . . .** is chosen as shown in Fig. 4.13, a new window will appear. A file explorer will appear upon clicking the file path in the new window with the right button of the mouse, and then the arbitrary name of the UDF file is inputted. Selecting *OK* outputs the UDF file. In this example, the UDF file name was "BS_n100m50_tmpin.udf".

4.12.2 Step 1': Generation of a COGNAC Input UDF File Using SILK Script

A simple molecular architecture such as a linear homopolymer and block copolymer can be defined using the menu of Action SILK. However, in the case of modeling a complex architecture such as highly branched chains, the user must edit SILK script for their own purpose.

Here, the use of SILK is introduced by modeling a four-armed star, as shown in Fig. 4.14.

1. *Loading of "potential_map.udf" and selection of record*
   The template file "potential_map.udf" was loaded on GOURMET and record 4 was selected, the same as in Step 1 above.
2. *Loading and editing a SILK sample script*
   Because programming SILK script from scratch is difficult, it is recommended that one of the sample SILK scripts is used as a template.

   By selecting the **Load . . .** button, which is located just above the Python window of GOURMET Editor, the sample script "silk_use.py" in the same folder as "potential_map.udf" (C:\OCTA8\ENGINES\COGNAC90\python\silk\sample in the case of the default installation on Windows) is loaded. Then, the script presented in Fig. 4.15 will appear.

   This SILK script consists of two parts: specifications of the output file and modeling of molecular architecture. Figure 4.16 shows the edited script. Some descriptions have been inserted into the script.

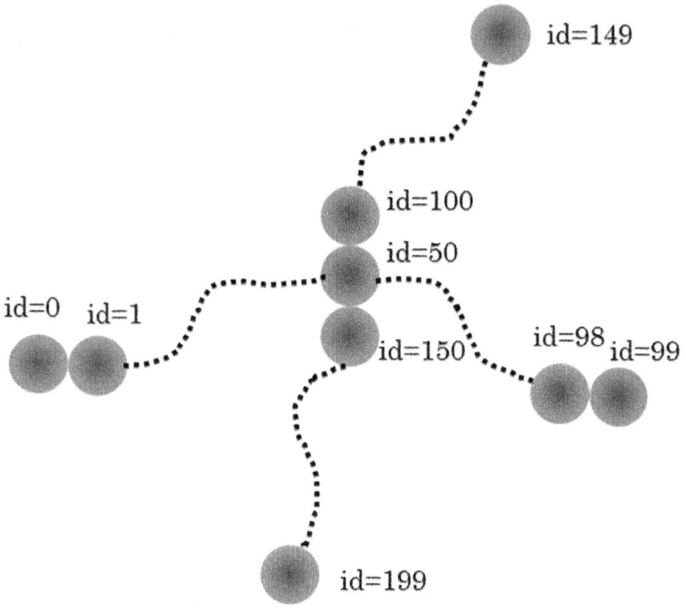

Fig. 4.14 Structure of a four-armed star with atom labels

Defining molecular topology using SILK script looks complex but is very flexible, and arbitrary molecular topology can be modeled using this script.

3. *Output to UDF file*
   UDF data will be output to a specified UDF file when the script is invoked with the *Run* button in the Python window.

4.12.3 Step 2: Editing of Simulation Conditions

In this section, various simulation conditions are edited to suit the objective of a simulation. Typical variables for conditions that are usually edited are as follows:

1. *Simulation_Conditions.Dynamics_Conditions.Time*
   Total steps, step width, and interval steps to be outputted are specified. In this example, *delta_T* was set as 0.012 [τ] following the original paper of Kremer and Grest. *Total_Steps* must be determined by the time scale of observation and was set to 100,000 [τ] here.
2. *Simulation_Conditions.Dynamics_Conditions.Temperature*
   Temperature was set to 1.0 [ε/k_B]. Because the temperature is controlled by a thermal bath with Langevin dynamics, the velocity scale was not executed by setting *Interval_of_Scale_Temp* to zero.

```
############## SECTION "USER DEFINITION" ##############
    ####### SUBSECTION "outputpath" ######
    def setOutParam(self):
    #output Directory (ex. outDir="c:/OCTA8/***"  (dos), outDir="/home/yourdir/***")
            self.engine.outDir="C:/OCTA8/ENGINES/COGNAC90/python/silk/sample"
    #filename without  suffix(.udf)
            self.engine.cognacFileName="test_in"
    #project name
            self.engine.prjName="DEVELOP"
    #output file is diveded to Structure_data and other Parameters when "TWO_FILES" is chosen.
            self.engine.fileNumCom="ONE_FILE"
            #self.engine.fileNumCom="TWO_FILES"
    ####### SUBSECTION "system"  ######
    def userDef(self):
    ##### BuildSytem(Make sure to execute at "record 6") #####
            name="mol"
            numMol=64
            self.engine.createMolecule(name)
            for i in range(0, 4):
                    self.engine.addAtoms(name, "UA", "UA_Kuwajima")
            for i in range(0, 3):
                    self.engine.addBonds(name,  i, i+1, "BOND_PE_Kuwajima")
            for i in range(0, 2):
                    self.engine.addAngles(name, i, i+1, i+2, "ANGLE_PE_Kuwajima")
            for i in range(0, 1):
                    self.engine.addTorsions(name, i, i+1, i+2, i+3, "TORSION_PE_Kuwajima")
            for i in range(0, 4):
                    self.engine.addInteractionSites(name, [i], "NB_PE_Kuwajima", "PAIR")
            self.engine.setSystem(name, numMol)
```

Fig. 4.15 Contents of silk_use.py (partially shown)

3. *Simulation_Conditions.Solver.Dynamics.Dynamics_Algorithm*
   The algorithm of temperature and pressure control is specified here. If temperature is controlled by velocity scaling, the ensemble must be set to NVE, and the interval of scale is set in *Interval_of_Scale_Temp* (normally 10–100 steps).

This example uses the Langevin dynamics (NVT_Kremer_Grest) for temperature control. The friction constant was set as 0.5 following the original paper [1].

4. *Initial_Structure.Initial_Unit_Cell*
Unit cell is specified here. Unit cell size can be specified by the combination of density and length of the unit cell. If the density is set and cell lengths of all axes are set to zero, a cubic cell is specified and the lengths are calculated automatically from the density and total mass. Conversely, if cell lengths of all axes are set, unit cell size is set as the specification without reading density.
In this example, the density was set to 0.85 [m/σ^3], which is usually used in the bead–spring model.
5. *Initial_Structure.Generate_Method*
A method to generate initial coordinates is specified here. This example uses random generation without angle constraint by setting *Initial_Structure.Generate_Method.Random.Fix_Angle* to zero, which corresponds to a false Boolean variable.

a

```
############## SECTION "USER DEFINITION" ##############
  ####### SUBSECTION "outputpath" ######
  def setOutParam(self):
  #output Directory (ex. outDir="c:/OCTA8/***" (dos), outDir="/home/yourdir/***")
        self.engine.outDir="C:/OCTA_Textbook/4/Sample"
```

⇒ **Output directory (arbitral name)**

```
  #filename without suffix(.udf)
        self.engine.cognacFileName="BS_n200_m25_4arm_in"
```

⇒ **File name of COGNAC input UDF (arbitral name)**

```
  #project name
        self.engine.prjName="DEVELOP"
```

⇒ **Project name appeared in hearer part of UDF (arbitral name)**

```
  #output file is diveded to Structure_data and other Parameters when "TWO_FILES" is
chosen.
        self.engine.fileNumCom="ONE_FILE"
        #self.engine.fileNumCom="TWO_FILES"
```

⇒ **Output option. The option "TWO_FILES" output "Set_of_Molecules" as a separated file.**

Fig. 4.16 SILK script to model four-armed star, (**a**) a part of specification of output file, (**b**) modeling of molecular architecture

b

```
####### SUBSECTION "system" ######
  def userDef(self):
  #####BuildSytem(Make sure to execute at "record 6")        #####
        name="star"
⇒ Molecular name (arbitral name)

        numMol=25
⇒ Number of molecules. 25 is set to adjust the number of total atoms to those of Step 1.

        self.engine.createMolecule(name)
        for i in range(0, 200):
              self.engine.addAtoms(name, "A",  "atom1")
⇒Registration of 200 atoms. Molecular name, atom name and atom type are specified
sequentially

        for i in range(0, 99):
              self.engine.addBonds(name,  i, i+1, "bond1")
⇒  Defining bond from atom id 0 to 99. Molecular name, atom ids at both end of bond
and bond type are specified sequentially

        self.engine.addBonds(name, 50, 100, "bond1")
⇒  Defining bond between the center (id =50) and an end of first branch (id=100)

        for i in range(100, 149):
              self.engine.addBonds(name,  i, i+1, "bond1")
⇒  Defining bond of first branch sequentially

        self.engine.addBonds(name, 50, 150, "bond1")
⇒  Defining bond between the center (id =50) and an end of second branch (id=150)

        for i in range(150, 199):
              self.engine.addBonds(name,  i, i+1, "bond1")
⇒  Defining bond of second branch sequentially
```

Fig. 4.16 (continued)

6. *Molecular_Attributes.Bond_Potential*[]
   Bond-stretching potential is set here. Because this example contains only a homopolymer of the bead–spring model, there is only one type of bond potential. FENE+LJ potential was used following the original paper [1].
   If angle bending and torsion potential are implemented in the coarse-grained model, the user must edit *Molecular_Attributes.Angle_Potential*[] and *Molecular_Attributes.Torsion_Potential*[].

7. *Interactions.Pair_Interaction*[]
   Nonbonding interactions are specified here. COGNAC usually uses reduced units, in which the Lennard–Jones parameters *sigma* and *epsilon* are set to 1.0.
   The cutoff distance affects the force acting on each atom. If the cutoff distance is set to $2^{1/6}\sigma$, only repulsive forces act on each atom. This short cutoff gives a reasonable computational time and is accurate enough to study polymer melts of constant volume.
   In contrast, attractive force with long cutoff is needed to study the dynamics of a polymer with volume change by using a constant pressure algorithm and cell deformation.
   Because this example studies molecular structure and dynamics using an NVT ensemble, the cutoff distance was set to $2^{1/6}\sigma$.
   The value of *Molecular_Attributes.Interaction_Site_Type*[].*Range* must be carefully considered if the cutoff distance is changed. This value corresponds to the virtual size of each interaction site, and nonbonding potential is calculated only for the distances smaller than the sum of the range of two sites. COGNAC shows an error message if the range is too short.
   Here, the edited file was saved as "BS_n100m50_in.udf".

4.12.4 Step 3: Execution of COGNAC

COGNAC can be executed by following Sect. 5.1 of the COGNAC manual. It strongly recommended executing OCTA engines including COGNAC with command line interface. Using GourmetTerm or Cygwin is convenient in Windows. Examples of environmental variables for bash on Cygwin are as follows:

```
export PF_FILES=/cygdrive/c/OCTA8/GOURMET
export PF_ENGINE=/cygdrive/c/OCTA8/ENGINES
export UDF_DEF_PATH=/cygdrive/c/OCTA8/ENGINES/udf
export PF_ENGINEARCH=win64

export PATH=$PATH:$PF_ENGINE/bin/$PF_ENGINEARCH
```

These paths of variables depend on the specifications at installation.
An example of the execution of COGNAC is:

```
> cognac90 -IBS_n100m50_in.udf -OBS_n100m50_out.udf >
BS_n100m50_out.log
```

This example will take 10–20 min with a single core of middle-range Intel CPU. Parallel computing with the –n option will shorten the throughput time.

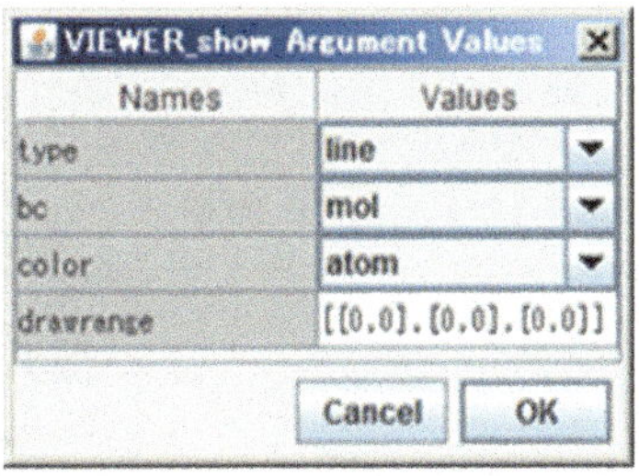

Fig. 4.17 Option of VIEWER_show . . .

4.12.5 *Step 4: Visualization and Analysis of the Simulation Results*

Visualization of Molecular Structure

After the output file of COGNAC "BS_n100m50_out.udf" is loaded on GOURMET, molecular structure can be visualized using the Action tool.

When **VIEWER_show . . .** is selected from the list of commands, which appear by clicking the icon of "BS_n100m50_out.udf" with the right button of the mouse, as shown in Fig. 4.17, each parameter can then be specified.

Here, the default values were chosen by selecting **OK**. Further details of the parameters are provided in the COGNAC manual. Then, a Viewer window will appear as illustrated in Fig. 4.18 (molecule and background colors may be different depending on the default settings.)

Because the default setting of this Viewer is a perspective view, the cubic cell looks distorted. To avoid this view, the user can turn the perspective view off within the menu **Display**.

Randomly generated initial structure is displayed if this procedure is executed on Record 0. More relaxed structure is displayed by sliding the bar on the bottom of the window to the right, as indicated in Fig. 4.19.

Calculation of $<R^2>$ and $<Rg^2>$

This section calculates the averages of the squares of end-to-end distance $<R^2>$ and radius of gyration $<Rg^2>$.

When the Action tool **ANALYSYS_R2_Rg2 . . .** is selected, a new window like that shown in Fig. 4.20 will appear.

If the default value is selected at *molecular_name*, the molecular name of index zero of *Set_of_Molecules* is used for the input. This example has only one type of molecules, so the default value can be used.

When *item* is changed to "both" and **OK** is selected with keeping *use_record* "current", results of the analysis are output to the Python output window as shown in Fig. 4.21.

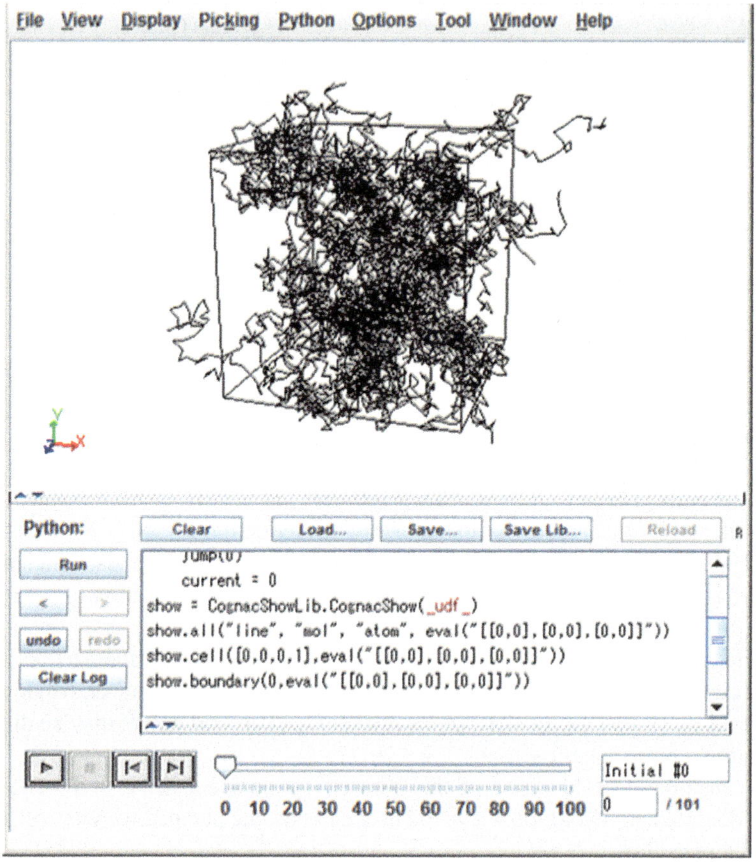

Fig. 4.18 Visualization of molecular structure

The output contains $<R^2>$ and $<Rg^2>$ and their components of the *x*, *y*, and *z* axes of every molecule named "linear_homo". The average $<R^2>$ and $<Rg^2>$ of molecules can be calculated using this output.

When *use_record* is changed to "range", the analysis is conducted with the range of records from *start_record* to *end_record*. Only averages of molecules in each record are plotted. Figure 4.22 shows $<R^2>$ and its components as a function of time for the current example.

Figure 4.23 presents $<R^2>$ and its components of the four-armed star that was modeled in Step 1'. Compared with a linear polymer, $<R^2>$ does not change very much when the degree of polymerization doubles. This result indicates that $<R^2>$ is determined by the degree of polymerization between the ends of a chain and not by the total degree of polymerization of a star polymer.

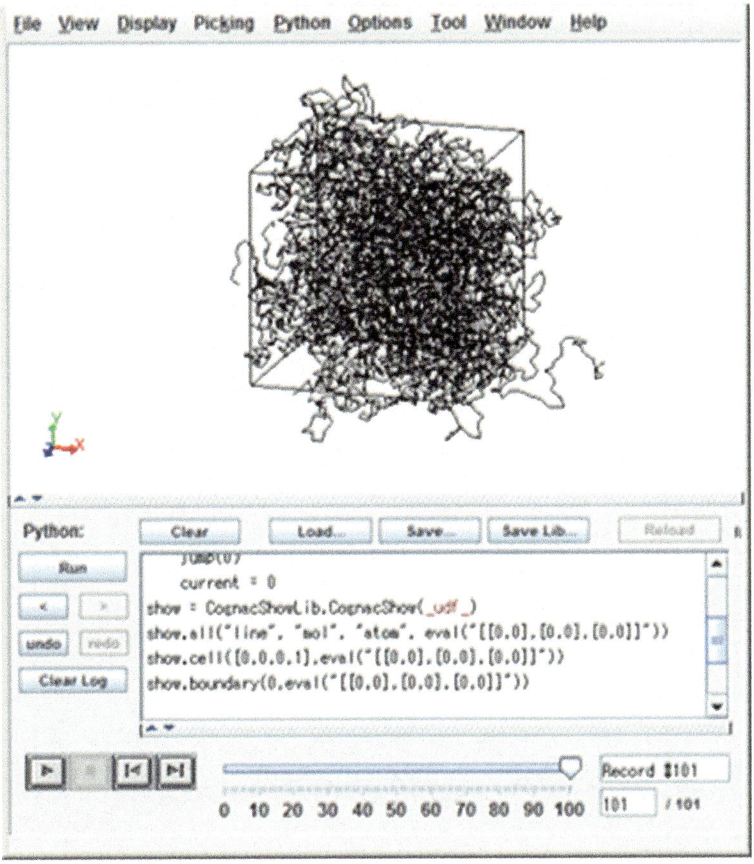

Fig. 4.19 Snapshot structure for the final record (perspective is turned off)

Fig. 4.20 Parameters for ANALYSIS_R2_Rg2

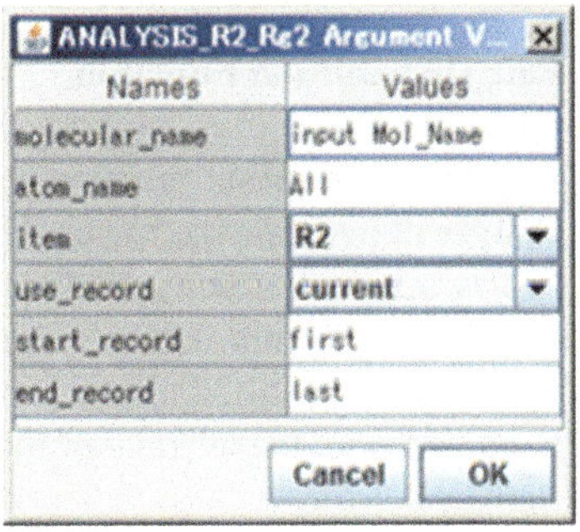

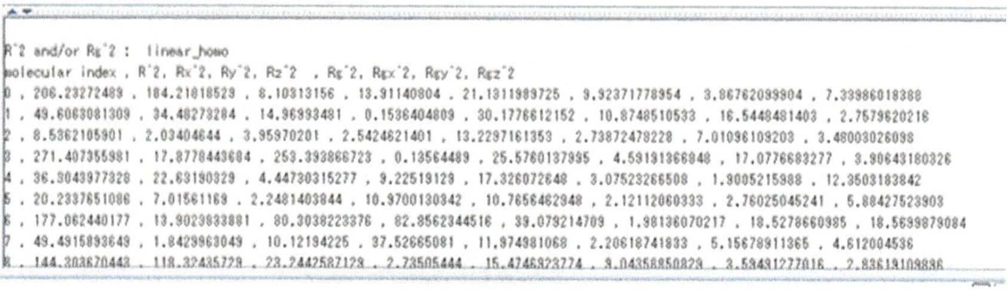

```
R^2 and/or Rg^2 :  linear_homo
molecular index , R^2, Rx^2, Ry^2, Rz^2  , Rg^2, Rgx^2, Rgy^2, Rgz^2
0 , 206.23272489 , 184.21818529 , 8.10313156 , 13.91140804 , 21.1311989725 , 9.92371778954 , 3.86762099904 , 7.33986018388
1 , 49.6063081309 , 34.48273284 , 14.96993481 , 0.1536404809 , 30.1776612152 , 10.8748510533 , 16.5448481403 , 2.7579620216
2 , 8.5362105901 , 2.03404644 , 3.95970201 , 2.5424621401 , 13.2297161353 , 2.73872478228 , 7.01096109203 , 3.48003026098
3 , 271.407355981 , 17.8778443684 , 253.393866723 , 0.13564489 , 25.5760137995 , 4.59191366848 , 17.0776683277 , 3.90643180326
4 , 36.3043977328 , 22.63190329 , 4.44730315277 , 9.22519129 , 17.326072648 , 3.07523266508 , 1.9005215988 , 12.3503183842
5 , 20.2337651086 , 7.01561169 , 2.2481403844 , 10.9700130342 , 10.7656462948 , 2.12112060333 , 2.76025045241 , 5.88427523903
6 , 177.062440177 , 13.9023933881 , 80.3038223376 , 82.8562344516 , 39.079214709 , 1.98136070217 , 18.5278660985 , 18.5699879084
7 , 49.4915893649 , 1.8429963049 , 10.12194225 , 37.52665081 , 11.974981068 , 2.20618741833 , 5.15678911365 , 4.612004536
8 , 144.203670443 , 118.32435729 , 23.2442587129 , 2.73505444 , 15.4746923774 , 3.04358850829 , 3.59491277016 , 2.83619109886
```

Fig. 4.21 Output of ANALYSIS_R2_Rg2

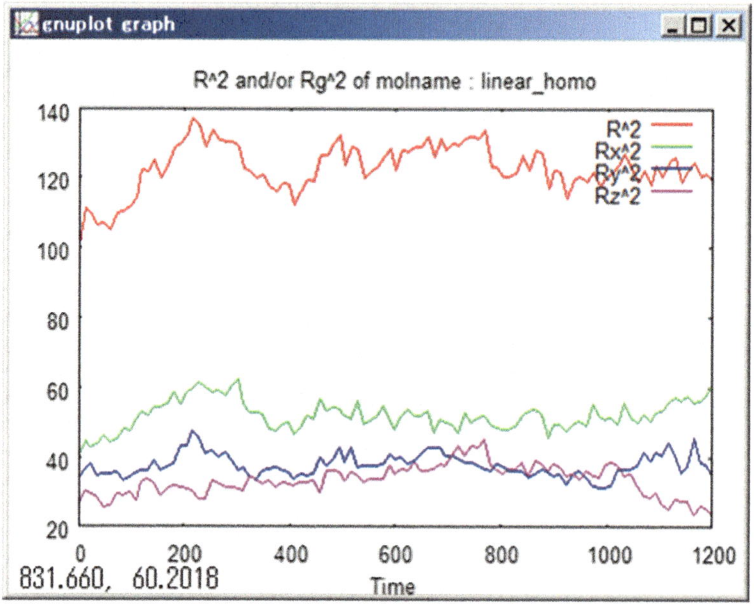

Fig. 4.22 $<R^2>$ and its components as a function of time

Pair Distribution Function

The pair distribution function analyzes the order of beads, while $<R^2>$ and $<Rg^2>$ analyze molecular size. When the Action command **ANALYSIS_pair_distribution . . .** is selected, a new window will appear as shown in Fig. 4.24. This example contains only atom name "A"; *name_list* should be modified as illustrated on the right side of Fig. 4.24.

In addition, the average of the record from No. 91 to 101 is calculated to reduce statistical error.

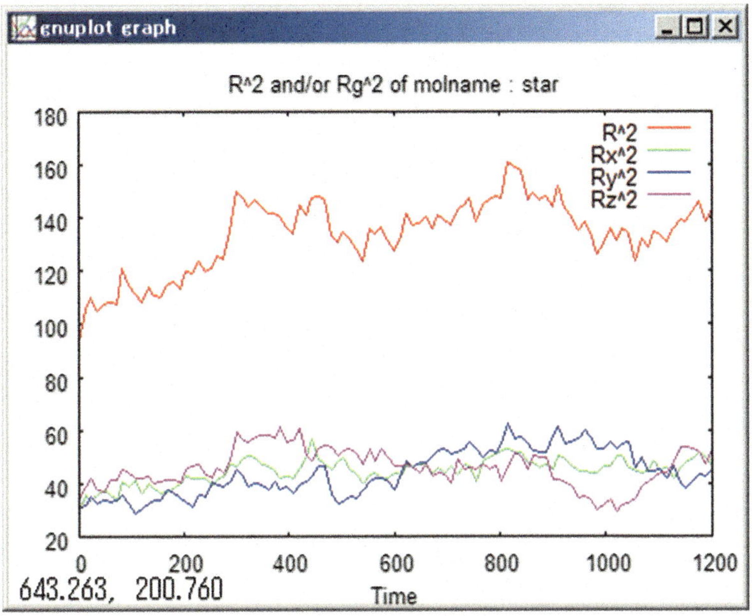

Fig. 4.23 $<R^2>$ and its components of a four-armed star

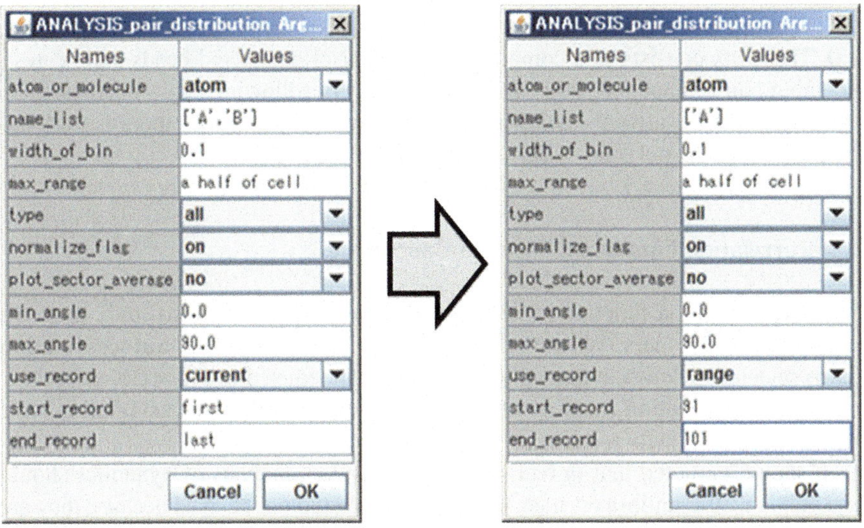

Fig. 4.24 Setup of ANALYSIS_pair_distribution

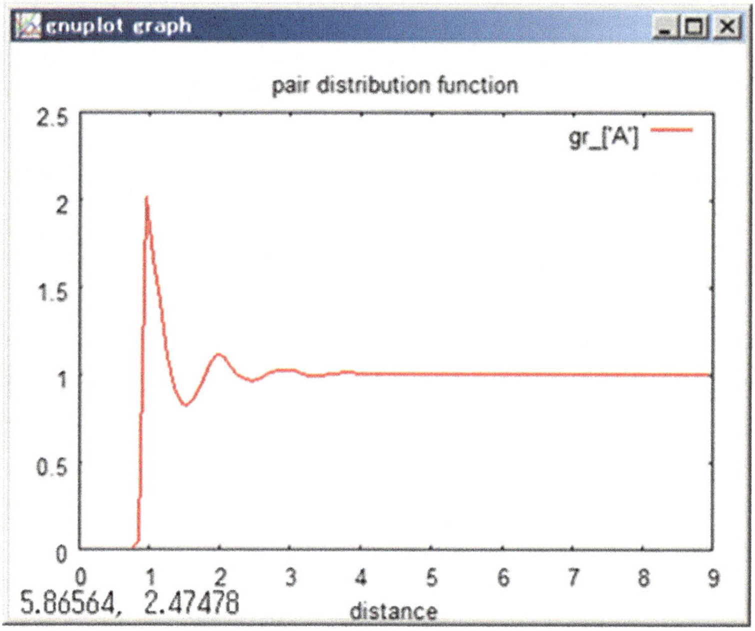

Fig. 4.25 Plot of the results of ANALYSIS_pair_distribution

The result is displayed in Fig. 4.25. The highest peak appears around the distance 1.0. This peak consists of a combination of a pair of connected beads and nearest-neighbor pair of unconnected beads. The intensity of following peaks decreases and almost disappears after the third-nearest peak, which means that there is no long-range order in the polymer system.

Autocorrelation Function of Normal Coordinates

The autocorrelation function of normal coordinates is calculated to study dynamic properties of polymers. When the Action command **ANALYSIS_autocorrelation ...** is selected, a new window will appear like that depicted in Fig. 4.26.

The autocorrelation function of first-order normal coordinates using all records will be calculated with the default parameters.

It should be noted that in real studies, this type of analysis of dynamics should use only a well-equilibrated state. Early records should not be used because they are not equilibrated enough.

Figure 4.27 shows the results of the autocorrelation function of the normal coordinates with order $p = 1$, which corresponds to the largest and slowest dynamics.

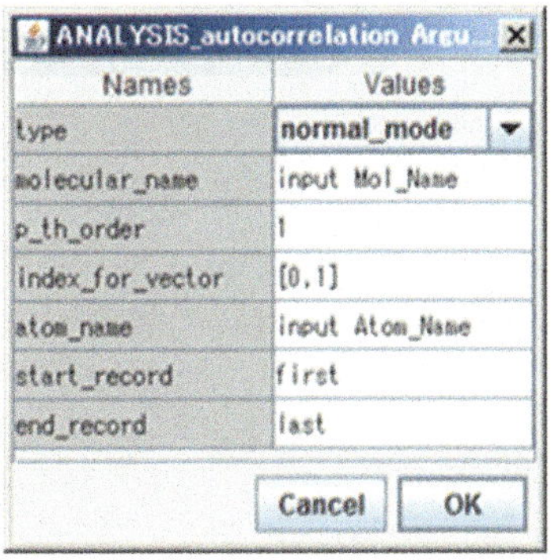

Fig. 4.26 Parameters of ANALYSIS_autocorrelation

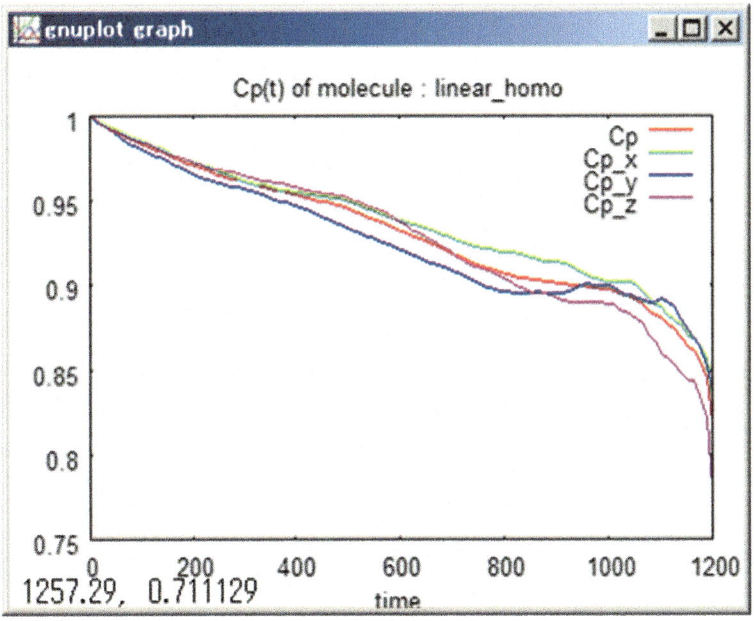

Fig. 4.27 Plot of the results of ANALYSIS_autocorrelation

The autocorrelation function decreases from only 1.0–0.9 at time 1000τ. Thus, much longer time steps than thosc uscd hcrc arc nccdcd to study thc rclaxation of this system.

Figure 4.28 displays the results of a longer run to 12,000τ as a semilog plot. The autocorrelation function decreases to about 0.5 at 10,000τ, which means the relaxation time of the $N = 100$ chains in this example is longer than 10,000τ.

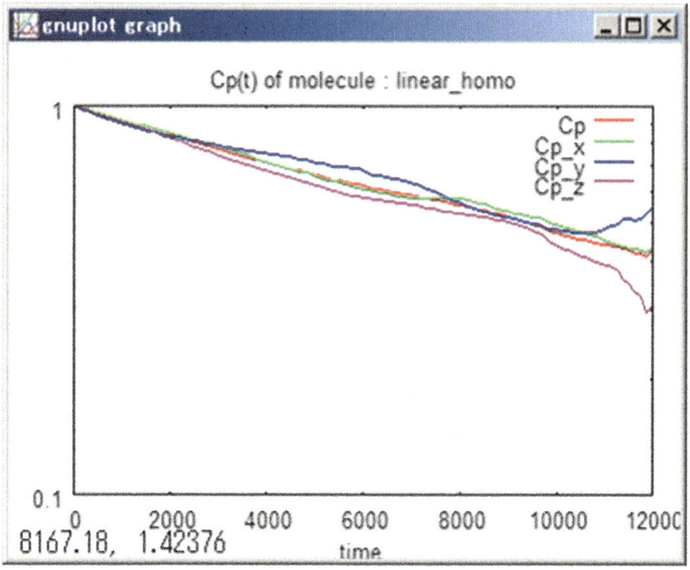

Fig. 4.28 Semilog plot of the results of ANALYSIS_autocorrelation (time = 0–12,000τ)

4.13 Download Instruction

Sample files are available from the OCTA website. See the instruction on http://octa.jp/download_bookdata.html. Password for the sample(zip file) of this chapter is "XYNzcXLF".

References

1. G.S. Grest, K. Kremer, Phys. Rev. A **33**, 3628 (1986)
2. R.D. Groot, P.B. Warren, J. Chem. Phys. **107**, 4423 (1997)
3. http://ambermd.org/
4. S.L. Mayo, B.D. Olafson, W.A. Goddard, J. Phys. Chem. **94**, 8897 (1990)
5. D. Bedrov, G.D. Smith, J.S. Smith, J. Chem. Phys. **119**, 10438 (2003)
6. J.G. Gay, B.J. Berne, J. Chem. Phys. **74**, 3316 (1981)
7. G.S. Grest, J. Chem. Phys. **105**, 5532 (1996)
8. T. Aoyagi, F. Sawa, T. Shoji, H. Fukunaga, J. Takimoto, M. Doi, Comput. Phys. Commun. **145**, 267 (2002)
9. M.P. Allen, D.J. Tildesley, *Computer Simulation of Liquids* (Clarendon Press, Oxford, 1987)
10. J.V.L. Beckers, C.P. Lowe, S.W.D. Leeuw, Mol. Simul. **20**, 369 (1998)
11. H.J.C. Berendsen, J.P.M. Postma, W.F. van Gunsteren, A. Dinola, J.R. Haak, J. Chem. Phys. **81**, 3684 (1984)
12. W.G. Hoover, Phys. Rev. A **31**, 1695 (1985)

13. D. Brown, J.H.R. Clarke, Macromolecules **24**, 2075 (1991)
14. H.C. Andersen, J. Chem. Phys. **72**, 2384 (1980)
15. M. Parrinello, A. Rahman, J. Appl. Phys. **52**, 7182 (1981)
16. A.W. Lees, S.F. Edwards, J. Phys. C Solid State Phys. **5**, 1921 (1972)
17. http://lammps.sandia.gov/

Chapter 5
SUSHI: Density Functional Theory Simulator

Takashi Honda

5.1 Introduction

In the field of multi-scale simulations of polymeric materials, theory is needed to bridge the particle picture treated with coarse-grained molecular dynamics (as in the case of OCTA/COGNAC) and the continuous-field picture treated with the finite element method and other pictures (as in the case of OCTA/MUFFIN). The theory must allow the reduction of the many-body problem to the one-body approximation. OCTA includes a simulator named Simulation Utilities for Soft and Hard Interfaces (SUSHI), which solve the bridging problem. SUSHI has implemented several types of density functional theory (DFT) of polymers. These DFTs address problems relating to interfacial structures and phase-separated structures of polymer melts and blends at scales ranging from several nanometers to several hundred nanometers, e.g., the macrophase separation of polymer blends and the microphase separation of block polymers.

This section introduces the basics of DFT and several free energy models for the DFT, namely, phenomenological density functional theories, theories that introduce polymer conformation by linear approximation employing the random phase approximation (RPA) and self-consistent field (SCF) theory that introduces polymer conformation accurately.

T. Honda (✉)
Foundation Technology Laboratory, Research & Development Center, Zeon Corporation, 1-2-1 Yako, Kawasaki-ku, Kawasaki, Kanagawa, 210-9507, Japan
e-mail: T5.Honda@zeon.co.jp

Japan Association for Chemical Innovation, *Computer Simulation of Polymeric Materials*, DOI 10.1007/978-981-10-0815-3_5

5.2 Overview of DFT for Polymer Blends

In this section, we define DFT for polymer blends where the system is composed of polymeric materials and solvents, and these components are made of segments. A segment is a coarse-grained object of repeating chemical units or solvents, and, for a polymer, the Gaussian statistics of the distribution of the segments are satisfied when the number of repeating units in the segment exceeds a critical threshold, where exclude volume effects in segment–segment interactions are ignored. We recognize that a system is described by continuum fields of segment densities under the incompressibility condition because the system is a melt or solution. In this case, the free energy of the system can be written according to the functional of densities of the components as the Helmholtz energy

$$\mathscr{F}_{\text{Helmholtz}} = \mathscr{F}[\{\phi_i(\mathbf{r})\}], \tag{5.1}$$

where $\phi_i(\mathbf{r})$ is the segment density of the i-th component, and the bracket represents a set of field variables.

We can search for a state with low free energy for the evaluation of $\{\phi_i(\mathbf{r})\}$. The state becomes the phase-separated structure. The evaluation refers to the optimization of the phase-separated structure. If we can directly minimize the free energy by optimizing $\{\phi_i(\mathbf{r})\}$, we get the equilibrated structure; such a method is described in a later section as the static method of SCF theory. Meanwhile, the thermodynamics of the free energy model give the chemical potential of the i-th component in $\{\phi_i(\mathbf{r})\}$ according to the functional derivative equation

$$\mu_i(\mathbf{r}) = \frac{\delta \mathscr{F}[\{\phi_i(\mathbf{r})\}]}{\delta \phi_i(\mathbf{r})}. \tag{5.2}$$

Using the chemical potential $\mu_i(\mathbf{r})$, we get the time evolution equation of $\{\phi_i(\mathbf{r})\}$ as

$$\frac{\partial}{\partial t}\phi_i(\mathbf{r},t) = \nabla \cdot [L_i(\mathbf{r},t)\nabla \mu_i(\mathbf{r},t)], \tag{5.3}$$

where t is time, $L_i(\mathbf{r},t)$ expresses the mobility of the i-th component, and the effects of the incompressibility condition and thermal fluctuation can be introduced in calculations by SUSHI but we neglect these description for simplicity [1–3].

We refer to this time evolution method as "dynamic DFT." Several free energy models of polymer melts have been proposed. In the following sections, typical free energy models are introduced.

5.3 Flory–Huggins Free Energy Model

The most basic free energy model of polymer mixtures is the Flory–Huggins free energy model [4, 5], which is derived using the lattice model on which segments of polymer are placed. This theory is applied to the system of homopolymer blends. The general expression of the Flory–Huggins free energy model is

$$\mathscr{F}[\{\phi_i(\mathbf{r})\}] = \frac{1}{\beta}\sum_i \int d\mathbf{r}\frac{\phi_i(\mathbf{r})}{N_i}\ln\phi_i(\mathbf{r}) + \frac{1}{\beta}\sum_i^n\sum_{j>i}^n \int d\mathbf{r}\chi_{ij}\phi_i(\mathbf{r})\phi_j(\mathbf{r}), \quad (5.4)$$

where $\beta = 1/k_BT$, $\phi_i(\mathbf{r})$ is the segment density of the i-th chain, N_i is the length of the i-th chain, and χ_{ij} is an interaction parameter of adjacent segments of i-th and j-th chains defined by

$$\chi_{ij} \equiv z\beta\left\{\epsilon_{ij} - \frac{1}{2}(\epsilon_{ii} + \epsilon_{jj})\right\}, \quad (5.5)$$

where z is the coordinate number of segments around a segment and ϵ_{ij} is the energy of interaction between segments of i-th and j-th chains.

This free energy model is appropriate for the comparison of stability between two uniform states of polymer blends, of which the segment densities of the polymers are different. This theory has thus been used to predict phase diagrams of homopolymer blends [6].

5.4 Phenomenological Flory–Huggins Free Energy Model

The Flory–Huggins free energy model, however, cannot be used directly for dynamic DFT because the model is derived to obtain the free energy model of a uniform state and ignores the effect of interfaces. This disadvantage is overcome by adding a term for interfacial energies, which is the so-called square gradient term. This term is derived from employing the phase field method [7] as follows.

We focus on the interface of a phase-separated structure, where energy is accumulated through segment–segment interactions and a decrease in the conformational entropy of polymers. We therefore assume that the interface energy at position $\mathbf{r}$ can be expressed using the phenomenological function given by

$$\mathscr{I}_i(\mathbf{r}) = f(\phi_i(\mathbf{r}), \nabla\phi_i(\mathbf{r}), \nabla^2\phi_i(\mathbf{r})) \quad (5.6)$$

$$= f(\phi_i(\mathbf{r}), 0, 0) + K_0\nabla\phi_i(\mathbf{r}) + K_1|\nabla\phi_i(\mathbf{r})|^2 + K_2\nabla^2\phi_i(\mathbf{r}) + \ldots, \quad (5.7)$$

where parameters $\nabla\phi_i(\mathbf{r})$ and $\nabla^2\phi_i(\mathbf{r})$ correspond to the effects of the segment–segment interaction and the curvature of the interface, respectively. From (5.6) to (5.7), the equation is expanded for each of the parameters, where "...." indicates terms of higher order, and K_0, K_1 and K_2 are functions of $\phi_i(\mathbf{r})$. The overall accumulated energy in a local region is obtained by integrating $\mathscr{I}_i(\mathbf{r})$ over volume as

$$\mathscr{I}_i = \int dVf(\phi_i(\mathbf{r}), \nabla\phi_i(\mathbf{r}), \nabla^2\phi_i(\mathbf{r})) \tag{5.8}$$

$$\simeq C + \int dV\{K_1|\nabla\phi_i(\mathbf{r})|^2 + K_2\nabla^2\phi_i(\mathbf{r})\}, \tag{5.9}$$

where C is the constant term and the K_0 term with $\phi_i(\mathbf{r})$ on the right-hand side of (5.9) is neglected according to the isotropy of phase separation and terms of higher order are neglected. To apply the divergence theorem of Gauss to the last term in (5.9), its expression is rewritten as

$$\int dVK_2\nabla^2\phi_i(\mathbf{r}) = \int dSK_2\nabla\phi_i(\mathbf{r}) \cdot \mathbf{n} - \int dV\nabla K_2 \cdot \nabla\phi_i(\mathbf{r}) \tag{5.10}$$

$$= -\int dV\frac{\partial K_2}{\partial\phi_i(\mathbf{r})}|\nabla\phi_i(\mathbf{r})|^2, \tag{5.11}$$

where the first term in (5.10) is the surface integration, which is neglected according to the isotropy of phase separation in (5.11). By replacing $K_2\nabla^2\phi_i(\mathbf{r})$ in (5.9) with (5.11) and subtracting the constant term C from (5.9), we get the phenomenological interface energy expressed as

$$\Delta\mathscr{I}_i \simeq \int dV\left\{K_1 - \frac{\partial K_2}{\partial\phi_i(\mathbf{r})}\right\}|\nabla\phi_i(\mathbf{r})|^2. \tag{5.12}$$

If we assume that $\{K_1 - \partial K_2/\partial\phi_i(\mathbf{r})\}$ is a constant parameter, we get the local interface energy in square-gradient form:

$$\Delta\mathscr{I}_i(\mathbf{r}) = \frac{\kappa_i}{2}|\nabla\phi_i(\mathbf{r})|^2, \tag{5.13}$$

where κ_i is the constant parameter of the interfacial energy for $\phi_i(\mathbf{r})$. The generalized Flory–Huggins equation (5.4) can therefore be modified to add this interface energy as

$$\mathscr{F}[\{\phi_i(\mathbf{r})\}] = \frac{1}{\beta}\sum_i\int d\mathbf{r}\frac{\phi_i(\mathbf{r})}{N_i}\ln\phi_i(\mathbf{r}) + \frac{1}{\beta}\sum_i^n\sum_{j>i}^n\int d\mathbf{r}\chi_{ij}\phi_i(\mathbf{r})\phi_j(\mathbf{r})$$
$$+\frac{1}{2\beta}\int d\mathbf{r}\sum_i\kappa_i|\nabla\phi_i(\mathbf{r})|^2. \tag{5.14}$$

Equation (5.14) is a conventional free energy model based on the Flory–Huggins free energy model.

5.5 Gaussian Chain Model

κ_i in the Flory–Huggins free energy model equation (5.14) is obtained with theory based on the RPA. We will explain the RPA for polymer blends briefly. For this purpose, this section introduces the Gaussian chain model, which is a basic concept of the theory of polymer physics.

A Gaussian chain is also called an ideal chain, which means that the segment–segment interactions are negligible. Modeling a polymer chain on a coarse-grained scale with a Gaussian chain is called the Gaussian approximation [3]. To explain the nature of a Gaussian chain and following the RPA and the SCF theory, we consider a system consisting of subchains that are parts of the polymer architecture; i.e., we recognize that homopolymers and blocks of block polymers are subchains.

We focus on a single subchain. As shown in Fig. 5.1, it is convenient to introduce a lattice model of the polymer chain where the lattice constant corresponds to the effective bond length b of the ideal chain.

Let us calculate the statistical weight of a subchain whose end segments are specified by indices i' and i ($i' < i$) and are fixed at sites $\mathbf{r}'$ and $\mathbf{r}$, respectively. We denote the statistical weight of this subchain by $Q(i', \mathbf{r}'; i, \mathbf{r})$. Owing to the chain connectivity, the statistical weight $Q(i', \mathbf{r}'; i+1, \mathbf{r})$ is related to $Q(i', \mathbf{r}'; i, \mathbf{r})$ through the recurrence formula

$$Q(i', \mathbf{r}'; i+1, \mathbf{r}) = \frac{1}{z} \sum_{\mathbf{r}''} Q(i', \mathbf{r}'; i, \mathbf{r}''), \tag{5.15}$$

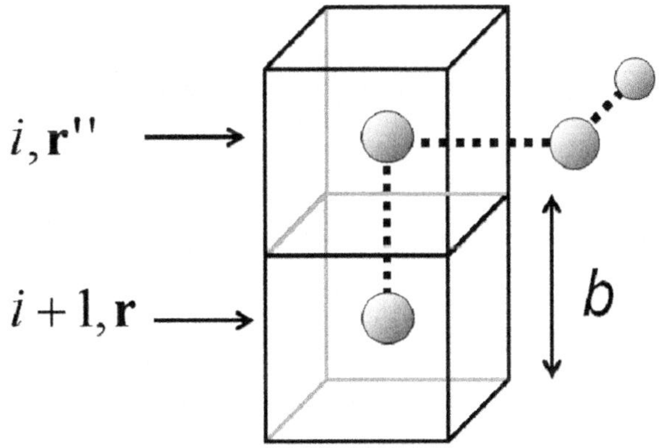

Fig. 5.1 Possible positions of the nearest-neighbor segment in a lattice model of a polymer chain

where z is the number of nearest-neighbor sites ($z = 6$ in the case shown in Fig. 5.1) and $\mathbf{r}''$ is one of the possible positions of the nearest-neighbor sites of $\mathbf{r}$.

From the discretized expression of the recurrence formula, Eq. (5.15), we derive a continuum formula. Subtracting $Q(i', \mathbf{r}'; i, \mathbf{r})$ from both sides of Eq. (5.15) leads to

$$Q(i', \mathbf{r}'; i+1, \mathbf{r}) - Q(i', \mathbf{r}'; i, \mathbf{r}) = \frac{1}{z} \sum_{\mathbf{r}''} Q(i', \mathbf{r}'; i, \mathbf{r}'') - Q(i', \mathbf{r}'; i, \mathbf{r}). \quad (5.16)$$

In the continuum limit, the left-hand side becomes the derivative of $Q(i', \mathbf{r}'; i, \mathbf{r})$ with respect to the index i and the right-hand side becomes the Laplacian of $Q(i', \mathbf{r}'; i, \mathbf{r})$. Thus, Eq. (5.16) reduces to the three-dimensional ($z = 6$) continuum equation

$$\frac{\partial}{\partial s} Q(s', \mathbf{r}'; s, \mathbf{r}) = \frac{b^2}{6} \nabla^2 Q(s', \mathbf{r}'; s, \mathbf{r}), \quad (5.17)$$

where s is used as a continuous parameter instead of the discrete index i and the b^2 factor appears when the finite difference is replaced by the Laplacian.

The solution to (5.17) is given by [8]

$$Q(s', \mathbf{r}'; s, \mathbf{r}) = \left(\frac{3}{2\pi |s - s'| b^2}\right)^{3/2} \exp\left(-\frac{3|\mathbf{r} - \mathbf{r}'|^2}{2|s - s'|b^2}\right). \quad (5.18)$$

As this statistical weight distribution is Gaussian, the ideal chain obeys Gaussian statistics.

The Fourier transformation of the distribution of the end-to-end vector of a subchain is given by

$$\begin{aligned} \int_{-\infty}^{-\infty} d\mathbf{r} \exp\,(i\mathbf{q} \cdot \mathbf{r}) \left(\frac{3}{2\pi |s - s'| b^2}\right)^{3/2} \exp\left(-\frac{3\mathbf{r}^2}{2|s - s'|b^2}\right) \\ = \exp\,\left(-\frac{b^2}{6}|s - s'||\mathbf{q}|^2\right). \end{aligned} \quad (5.19)$$

This expression gives the intensity of scattering from a pair of segments s and s'. Thus, the function of total scattering from the whole chain (i.e., the structure factor for a Gaussian chain) is given by

$$\begin{aligned} S^0(\mathbf{q}) &= \frac{1}{N} \int_0^N ds' \int_0^N ds \,\exp\left(-\frac{b^2}{6}|s - s'||\mathbf{q}|^2\right) \\ &= \frac{2N}{x^2}(e^{-x} - 1 + x), \end{aligned} \quad (5.20)$$

where $x \equiv (1/6)Nb^2|\mathbf{q}|^2 = R_G^2|\mathbf{q}|^2$. The function $2(e^{-x} - 1 + x)/x^2$ in (5.20) is called the Debye function, which is used in the RPA explained in the following section [5, 9].

5.6 Linear SCF Theory with RPA

If we get the relation between an external field imposed to a segment and a fluctuation of the segment density, we can extract useful information of polymer blends. Obtaining the relation is a problem of linear response theory, and the RPA plays an important role in solving this problem for polymer blends [9–11].

To obtain the solution, we prepare several vectors and matrices in Fourier space. We consider a polymer melt system consisting of several subchain segment density distributions $\phi_i(\mathbf{r})$ and fields $u_i(\mathbf{r})$ conjugate to each $\phi_i(\mathbf{r})$. The fluctuation of $\phi_i(\mathbf{r})$ is defined by

$$\delta\phi_i(\mathbf{r}) \equiv \phi_i(\mathbf{r}) - \bar{\phi}_i, \tag{5.21}$$

where $\bar{\phi}_i$ is the average density of the i-th subchain. We define vectors $u_i(\mathbf{q})$ and $d_i(\mathbf{q})$ in Fourier space as

$$\mathbf{u} \equiv \{u_i(\mathbf{q})\}, \tag{5.22}$$

$$\mathbf{d} \equiv \{\delta\phi_i(\mathbf{q})\}. \tag{5.23}$$

Next, we define the segment–segment energy matrix and the "intra" chain scattering function matrix as

$$\mathbf{C} \equiv \{z\epsilon_{ij}\}, \tag{5.24}$$

$$\mathbf{S}^0 \equiv \{S^0_{ij}(\mathbf{q})\}. \tag{5.25}$$

The formula for $S^0_{ij}(\mathbf{q})$ is explained later. We then prepare the pressure for the incompressibility condition as a function of $\mathbf{u}, \mathbf{S}^0$, and $\mathbf{C}$ as

$$u^* = f(\mathbf{u}, \mathbf{S}^0, \mathbf{C}). \tag{5.26}$$

Using these vectors and matrices, we can write the self-consistent equation of linear response theory [10] as

$$\mathbf{d} = -\beta\mathbf{S}^0(\mathbf{u} + \mathbf{C}\mathbf{d} + u^*\mathbf{e}), \tag{5.27}$$

where $\mathbf{e} \equiv \{1\}$.

The general solution of (5.27) under the incompressibility condition has the linear form

$$\mathbf{u} = -\frac{1}{\beta}\mathbf{S}^{-1}\mathbf{d}, \tag{5.28}$$

where $\mathbf{S}$ is a matrix for which the ij element is the function of scattering between the segments of i-th and j-th subchains in the system.

5.6.1 Application of the RPA to a Polymer Blend

We derive a practical solution for a polymer blend using (5.27). Here, we consider a simple polymer blend of two types of subchains with one segment–segment interaction energy $\epsilon = \epsilon_{12} = \epsilon_{21}$. The matrices $\mathbf{C}$ and $\mathbf{S}^0$ are written as

$$\mathbf{C} = \begin{pmatrix} 0 & z\epsilon \\ z\epsilon & 0 \end{pmatrix}, \tag{5.29}$$

$$\mathbf{S}^0 = \begin{pmatrix} S^0_{11} & 0 \\ 0 & S^0_{22} \end{pmatrix}. \tag{5.30}$$

An element of S^0_{ii} is given by the product of $\bar{\phi}_i$ and the structure factor for a Gaussian chain (i.e., the Debye function (5.20)):

$$S^0_{ii} = \bar{\phi}_i \frac{2N_i}{x_i^2}(e^{-x_i} - 1 + x_i). \tag{5.31}$$

We substitute (5.29) and (5.30) into (5.27) and eliminate u^* using the incompressibility condition $d_1 + d_2 = 0$. We obtain the linear relation between d_i and u_i as

$$u = -\frac{1}{\beta}\left(\frac{1}{S^0_{11}} + \frac{1}{S^0_{22}} - 2\chi\right) d, \tag{5.32}$$

where $d = d_1 = -d_2$, $u = u_1 - u_2$, and $\chi = z\epsilon$. Under the incompressibility condition, d_2 and u_2 are dependent on d_1 and u_1, and the effect of the external potential is thus provided by u.

5.7 Ginzburg–Landau Theory

A linear relation like that mentioned in the preceding section can be seen in terms of the equation of Ginzburg–Landau theory, which is justified for the weak segregation region of the polymer blend [3]. We explain Ginzburg–Landau theory to clarify the phenomenological DFT of polymer blends.

Using the Ginzburg–Landau expansion, the free energy functional is given by

$$\mathscr{F}[\{\phi_i(\mathbf{r})\}] = \mathscr{F}_0[\{\bar{\phi}_i\}] + \frac{1}{2\beta}\sum_{ij}\int\int d\mathbf{r}\, d\mathbf{r}' S^{-1}_{ij}(\mathbf{r}-\mathbf{r}')\delta\phi_i(\mathbf{r})\delta\phi_j(\mathbf{r}') + \ldots . \tag{5.33}$$

$$= \mathscr{F}_0[\{\bar{\phi}_i\}] + \frac{1}{2\beta}\sum_{ij}\int d\mathbf{q} S^{-1}_{ij}(\mathbf{q})\delta\phi_i(\mathbf{q})\delta\phi_j(-\mathbf{q}) + \ldots, \tag{5.34}$$

where $\delta\phi_i(\mathbf{r})$ is the local segment density fluctuation of the i-th subchain at position $\mathbf{r}$. The first term $\mathscr{F}_0[\{\bar{\phi_i}\}]$ is the free energy of the uniformly mixed state that is used as the reference state for the expansion. Equations (5.33) and (5.34) are the Taylor series expansions of the free energy with respect to the local segment density fluctuations of the subchains. The expansion coefficients, $S_{ij}^{-1}(\mathbf{r}-\mathbf{r}')$, in Eq. (5.33) are the inverse of the density–density autocorrelation functions between the segment density fluctuations belonging to the i-th and j-th subchains at positions $\mathbf{r}$ and $\mathbf{r}'$, respectively. The Fourier representation is $S_{ij}^{-1}(\mathbf{q})$ in Eq. (5.34), where $\mathbf{q}$ is the scattering wave vector. The dots at the ends of (5.33) and (5.34) indicate the terms of higher order in the Taylor series expansion in $\delta\phi_i(\mathbf{r})$ and $\delta\phi_i(\mathbf{q})$, respectively.

The second term in (5.34) corresponds to the linear relation (5.32) for the polymer blend mentioned in the above section.

5.8 Flory–Huggins–de Gennes Model

Using the theories described in Sects. 5.6 and 5.7, we can theoretically set the value of the phenomenological parameter κ_i in the phenomenological Flory–Huggins free energy model equation (5.14).

The linear relation (5.32) includes both the Gaussian chain conformation effect through S_{ii} and the segment–segment interaction effect through χ. We can add only the Gaussian chain conformation effect to the Flory–Huggins equation because the segment–segment interaction effect is already included in the Flory–Huggins free energy model as the enthalpy approximate term with χ.

We neglect χ in (5.32) and modify the Debye function in S_{ii}^0 with the Taylor series expansion of the exponential term as

$$\frac{2}{x_i^2}(e^{-x_i}-1+x_i) \simeq \frac{2}{x_i^2}\left(\frac{x_i^2}{2!}-\frac{x_i^3}{3!}\right) \tag{5.35}$$

$$= \left(1-\frac{x_i}{3}\right) \tag{5.36}$$

$$\simeq e^{-x_i/3}. \tag{5.37}$$

These modifications and the inverse Fourier transformation change (5.32) to an equation in the real space:

$$(u_1-u_2)|_{\text{real space}} = -\frac{1}{\beta}\frac{b^2}{18\bar{\phi}_1\bar{\phi}_2}|\nabla\phi_1(\mathbf{r})|^2. \tag{5.38}$$

By substituting (5.38) for the square-gradient term of (5.14), we get a free energy model of the polymer blend:

$$\beta\mathscr{F}[\{\phi(\mathbf{r})\}] = \frac{\phi_1(\mathbf{r})}{N_1}\ln\phi_1(\mathbf{r}) + \frac{\phi_2(\mathbf{r})}{N_2}\ln\phi_2(\mathbf{r}) + \chi\phi_1(\mathbf{r})\phi_2(\mathbf{r}) + \frac{b^2}{36\bar{\phi}_1\bar{\phi}_2}|\nabla\delta\phi_1(\mathbf{r})|^2. \tag{5.39}$$

This free energy model is called the Flory–Huggins–de Gennes model and it can be used in the simulations of homopolymer blends [12].

5.9 Combination of Ginzburg–Landau Theory and the RPA

The linear relation mentioned in Sect. 5.6 can be derived from any polymer blend of almost any polymer architecture (excluding a looped structure), and the combination of Ginzburg–Landau theory and the RPA produce the generalized free energy model of polymer blends [10, 11]. We refer to the method as GRPA (as in either Ginzburg–Landau + RPA or generalized RPA).

For this generalization, we use the free energy model developed by Bohbot-Raviv and Wang [13]. They replaced the second-order contributions $\delta\phi_i(\mathbf{r})$ on the right-hand side of the Flory–Huggins free energy equation with the second term of (5.34) and obtained the expression

$$\mathscr{F}[\{\phi_i(\mathbf{r})\}] = \frac{1}{\beta}\sum_i\int d\mathbf{r}\frac{\phi_i(\mathbf{r})}{N_i}\ln\phi_i(\mathbf{r}) - \frac{1}{2\beta}\sum_i\int d\mathbf{r}\frac{1}{N_i\bar{\phi}_i}\delta\phi_i(\mathbf{r})\delta\phi_i(\mathbf{r}) + \frac{1}{2\beta}\sum_{ij}\int d\mathbf{q}S_{ij}^{-1}(\mathbf{q})\delta\phi_i(\mathbf{q})\delta\phi_j(-\mathbf{q}), \tag{5.40}$$

where N_i is the length of the i-th subchain. In (5.40), the first term is the Flory–Huggins mixing entropy of the centers of mass of the subchains. The subtracted second term is given by the second-order term of the Taylor series expansion of the first term in $\delta\phi_i(\mathbf{r})$, which is needed to avoid a double count of the second-order segment–segment interaction given by the third term. The third term is obtained with the RPA in the same manner described in Sect. 5.6.

However, the analytical solution of a many-chain system is complicated, and it is thus practical to get numerical solutions for each $\mathbf{q}$. In SUSHI, the method is implemented as a GRPA engine.

5.10 SCF Theory

The SCF theory for polymer blends described in the following section is the most accurate DFT described in this book. SCF theory introduces the conformational entropy of subchains using the path integral, and it can solve special problems with a wall in the system. The detail of SCF theory is given in the literature [11].

5.10.1 Path Integral

We consider the i-th subchain composed of K-type segments in the p-type polymer. The total number of segments in this subchain is $N_i^{(p)}$. The K-type segment has two physical characteristics: an effective bond length b_K and specific volume ρ_K.

We denote the statistical weight of the i-th subchain as $Q_i(s', \mathbf{r}'; s, \mathbf{r})$ and the external potential acting on the segments of the i-th subchain as $V_i(\mathbf{r})$. When the values of b_K, ρ_K, and $V_i(\mathbf{r})$ are given, (5.17) is modified as

$$\frac{\partial}{\partial s} Q_i(s', \mathbf{r}'; s, \mathbf{r}) = \left[\frac{b_K^2}{6}\nabla^2 - \beta \rho_K V_i(\mathbf{r})\right] Q_i(s', \mathbf{r}'; s, \mathbf{r}). \tag{5.41}$$

This expression is a sum of Boltzmann weights for all possible chain conformations, which are identified as paths of a diffusing particle in (5.41). Therefore, $Q_i(s', \mathbf{r}'; s, \mathbf{r})$ is called the path integral.

$V_i(\mathbf{r})$ in (5.41) can be divided into two contributions: a segment–segment interaction potential and an external potential due to the constraints on the segment density distributions, such as the incompressibility condition. These potentials are derived as follows.

The contribution from the segment–segment interaction potential is given by

$$W_K(\mathbf{r}) = \sum_{K'} \epsilon_{KK'} \phi_{K'}(\mathbf{r}), \tag{5.42}$$

where $\phi_{K'}(\mathbf{r})$ is the sum of the segment densities of the subchains composed of the K'-type segment, written as

$$\phi_{K'}(\mathbf{r}) = \sum_{j \in K'} \phi_j(\mathbf{r}). \tag{5.43}$$

The external potential for the constraint on the segment density distribution maintains a given profile of the segment density distribution. This constraining potential should therefore balance with the chemical potential of the segment $\mu_i(\mathbf{r})$, and the constraining potential is given by $-\mu_i(\mathbf{r})$.

The self-consistent potential $V_i(\mathbf{r})$ is thus given by

$$V_i(\mathbf{r}) = W_K(\mathbf{r}) - \mu_i(\mathbf{r}). \tag{5.44}$$

The potential $V_i(\mathbf{r})$ and $\phi_i(\mathbf{r})$ are related to each other under the constraining and incompressibility conditions [3, 11, 14–17].

The segment density at position $(s, \mathbf{r})$ of the i-th subchain is calculated using two path integrals. When we denote initial statistical weights at both ends of the subchain $s = 0$ and $s = N_i^{(p)}$ as $q_i^0(0, \mathbf{r}')$ and $q_i^0(N_i^{(p)}, \mathbf{r}')$, we can introduce integrated path integrals from both ends of the subchain expressed as

$$q_i(s, \mathbf{r}) \equiv \int d\mathbf{r}' q_i^0(0, \mathbf{r}') Q_i(0, \mathbf{r}'; s, \mathbf{r}), \tag{5.45}$$

$$\tilde{q}_i(s, \mathbf{r}) \equiv \int d\mathbf{r}' q_i^0(N_i^{(p)}, \mathbf{r}') Q_i(0, \mathbf{r}'; s, \mathbf{r}). \tag{5.46}$$

It is easy to confirm that $q_i(s, \mathbf{r})$ and $\tilde{q}_i(s, \mathbf{r})$ also satisfy (5.41). We refer to these $q_i(s, \mathbf{r})$ and $\tilde{q}_i(s, \mathbf{r})$ as the normal-direction integrated path integral and the reverse-direction integrated path integral, respectively.

When an end of a subchain is free and this free end is everywhere uniform, we have $q_i^0(0, \mathbf{r}') = 1$ or $q_i^0(N_i^{(p)}, \mathbf{r}') = 1$. Otherwise, the end is a junction point, where the initial statistical weight of the end must be a product of all statistical weights of the other subchains connected to this junction point [11].

5.10.2 Calculation of Segment Density

When the path integrals are evaluated, the segment density of the i-th subchain at position $\mathbf{r}$ is given by

$$\phi_i(\mathbf{r}) = C^{(p)} \int_{s \in i\text{-th subchain}} ds\, q_i(s, \mathbf{r}) \tilde{q}_i(N_i^{(p)} - s, \mathbf{r}), \tag{5.47}$$

where $C^{(p)}$ is a normalization constant and the integral over s is taken over the whole i-th subchain. To accumulate the contributions from all segments of the i-th subchain, the integrand is integrated over s in the subchain.

The normalization constant $C^{(p)}$ is evaluated by the kind of ensemble as follows.

Canonical Ensemble

The normalization constant $C^{(p)}$ is given by

$$C^{(p)} = \frac{M^{(p)}}{\int d\mathbf{r}\, q_i(s, \mathbf{r}) \tilde{q}_i(N_i^{(p)} - s, \mathbf{r})} = \frac{M^{(p)}}{\mathscr{Z}^{(p)}}, \tag{5.48}$$

where $M^{(p)}$ is the total number of p-type polymer chains to which the i-th subchain belongs and $\mathscr{Z}^{(p)}$ is the partition function of a p-type chain. Note that this partition function is independent of the subchain index i in the p-type chain.

Grand Canonical Ensemble

The normalization constant $C^{(p)}$ is given by

$$C^{(p)} = \frac{m^{(pb)}}{\exp[-\beta \sum_{i \in p} \rho_K N_i^{(p)} W_K(b)]} = \frac{m^{(pb)}}{\mathscr{Z}^{(pb)}}, \tag{5.49}$$

where $m^{(pb)}$ is the total number of p-type polymers per volume in the bulk reservoir that is considered to be a set of mean fields of components [14] and $\mathscr{Z}^{(pb)}$ is the partition function of the p-type polymer per volume in the bulk reservoir.

5.10.3 *Free Energy*

The Helmholtz energy of the system can be given as a functional of the segment densities of subchains $\{\phi_i\}$ [3, 11, 14]:

$$\begin{aligned} \mathscr{F}[\{\phi_i\}] = & -\frac{1}{\beta}\sum_p M^{(p)} \ln \mathscr{Z}^{(p)} + \frac{1}{\beta}\sum_p M^{(p)} \ln M^{(p)} \\ & + \frac{1}{2}\sum_K \sum_{K'} \int d\mathbf{r} \varepsilon_{KK'} \phi_K(\mathbf{r}) \phi_{K'}(\mathbf{r}) \\ & - \sum_i \int d\mathbf{r} V_i(\mathbf{r}) \phi_i(\mathbf{r}). \end{aligned} \tag{5.50}$$

When a system includes solvents, solvent distributions are calculated as a Boltzmann factor under the external field for solvents, and solvent contributions are added to the SCF calculation [11].

5.10.4 *Practical Method of Calculating the Path Integral*

Equation (5.41) makes the approximation that the effect of the external field is sufficiently weak and the Boltzmann factor can be given by the linear approximation. However, the method implemented in SUSHI calculates the Boltzmann factor accurately as

$$q_i(s+ds, \mathbf{r}) = \exp\left[-\beta \frac{\rho_K V_K(\mathbf{r}) ds}{2}\right]\left(1 + \frac{b_K^2}{6}\nabla^2 ds\right)\left\{\exp\left[-\beta \frac{\rho_K V_K(\mathbf{r}) ds}{2}\right] q_i(s, \mathbf{r})\right\}. \tag{5.51}$$

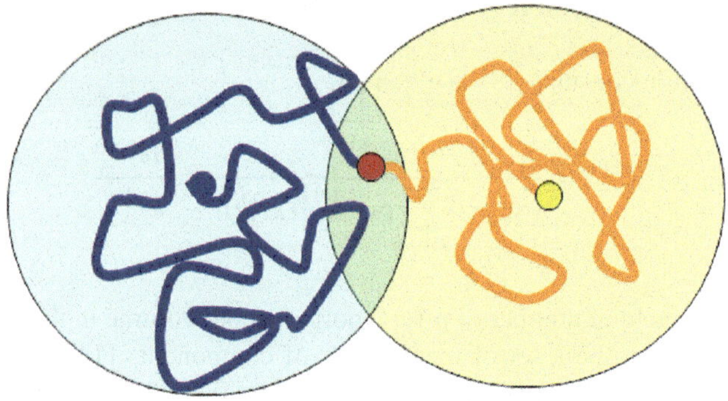

Fig. 5.2 Schematic figure of the SCF calculation. The segment density of the middle segment is obtained as the product of the path integrals from both ends of the polymer

This is the discretized path integral scheme for the finite difference method. SUSHI uses the finite difference method to obtain a numerical solution. In (5.51), the factor $1/2$ in exp[] is needed to discretize the path integral starting from a mesh grid point and ending at a neighboring mesh grid point. Through such a one-step path integral, a segment feels the average external field at the start and end grid points. The SCF scheme in SUSHI therefore uses ∇^2 and an exponential calculation many times.

5.10.5 SCF Calculation

As described in the above section and shown in Fig. 5.2, the calculation of $\{\phi_i(\mathbf{r})\}$ requires path integrals. The path integrals require external potentials $\{V_i(\mathbf{r})\}$. However, part of the external potential is the segment–segment interaction energy $\{W_i(\mathbf{r})\}$, which is calculated with $\phi_i(\mathbf{r})$.

Therefore, this calculation cannot be solved analytically and a nonlinear calculation (i.e., SCF calculation) is needed as shown in Fig. 5.3. These calculation methods using the path integral are thus referred to as SCF theory.

In SUSHI, two different SCF methods are implemented as follows.

(1) Static SCF method
All segments of the same type are subjected to the same SCF irrespective of the subchains they belong to. To get the equilibrium state, $\{W_i(\mathbf{r})\}$ and $\{\mu_i(\mathbf{r})\}$ are simultaneously and independently updated, starting from an appropriate initial distribution [11, 18].

(2) Dynamic SCF method
First, we give the initial $\{\phi_i(\mathbf{r})\}$ that satisfies the incompressibility condition. The segments of the different types of subchains are subjected to different SCFs. To get $\{\mu_i(\mathbf{r})\}$ for given $\{\phi_i(\mathbf{r})\}$, $\{\mu_i(\mathbf{r})\}$ is simultaneously and independently updated starting from an appropriate initial distribution. After the convergence of an SCF calculation, a new $\{\mu_i(\mathbf{r})\}$ is used for dynamic of $\{\phi_i(\mathbf{r})\}$ DFT. We return to the SCF calculation and repeat the time evolution calculation [19–27].

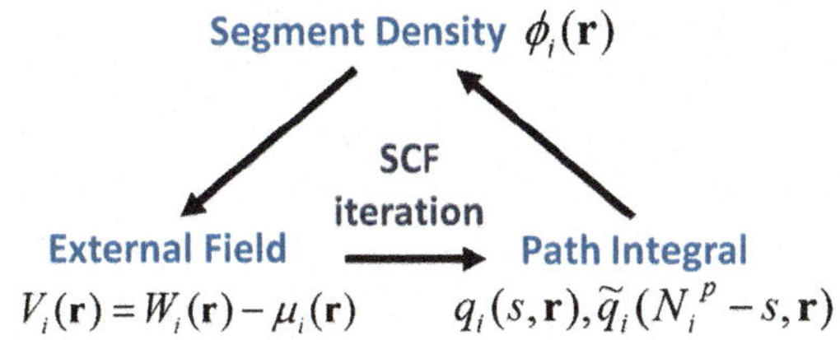

Fig. 5.3 Basic scheme of SCF theory

5.11 Hydrodynamics Effect

We can introduce the hydrodynamics effect to dynamic DFT, which is important in accelerating phase separations [28]. First, velocity fields $\{\mathbf{v}_i(\mathbf{r})\}$ are introduced to the dynamic DFT equation:

$$\frac{\partial}{\partial t}\phi_i(\mathbf{r}, t) = \nabla \cdot \{L_i(\mathbf{r}, t)\nabla\mu_i(\mathbf{r}, t)\} - \nabla \cdot \{\phi_i(\mathbf{r})\mathbf{v}_i(\mathbf{r})\}. \tag{5.52}$$

5.11.1 *Coupling with the Navier–Stokes Equation*

The velocity fields $\{\mathbf{v}_i(\mathbf{r})\}$ are obtained with the Navier–Stokes equation. Under the incompressibility condition $\nabla \cdot \mathbf{v}(\mathbf{r}, t) = 0$, the Navier–Stokes equation is given by [28]

$$\rho\frac{\partial \mathbf{v}(\mathbf{r}, t)}{\partial t} = -\rho\{\mathbf{v}(\mathbf{r}, t)\cdot\nabla\}\mathbf{v}(\mathbf{r}, t) - \nabla p(\mathbf{r}, t) + \nabla\{\eta(\mathbf{r}, t)\nabla\cdot\mathbf{v}(\mathbf{r}, t)\} - \sum_i \phi_i(\mathbf{r}, t)\nabla\mu_i(\mathbf{r}, t), \tag{5.53}$$

where $p(\mathbf{r}, t)$ is the pressure, $\eta(\mathbf{r}, t)$ is the local viscosity, and the last term is the volume force introduced to couple the dynamic DFT. We assume that the local viscosity is linearly dependent on the local segment density according to

$$\eta(\mathbf{r}, t) \equiv \sum \eta_K \phi_K(\mathbf{r}, t), \tag{5.54}$$

where η_K is the viscosity of a pure system composed of K-type segments.

We introduce the steady-state approximation that the characteristic relaxation time of the local velocity field is much shorter than that of the local volume force (i.e., the characteristic time scale of the change in the segment density fields). This assumption leads to the relation

$$\frac{\partial \mathbf{v}(\mathbf{r}, t)}{\partial t} = 0. \tag{5.55}$$

The incompressibility condition and the steady-state approximation deform the Navier–Stokes equation and give Poisson's equation with respect to $p(\mathbf{r}, t)$:

$$\nabla^2 p(\mathbf{r}, t) = \nabla \cdot [-\rho\{\mathbf{v}(\mathbf{r}, t) \cdot \nabla\}\mathbf{v}(\mathbf{r}, t) + \nabla\{\eta(\mathbf{r}, t)\nabla \cdot \mathbf{v}(\mathbf{r}, t)\} - \sum_i \phi_i(\mathbf{r}, t)\nabla\mu_i(\mathbf{r}, t)]. \tag{5.56}$$

By solving this equation and substituting the obtained $p(\mathbf{r}, t)$ into (5.53), we finally obtain the velocity field $\mathbf{v}(\mathbf{r}, t)$. This local velocity $\mathbf{v}(\mathbf{r}, t)$ is then substituted into (5.51) to calculate the time evolution of the segment density fields.

5.12 Example: Phase Diagram Generated with the Flory–Huggins Free Energy Model

From this section, we present examples using SUSHI with wild card and tools concerning DFTs. We denote the SUSHI+version and sushi+version as SUSHI* and sushi*, respectively.

We construct a phase diagram of an A/B polymer blend using the Flory–Huggins free energy model

$$\beta\mathscr{F} = \frac{\phi_A}{N_A} \ln \phi_A + \frac{\phi_B}{N_B} \ln \phi_B + \chi_{AB}\phi_A\phi_B. \tag{5.57}$$

For simplicity, we use the notation ϕ for ϕ_A ($\phi_A = \phi, \phi_B = 1 - \phi$ and χ for χ_{AB}. When χ exceeds the critical point of phase separation, the profile of (5.57) is that shown in Fig. 5.4, and the system is separated into two phases of which the segment densities are those of the contact points of the common tangential line on the free energy curve. The gradient of the common tangential line is the chemical potential of segments of polymers in these phases. The star-shaped points are the inflection points of the free energy curve, and the shape of the curve is convex within the region between the points. Therefore, the phase separation in the region is accelerated and the region is called the spinodal region. The inflection point is called the spinodal point. The phase separation in the region between the contact points of the common tangential line and the inflection point generates droplets, and is slow. This region is called the binodal region and the contact point is called the binodal point.

We can thus draw a phase diagram of polymer blends to plot χN values of all spinodal and binodal points versus the value of ϕ as shown in Fig. 5.5.

The critical point of the Flory–Huggins phase diagram, which is the bottom point of the spinodal and binodal curves, satisfies the equations

$$\frac{\partial^2 \Delta F}{\partial \phi^2} = 0, \tag{5.58}$$

$$\frac{\partial^3 \Delta F}{\partial \phi^3} = 0. \tag{5.59}$$

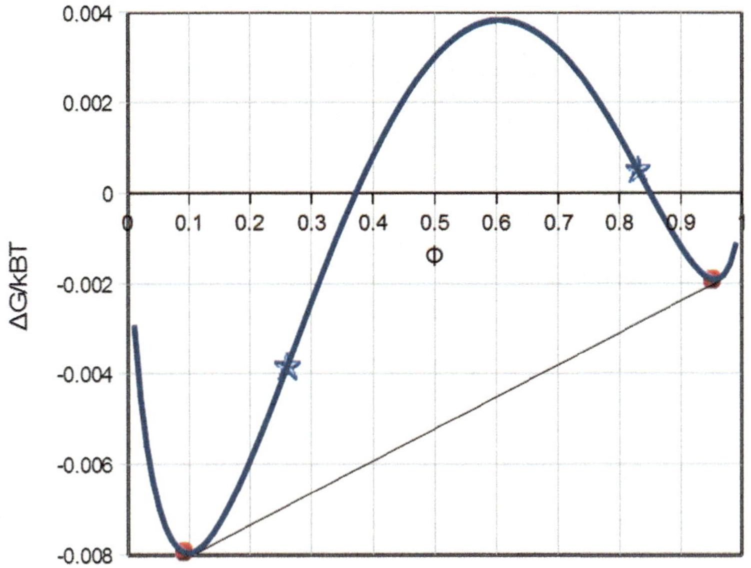

Fig. 5.4 Example of a profile of the Flory–Huggins free energy model: $N_A = 10, N_B = 20$, $\chi = 0.22$. The points of star shape are spinodal points, and the contact points of the common tangential line on the free energy curve are binodal points

We introduce a tool that can draw a phase diagram of a polymer blend with the Flory–Huggins free energy model, which can be obtained numerically as follows.

5.12.1 Critical Point

The value of ϕ_A at the critical point can be obtained by solving the equations (5.57), (5.58), and (5.59). The solution is given by

$$\phi_A = \frac{(N_A N_B)^{1/2} - N_B}{N_A - N_B}. \tag{5.60}$$

5.12.2 Spinodal Points

Spinodal points can be obtained by solving the equations (5.57) and (5.58). The solution is given by

$$\chi N_A N_B = \frac{1}{2}\left(\frac{N_A}{\phi_A} + \frac{N_B}{\phi_B}\right). \tag{5.61}$$

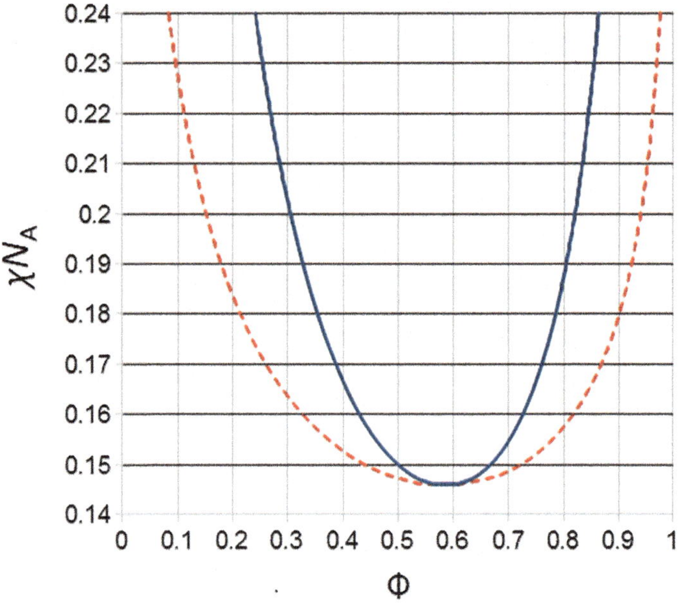

Fig. 5.5 Example of a phase diagram of the Flory–Huggins free energy model: $N_A = 10$, $N_B = 20$, the normal line is the spinodal line, and the dotted line is the bimodal line. The region above the spinodal line is the spinodal region and the region below the spinodal line and above the binodal line is the binodal region

5.12.3 Binodal Points

The problem of obtaining the binodal points corresponds to the problem of obtaining pairs of contact points of the common tangential line on the free energy curve. Thus, the pair of contact points of which ϕ values are ϕ_A and ϕ'_A satisfies the relation given by solving equations (5.57) and (5.58), written as

$$\frac{\partial F}{\partial \phi}|_{\phi=\phi_A} = \frac{\partial F}{\partial \phi}|_{\phi=\phi'_A} = \frac{F(\phi_A) - F(\phi'_A)}{\phi_A - \phi'_A}. \tag{5.62}$$

This equation can be solved under the incompressibility condition as

$$N_A \frac{\phi_B + \phi'_B}{2} \ln \frac{\phi_B}{\phi'_B} + N_B \frac{\phi_A + \phi'_A}{2} \ln \frac{\phi_A}{\phi'_A} + (N_B - N_A)(\phi'_A - \phi_A) = 0. \tag{5.63}$$

The procedure to obtain the pair ϕ_A and ϕ'_A is to fix the value of ϕ_A and search for the value of ϕ'_A numerically.

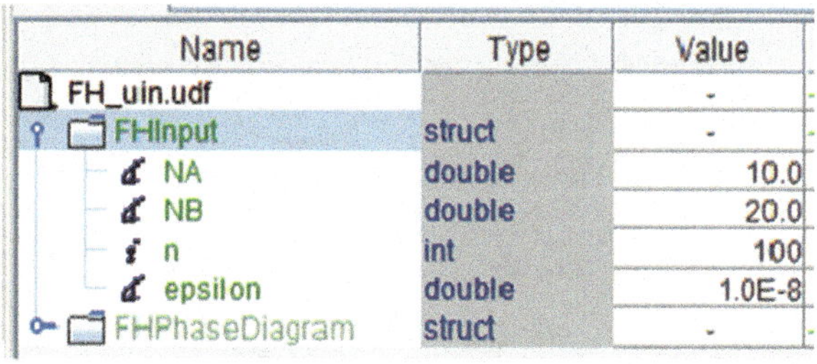

Fig. 5.6 UDF input sample for the Flory–Huggins phase diagram

5.12.4 Tool for the Flory–Huggins Phase Diagram

A tool named FHPhaseDiagram is used to draw the Flory–Huggins phase diagram for some pair of N_A and N_B. We copy the user definable format (UDF) file "SUSHI*/Potage/FHPhaseDiagram/sample/FH.udf" into our working directory and copy again the same file into the same directory with another name that is used for the input file. For example, we rename the file as "FH_uin.udf." We open the UDF file using GOURMET and write a few data into this file as shown in Fig. 5.6. There are only four data: NA, NB, n, and epsilon. NA and NB correspond to the chain lengths N_A and N_B, n is the number of plot points of ϕ_A, and epsilon is the threshold value of convergence of χN_A for the numerical calculation of binodal points. Default values of n and epsilon do not need to be changed. Thus, we only change the values of NA and NB. We set these values as 10 and 20, respectively. These conditions are the same conditions as those for Fig. 5.5.

We next open GourmetTerm and move to the working directory and invoke a program by entering

```
> FHPhaseDiagram -I FH_uin.udf -O FH_uot.udf
```

where the options "-I" and "-O" indicate the input and output UDF file names, respectively.

After finishing the run, we get the "FH_out.udf" file. As shown in Fig. 5.7, we open the file using GOURMET and set the View mode at the top to Table and then click the phae_diagram[] subfolder. We can see the data—*phi, spinodal, and binodal*—in a row; these are the data ϕ_A, χN_A at the spinodal point and χN_A at the binodal point. These data can be plotted in the following procedure. We click the Plot tab at the bottom and push the Make tab. Command lines for gnuplot appear in the Plot window. We then modify the command lines to plot χN_A vs. ϕ as shown in Fig. 5.7 and then push the Plot button. We can plot the phase diagram using gnuplot as shown in Fig. 5.5.

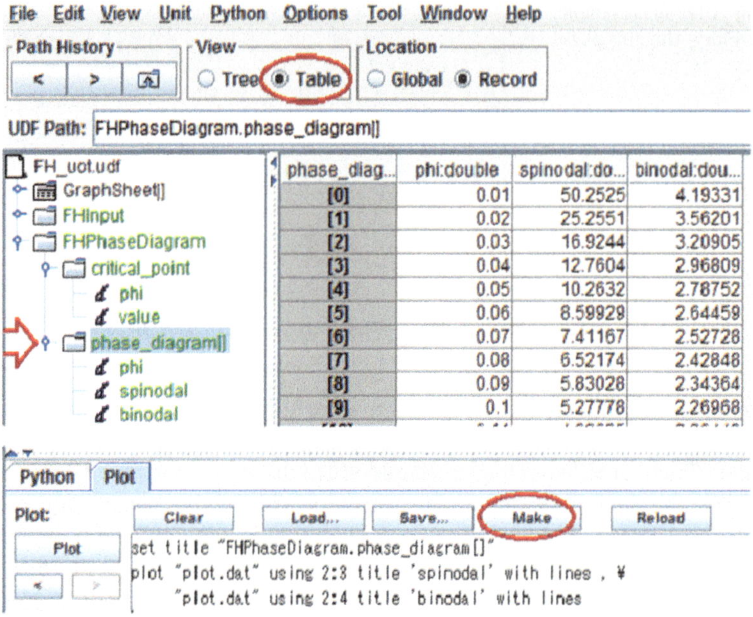

Fig. 5.7 UDF output sample for the Flory–Huggins phase diagram

5.13 Example: Estimation of the Critical Point of Spinodal Decomposition of a Diblock Copolymer

As mentioned above, the phase diagram of the polymer blend can be estimated using the Flory–Huggins free energy model. We now consider the estimation of the phase diagram of block copolymers. The most accurate estimation is to use SCF theory [18]. However, the calculation cost is high, and we need to predict the phase separation before an accurate calculation using SCF theory. The use of the RPA makes it easier to obtain information of the phase diagram of the diblock copolymer than is the case using the SCF method. The general outline of this method is as follows.

According to the temperature drop in the disordered melt composed of block copolymers, the critical point of the spinodal decomposition, where one of the scattering functions between the segments of subchains diverges at finite q, is reached [3, 9]. Mathematically, this method relates to the eigenvalue problem of the matrix $-\mathbf{S}^{-1}$ in (5.28), where one of the eigenvalues of $-\mathbf{S}^{-1}$ has a negative value at the spinodal point.

However, it is complicated to obtain the analytical equation of the scattering function of a system with many components, as mentioned in Sect. 5.6. The method is therefore implemented as a discretized numerical method of RPAEngine in SUSHI and RPAEngine in POTAGE2, which is a tool for constructing the phase diagrams of polymer blends. This eigenvalue problem can be solved using POTAGE2. We show an example of using POTAGE2 as follows.

Here we estimate the critical point of the symmetric A–B diblock polymer. The value is known to be $\chi N = 10.5$ according to the RPA. First, we copy the UDF file "SUSHI*/Potage/POTAG2_sample/A50_B50_A50B50.udf" to our working directory. As shown in Fig. 5.8, we change all structures of the three polymers in pd_data_for_RPA[0] to the same structure $A_{10}B_{10}$, which has the value $N = 20$. The input procedure for a polymer structure is the same as in SUSHI; refer to details in the SUSHI manual. Next, we set the value of chi_parameters[0] to 0.51, which sets the χN value to 10.2. We see only polymer1 in Fig. 5.8 but also set polymer2 and polymer3 the same. We save the UDF file as "A10B10_51.udf", and then, as shown in Fig. 5.8, click the right mouse button on the POTAGEInput subfolder and invoke the Potage2Run action in the appearing action window.

After finishing the run, a triangular phase diagram is automatically produced by gnuplot. In the phase diagram, we see grid marks "$*$" colored blue, the blue of which indicate out of spinodal regions for macrophase separations at the grid marks. We use just one kind of diblock copolymers and no macrophase separation occurred in this example.

We change all the values of chi_parameters[0] to 0.52 ($\chi N = 10.4$) in the UDF input file, save the file as "A10B10_52.udf", and restart POTAGE2. We see the same

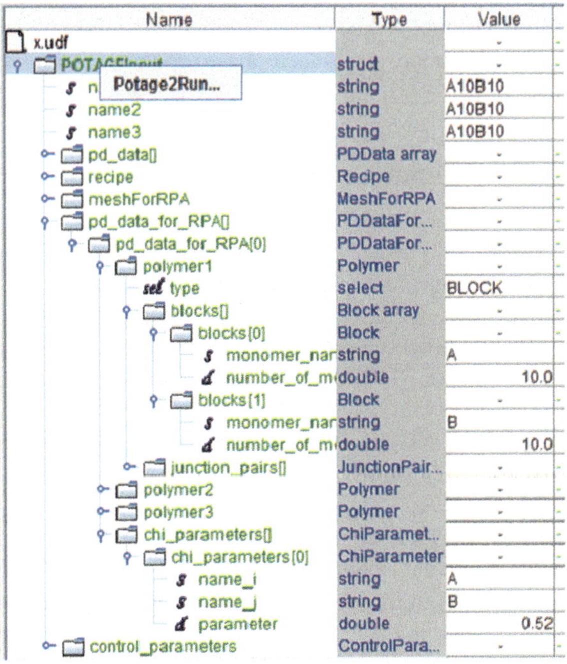

Fig. 5.8 UDF input file used to obtain the critical point of the diblock copolymer for POATGE2

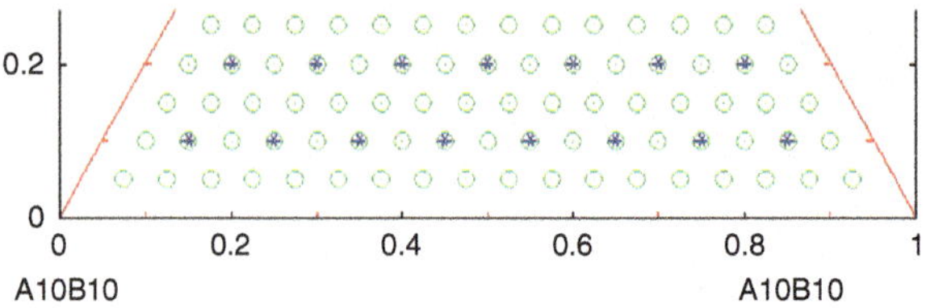

Fig. 5.9 Part of the phase diagram obtained in the estimation run of the critical point of the spinodal decomposition of a symmetric A–B diblock copolymer with POTAGE2. The many circles mark the spinodal region of the block polymer

triangular phase diagram, but many circular marks appear as shown in Fig. 5.9. The circular marks mean that spinodal region for microphase separation at the center of these circles on the phase diagram; this spinodal condition is estimated by solving the eigenvalue problem of $-\mathbf{S}^{-1}$ in (5.28).

Using the information of the two runs, we estimate the value of the critical point near the average value $\chi N = 10.3$ for symmetric diblock copolymers. The result has numerical error owing to the discretized calculation, but the calculation cost is low and the spinodal condition for microphase separation is easily predicted.

Furthermore, we can input up to three different structures of polymers into POTAGE2, and for such polymer blends, spinodal curves for macrophase separation automatically appear and the grid marks "$*$" are colored red in spinodal regions on the phase diagrams. The tool is thus useful for obtaining information of the spinodal decompositions for both micro- and macrophase separations of polymer blends.

5.14 Example: Estimation of χ Parameters

The DFT of polymer blends needs the Flory–Huggins χ parameters for calculations. In this section, we introduce a method of estimating χ parameters employing the group contribution. In the OCTA system, we offer several UDF files that contain useful data on physical properties of various kinds of polymers. These UDF files are stored in the POLYMERDATABASE subfolder and can be referred to using GOURMET.

First, we open "PF_ENGINE/POLYMERDATABASE/polymerdata.udf" with GOURMET, and then, as shown in Fig. 5.10, we open File/Header and add "potage.act" to Action File.

Next, as shown in Fig. 5.11, we run the POTAGEInputMaker action by clicking the right mouse button on the PolymerDatabase subfolder as shown in Fig. 5.11a. When the run is normally terminated, we see the message "normal end" in the Python windows as shown in Fig. 5.11b.

The "POTAGE_input_maker.udf" file is generated in the default working directory GOURMET/tmp. We open it using GOURMET. There are two subfolders

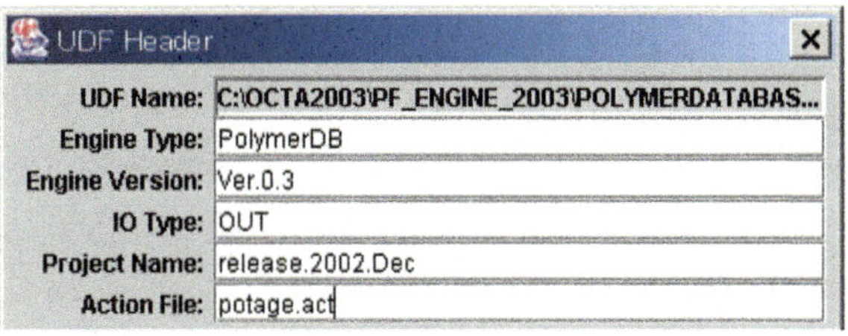

Fig. 5.10 UDF header file window of the polymer database

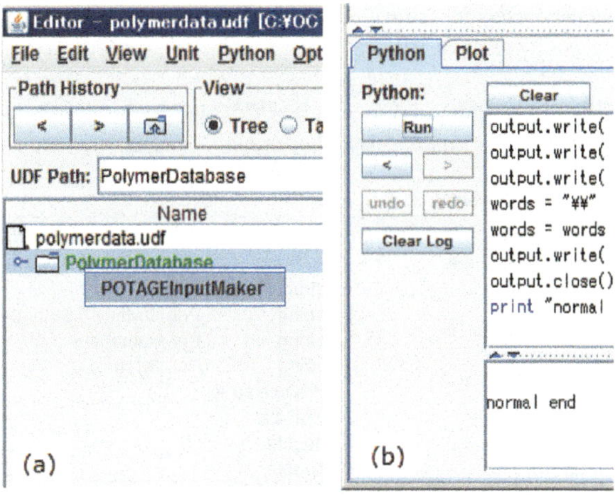

Fig. 5.11 Invoking the action script on the polymer database UDF

POTAGEInput and POTAGEInputMaker. We set the View mode at the top to Table and click the subfolder POTAGEInputMaker.polymer_parameter_for_potage[]. A table with SP values appears as shown in Fig. 5.12. There are "name," "use," and "ratio" parameters in a row. "name" is the polymer name and "use" is the ID of the polymer for the χ parameter we want. We thus input the values 1, 2, or 3 to "use" because the ID is defined as a positive number and the maximum number of components is three. Next, we input values to each "ratio"; the default value is 1.

If we need the χ parameter of a random copolymer, we input the same ID to "use" for all polymers of which monomers are used in the random copolymer and input values to "ratio" under the restriction that the sum of values is unity. We set the same parameters as for Fig. 5.12.

Next, we invoke the POTAGESetChi action by clicking the right mouse button on the POTAGEInputMaker subfolder. An input window opens and requires values of the molar volume of segment V_r[cm^3/mol], temperature T[°C], and constant term χ_s (default values are input as 100, 150, and 0). We change values of V_r[cm^3/mol] and T[°C] for our purpose. We use the default values as an example.

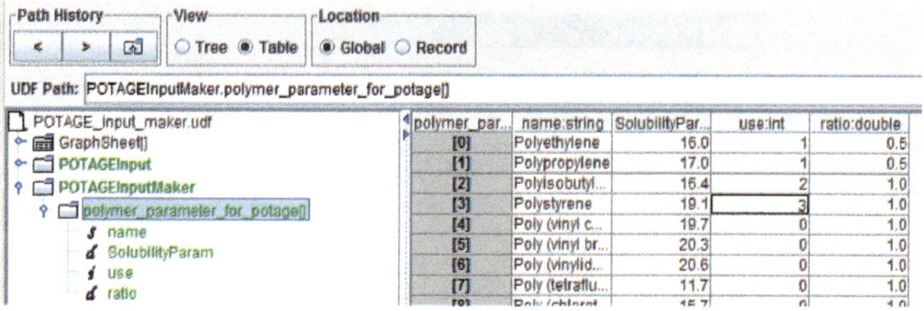

Fig. 5.12 Polymer parameters for estimating the χ parameter

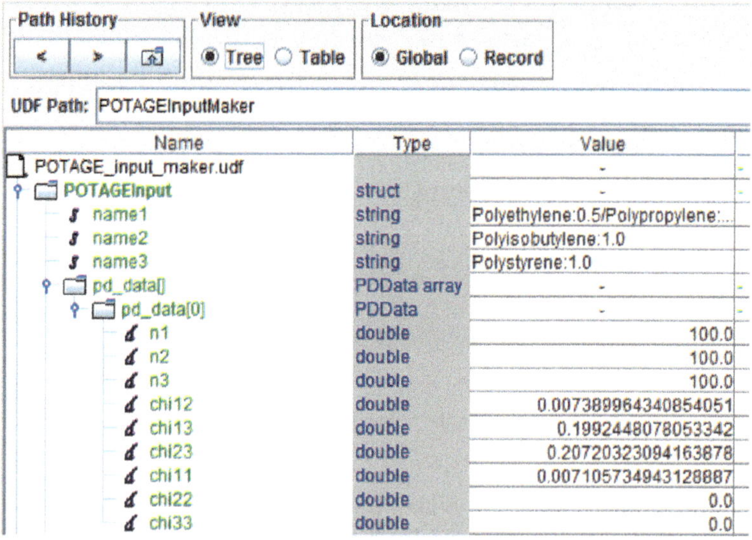

Fig. 5.13 Example of the estimation of χ parameters

After finishing the action, we see χ parameters for three components as shown in Fig. 5.13. Using the χ parameters, POTAGE2 can draw the triangular phase diagram with only spinodal curves by the action on the POTAGEInput subfolder. It is useful to estimate the phase separation roughly before DFT simulations.

As practical advice, the estimated χ mentioned above tends to exceed the stable limit of the SCF calculation. In such a case, we try to reduce the value of χ gradually until the SCF iteration converges because the domain morphologies for systems with large χ are in general unchanged even if we reduce the value of χ.

5.15 Example: Macrophase Separation and Microphase Separation

In this section, the static SCF calculation of a macrophase separation of an A/B polymer blend and the dynamic SCF calculation of a microphase separation of an A–B diblock copolymer are explained.

5.15.1 Macrophase Separation (Static SCF Calculation)

We open "SUSHI*/sample/Input/blend_uin.udf" using GOURMET. This sample file is for a static SCF calculation of an A/B blend in one dimension. We modify it for a two-dimensional calculation. As shown in Fig. 5.14, we open "mesh" subfolder and add an element to axes[] by clicking the left mouse button on axes[0] and clicking the left mouse button on "Edit/Add an Array Element," which means we change to two dimensions. Furthermore, we add three elements of values[] to added axes[2] by clicking the left mouse button on values[] in axes[2] and clicking the left mouse button on Edit/Add Array Elements. Please refer to the GOUEMT manual for the details of manipulation in GOURMET.

We input three values—0, 32, and 32—to each added values[]. These values are the initial value, end value, and number of divisions of the Y axis, respectively. In the same manner as adding elements to values[], we add one element to boundary_condition.conditions[] by clicking the left mouse button on boundary_condition.conditions[0] and clicking the left mouse button on "Edit/Add Array Elements" and input the character "PERIODIC" to the added element axis_conditions[]. These procedures set the system to two dimensions under the periodic boundary condition. We save this UDF file as "blend2D_uin.udf" and then open GourmetTerm and move to the directory of the saved file. We invoke SUSHI by entering

```
> sushi* -I blend2D_uin.udf -C blend2D_uin.ual
```

where the "-C" option redirects the log-on console to the blend2D_uin.ual file.

After finishing the run, the output file named "blend2D_uot.udf" is automatically generated. In the default operation of SUSHI, input and output file names are assumed as "*_uin.udf" and "*_uot.udf", respectively. We open the output file and move the record cursor at the bottom to the last record as shown in Fig. 5.15a. We move the cursor to the SUSHIOutput subfolder and clicking the right mouse button. A select window appears and we then select show_field in the window as shown in Fig. 5.15b. We see the interface of the A/B blend as shown in Fig. 5.15c. Furthermore, in the same manner of clicking the right mouse button on the SUSHIOutput subfolder and selecting plot_1D_field in the appearing window as shown in Fig. 5.15d, a graph of the one-dimensional segment profile at the interface is generated as shown in Fig. 5.15e.

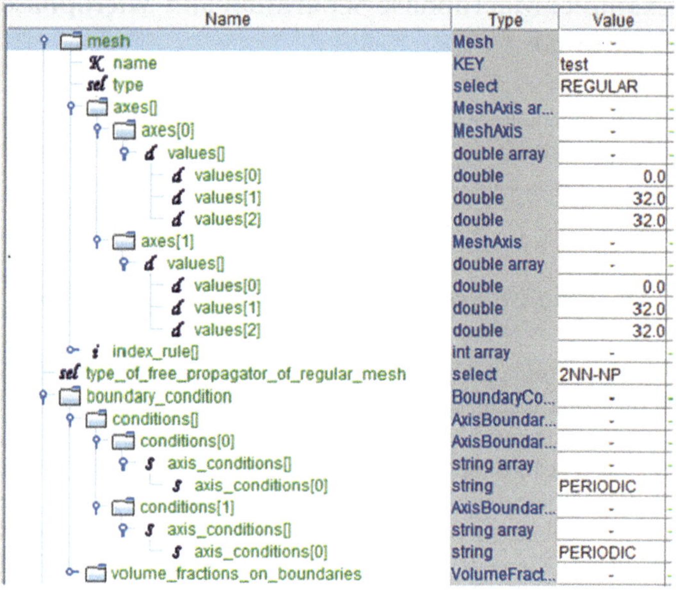

Name	Type	Value
mesh	Mesh	-
name	KEY	test
type	select	REGULAR
axes[]	MeshAxis ar...	-
axes[0]	MeshAxis	-
values[]	double array	-
values[0]	double	0.0
values[1]	double	32.0
values[2]	double	32.0
axes[1]	MeshAxis	-
values[]	double array	-
values[0]	double	0.0
values[1]	double	32.0
values[2]	double	32.0
index_rule[]	int array	-
type_of_free_propagator_of_regular_mesh	select	2NN-NP
boundary_condition	BoundaryCo...	-
conditions[]	AxisBoundar...	-
conditions[0]	AxisBoundar...	-
axis_conditions[]	string array	-
axis_conditions[0]	string	PERIODIC
conditions[1]	AxisBoundar...	-
axis_conditions[]	string array	-
axis_conditions[0]	string	PERIODIC
volume_fractions_on_boundaries	VolumeFract...	-

Fig. 5.14 UDF file for the input of macrophase separation

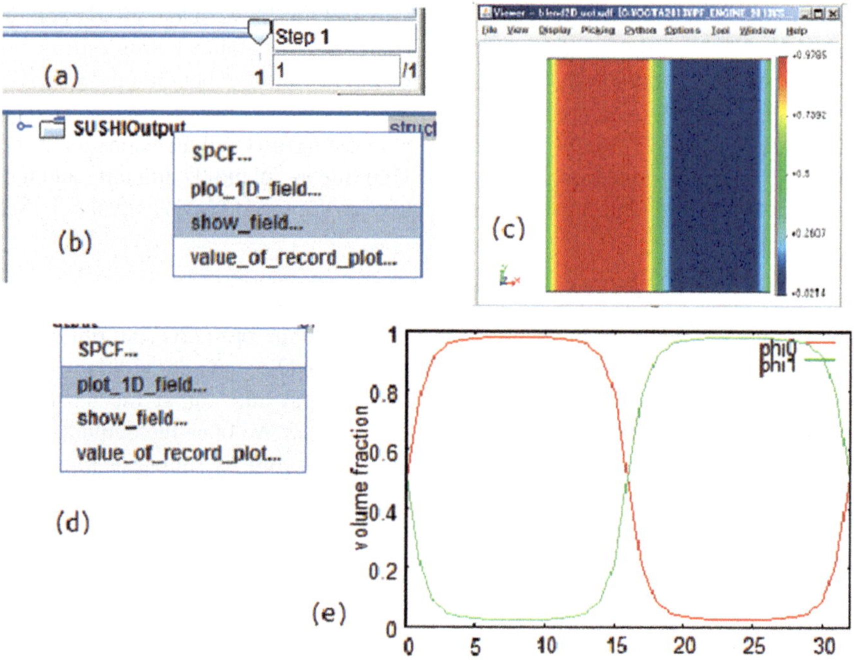

Fig. 5.15 UDF file of the output of macrophase separation

5.15.2 *Microphase Separation (Dynamic SCF Method)*

We copy and open "SUSHI*/sample/Input/blend2D_7_dy1_uin.udf" using GOURMET in our working directory. This sample file is for a dynamic SCF calculation of the A–B block copolymer melt in two dimensions. We add the hydrodynamics effect option and set the calculation to three dimensions.

First, we set the values 0.01, 100,000, and 10,000 for the parameters delta_t, max_dynamics_step and output_interval_step in the subfolder calculation_method.DYNAMICS. We then set the value of archives_interval_step the same as the value of output_interval_step on the same subfolder. We change to the three-dimensional calculation in the same manner described in the above section. As shown in Fig. 5.16, we add the elements of axes[] and increase the number of elements to three and then add the elements of axes[] and values[] and increase the number of elements to four. We input four values—0, 16, 32, and 2—to all axes[].values[]. The last value is the number of division of axis and is needed only for the parallel calculation using the Message Passing Interface (MPI), where the number of MPI processes (i.e., the usable number of CPUs for the parallel calculation) is $2^3 = 8$.

Next, as shown in Fig. 5.17, we change the parameters in external_conditions. dynamic_conditions, add one element to types_of_polymer_mobility[], and set the parameter 0 and "ROUSE" to component_ID and type, respectively. These settings provide stability to the dynamic SCF calculations. Additionally, we introduce the hydrodynamics calculation, set the value 1 for the density

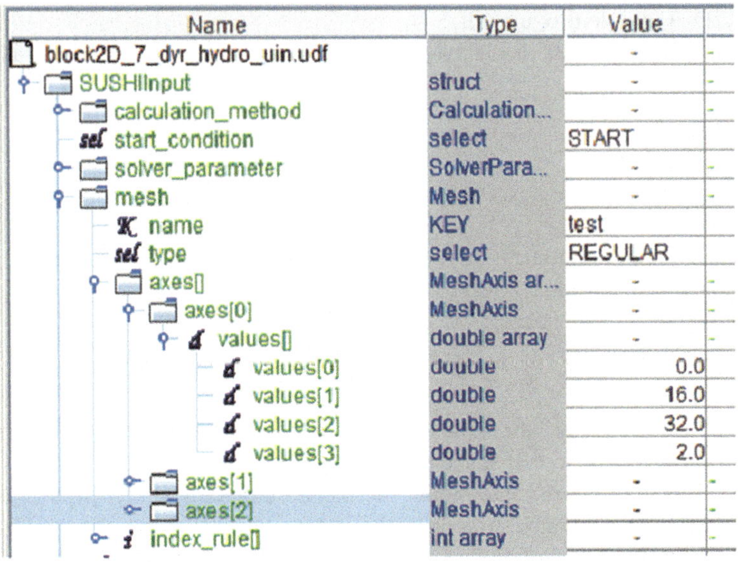

Fig. 5.16 Input UDF for a microphase separation 1/2

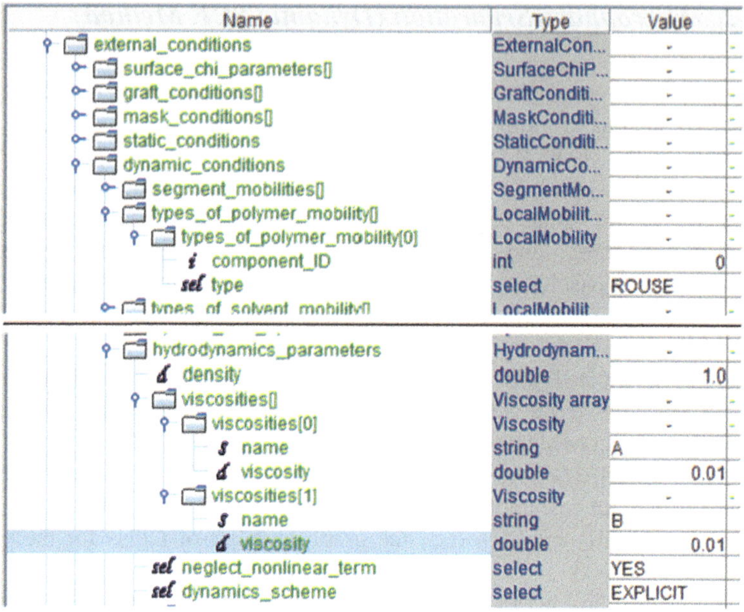

Fig. 5.17 Input UDF for a microphase separation 2/2

among the hydrodynamics_parameters, add two elements to viscosities[] in hydrodynamics_parameters, and set the names A and B for these elements and set both viscosities to 0.0001. We save the modified file as "block2D_7_dyr_hydro_uin.udf."

By invoking SUSHI in our environment, we can choose several run types as follows.

```
Message Passing Interface (MPI) parallel versions
Windows/MPICH2
> mpiexec -n 8 sushi*MPI -I block2D_7_dyr_hydro_uin.udf ....
Linux/OpenMPI
> mpirun -n 8 sushi*MPI -I block2D_7_dyr_hydro_uin.udf ....
Graphics processing unit (GPU) parallel version
> sushi*GPU -I block2D_7_dyr_hydro_uin.udf ....
One-core version
> sushi* -I block2D_7_dyr_hydro_uin.udf ....
```

We add the option "-t 1" for restart in the MPI calculation. In these types of runs, a GPU version may be fastest at such a scale.

After finishing the run, the output file named block2D_7_dyr_hydro_uot.udf is automatically generated. We open the output file and move the cursor to the SUSHIOutput subfolder, and clicking the right mouse button, a select window

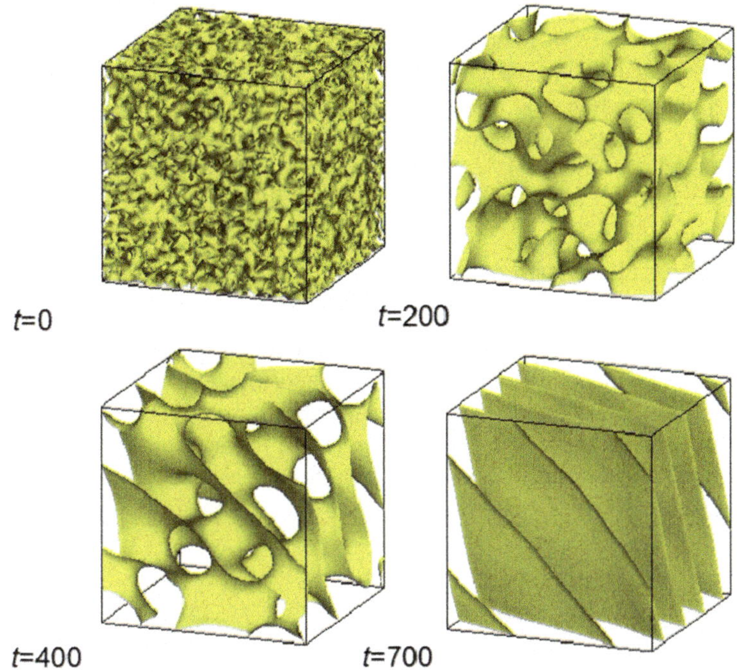

Fig. 5.18 Microphase separation structure of A–B diblock copolymer. Isosurfaces of the A segment with a system size of 16^3, mesh size of 32^3, $A_{10}B_{10}$ component, block ratio f of 0.5, χN of 20, and viscosity $\eta = 0.0001$

appears. We select show_field in the window and select the surface value ON. We see the isosurface three-dimensional graphic view window. We push the video "▷" button at the bottom left of the view window. We see an animation of microphase separation as shown in Fig. 5.18.

5.16 Example: Microphase Separated Structures of Diblock Copolymers

Block polymers generate domains of the length scales of 1–100 nm, and these phase-separated domains are promising candidates for functional materials in nanotechnologies [29, 30]. SCF theory has been noticed by the prediction of phase diagram of diblock copolymers in Fourier space by Matsen and Bates [18]. Instead of SCF theory in Fourier space, SUSHI uses SCF theory in real space. However, these two theories are based on the same SCF theory, and therefore, these theories give the same results. In this section, we optimize the microphase-separated structures of diblock copolymers and confirm the consistency of these two theories in the phase diagram of a diblock copolymer.

We use a diblock copolymer for which the chain length N is 20 and obtain all types of structures of microphase separation by changing the block ratio $f = N_A/(N_A + N_B)$ and χN. We start from $f = 0.5$ and $\chi N = 20$, which are the conditions for a lamellar structure, and move to a cylinder, BCC sphere, and gyroid structure condition by making epitaxial assumptions. The sample files can be downloaded from the website of this book.

5.16.1 Lamellar Structure

A lamellar structure can be simulated in one-dimensional space with the block ratio $f = 0.5$. The critical point of the spinodal decomposition of a lamellar structure is at $\chi N = 10.5$, as derived by the RPA calculation. We thus use $\chi N = 20$ and optimize the structure with system size optimization (SSO) in one dimension [31]. We get the structure shown in Fig. 5.19a. From this one-dimensional structure, we get the length of periodicity of the lamella D_L.

5.16.2 Cylinder Structure

A cylinder structure is packed hexagonally in a two-dimensional space. We assume that D_L is the same as length D_C, which is the cylinder spacing. This assumption satisfies

$$D_C = \frac{\sqrt{3}}{2} D_L. \tag{5.64}$$

In the actual calculation, we assume a two-dimensional rectangle cell for which the Y and X axis ratio is $\sqrt{3}$:1 for hexagonal packing. We set the block ratio and χN as $f = 0.35$ and $\chi N = 15$, respectively, for the SSO simulation. Furthermore, we set all boundary conditions of the cell to the reflective condition using the keyword "NEUMANN" for the UDF input of SUSHI. This condition satisfies the minimal space for the hexagonally packed cylinder. As shown in Fig. 5.19b, we get the minimal cut structure of the cylinder in two dimensions.

5.16.3 BCC Sphere Structure

We assume that D_C is the same as the length between BCC spheres. This assumption satisfies the relation

$$D_S = \frac{2}{\sqrt{3}} D_C, \tag{5.65}$$

where D_S is the edge length of the unit cubic cell of the BCC structure.

In the actual calculation, we use the three-dimensional cubic cell for which the edge length is set to $D_S/8$ with all boundary conditions set as "NEUMANN," which is the keyword that sets the reflective boundary condition in the SUSHI input UDF file, and set the block ratio and χN as $f = 0.25$ and $\chi N = 20$, respectively, for the SSO simulation. As shown in Fig. 5.19c, we get the cut structure of the BCC of one-eighth unit cell size.

5.16.4 Gyroid Structure

We make the assumption

$$D_G = \sqrt{6} D_L \tag{5.66}$$

for the gyroid structure, where D_G is the edge length of the gyroid unit cell. The gyroid structure has high spatial symmetry, and several assumptions are thus possible. Our assumption is one of those assumptions. Next, we set the block ratio and χN as $f = 0.35$ and $\chi N = 20$, respectively, for the SSO simulation and then assume the initial external field for the minor segment in the space given by

$$V(x, y, z) = V_0 \left(\cos\frac{2\pi x}{D_G} \sin\frac{2\pi y}{D_G} + \cos\frac{2\pi y}{D_G} \sin\frac{2\pi z}{D_G} + \cos\frac{2\pi z}{D_G} \sin\frac{2\pi x}{D_G} \right)^2, \tag{5.67}$$

where V_0 is a negative constant to accumulate under this field. This initial guess has been known empirically to produce a gyroid structure. The SSO under the condition gives the double gyroid structure as shown in Fig. 5.19d [31].

5.16.5 Fddd Structure

The Fddd structure is complicated, and we thus set the initial external field according to the result of Tyler and Morse [32] with the "domain specification" method implemented in SUSHI, which can set the initial external field properly, on the unit cell for which the edge length ratio of the lattice constant is given as $a{:}b{:}c = 1{:}2{:}\sqrt{3}$.

Figure 5.19e is the result of the run using those conditions. The Fddd structure can be obtained.

5.16.6 Phase Diagram of Microphase Separation of Diblock Copolymers Generated by SUSHI

We start with the structure of the diblock copolymer mentioned above. We continually restart by changing f and χN slightly. We get stable phases for wide ranges of

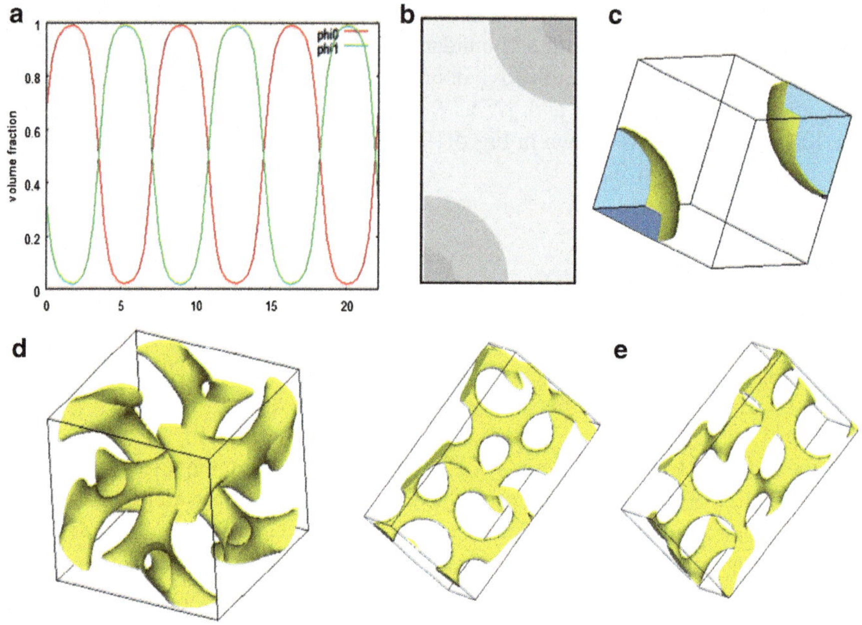

Fig. 5.19 Structures of a diblock copolymer obtained in an SCF calculation by SUSHI. (**a**) Lamella in one dimension, (**b**) part of a hexagonally packed cylinder in two dimensions, (**c**) isosurfaces of part of a BCC sphere in three dimensions, (**d**) isosurfaces of a double gyroid unit cell in three dimensions, and (**e**) isosurfaces of Fddd unit cells extracted in two different directions in three dimensions

f and χN because the SCF method gives the free energy of the optimized structure in SSO. We compare the result obtained with SUSHI with the phase diagram of the diblock copolymer obtained by Matsen and Bates [18] in Fig. 5.20. The calculation result of SUSHI well matches the phase diagram of Matsen and Bates.

5.17 Conclusion

SUSHI is a general purpose simulation program using DFTs for polymeric materials and covers from a small scale of polymer interfaces to a large scale of macro phase separations of polymer blends. SUSHI has been developing for large systems and can use MPI + GPU parallel calculations. We hope SUSHI be a good tool for researchers in the field of polymeric materials.

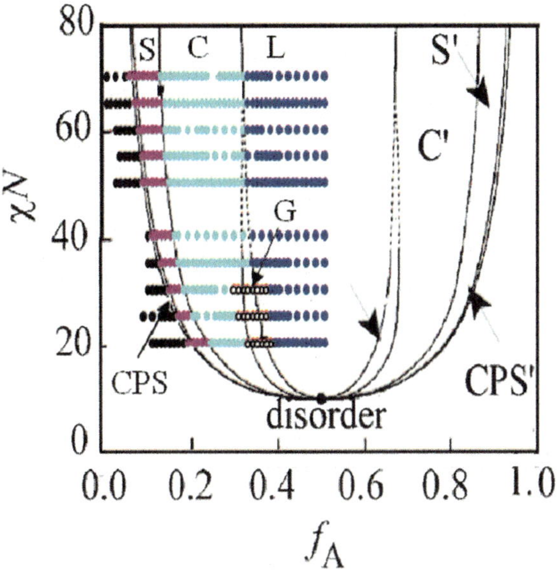

Fig. 5.20 Plot of the results of SUSHI on the phase diagram of the diblock copolymer obtained by Matsen and Bates [18]

5.18 Download Instruction

Sample files are available from the OCTA website. See the instruction on http://octa.jp/download_bookdata.html. Password for the sample(zip file) of this chapter is "CPAjRHtq."

Acknowledgment The author thank Prof. T. Kawakatsu in Tohoku University for many comments for this chapter and Prof. T. Koyama in Nagoya University for helpful information of phase field method. The implementation of parallel computation of SUSHI with MPI + GPU had been supported by the Global Scientific Information and Computing center in Tokyo Institute of Technology with supercomputer TSUBAME2.5. We thank Prof. T. Aoki, Mr. J. Sasaki, and Mr. Y. Matsumoto for giving us helpful information. The implementation of SUSHI with MPI library also had been supported by the Information Initiative Center in Hokkaido University with supercomputer SR16000/M1 and RIKEN Advanced Institute for Computational Science with K-computer. We thank Prof. M. Omiya in Hokkaido University, Dr. Hagita in National Defense Academy of Japan, and High Performance Computing Infrastructure in Japan. We also thank members of the technical seminar of polymer simulation organized by Japan Association for Chemical Innovation, for giving us many comments and opportunities of discussions.

References

1. J.G.E.M. Fraaije, J. Chem. Phys. **99**, 9202 (1993)
2. R. Hasegawa, M. Doi, Macromolecules **30**, 5490 (1997)
3. T. Kawakatsu, *Statistical Physics of Polymers—An Introduction* (Springer-Verlag, Berlin, 2004)

4. P.G. de Gennes, *Scaling Concepts in Polymer Physics* (Cornell University Press, Ithaca, 1979)
5. G.R. Strobl, *The Physics of Polymers* (Springer-Verlag, Berlin, 1997)
6. R. Koningsveld, W.H. Stockmayer, E. Nies, *Polymer Phase Diagrams* (Oxford University, New York, 2001)
7. H. Czichos, T. Saito, L. Smith, *Springer Handbook of Metrology and Testing* (Springer-Verlag, Berlin, 2011)
8. M. Doi, S.F. Edwards, *The Theory of Polymer Dynamics* (Oxford Science, Oxford, 1986)
9. L. Leibler, Macromolecules **13**, 1602 (1980)
10. T. Honda, T. Kawakatsu, Macromolecules **40**, 1227 (2007)
11. T. Honda, T. Kawakatsu, in *Nanostructured Soft Matter, Experiments, Theory and Perspectives: Computer Simulation of Nano-scale Phenomena based on the Dynamic Density Functional Theories Applications of SUSHI in the OCTA System*, ed. by A. Zvelindovsky (Springer-Verlag, Berlin, 2007)
12. P.G. de Gennes, J. Phys. (Paris) **31**, 235 (1970)
13. Y. Bohbot-Raviv, Z.-G. Wang, Phys. Rev. Lett. **85**, 3428 (2000)
14. G.J. Fleer, M.A. Cohen Stuart, J.M.H.M. Scheutjens, T. Cosgrove, B. Vincent, *Polymers at Interfaces* (Chapman & Hall, London, 1993)
15. E. Helfand, Z.R. Wasserman, Macromolecules **9**, 879 (1976)
16. E. Helfand, Z.R. Wasserman, Macromolecules **11**, 960 (1978)
17. E. Helfand, Z.R. Wasserman, Macromolecules **11**, 994 (1980)
18. M.W. Matsen, F.S. Bates, Macromolecules **29**, 1091 (1996)
19. A.V. Zvelindovsky, G.J.A. Sevink, B.A.C. van Vlimmeren, N.M. Maurits, J.G.E.M. Fraaije, Phys. Rev. E **57**, R4879 (1998)
20. A.V. Zvelindovsky, B.A.C. van Vlimmeren, G.J.A. Sevink, N.M. Maurits, J.G.E.M. Fraaije, J. Chem. Phys. **109**, 8751 (1998)
21. A.V. Zvelindovsky, G.J.A. Sevink, J.G.E.M. Fraaije, Phys. Rev. E **62**, R3063 (2000)
22. A.V. Zvelindovsky, G.J.A. Sevink, Europhys. Lett. **62**, 370 (2003)
23. H. Morita, T. Kawakatsu, Macromolecules **34**, 8777 (2001)
24. H. Morita, T. Kawakatsu, M. Doi, D. Yamaguchi, M. Takenaka, T. Hashimoto, Macromolecules **35**, 7473 (2002)
25. D.Q. Ly, T. Honda, T. Kawakatsu, A.V. Zvelindovsky, Macromolecules **40**, 2928 (2007)
26. D.Q. Ly, T. Honda, T. Kawakatsu, A.V. Zvelindovsky, Macromolecules **41**, 4501 (2008)
27. D.Q. Ly, T. Honda, T. Kawakatsu, A.V. Zvelindovsky, Soft Matter **5**, 4814 (2009)
28. T. Honda, T. Kawakatsu, J. Chem. Phys. **129**, 114904 (2008)
29. F.S. Bates, G.H. Fredrickson, Phys. Today **52**, 32 (1999)
30. I.W. Hamley, *Block Copolymers; Oxford* (Oxford University Press, Oxford, 1999)
31. T. Honda, T. Kawakatsu, Macromolecules **39**, 2340 (2006)
32. C.A. Tyler, D.C. Morse, Phys. Rev. Lett. **95**, 208302 (2005)

Chapter 6
PASTA and NAPLES: Rheology Simulator

Yuichi Masubuchi

6.1 Introduction

PASTA and NAPLES are simulation codes that have been developed for the long-time dynamics and rheology of entangled polymers.

Polymeric liquids have entangled dynamics when their concentrations and molecular weights are sufficiently high [1]. If the molecular weight of a polymer is low, the effect of surrounding molecules on the dynamics of the test chain can be described by the isotropic monomeric friction of a Rouse chain. In contrast, if the molecular weight and concentration exceed certain critical values (which depend on the chemistry of the molecule), appreciable retardation of the polymer dynamics is observed. This retardation is induced by a qualitative change in the molecular motion due to coupling between neighboring molecules, called entanglement. The mechanism of entanglement has not yet been fully elucidated; however, it has been established (by means of molecular simulations) that entangled dynamics of polymers can be reproduced if the polymer chains cannot cross one another.

Entangled polymer dynamics is difficult to simulate via atomic and coarse-grained molecular simulations [2]. As mentioned above, such dynamics can be reproduced if the uncrossability of polymer chains is guaranteed. An intuitive option is to attain the uncrossability via the excluded volume interactions. With recent advances in computer technologies, a few papers have conducted atomic [3] and coarse-grained [4] simulations of entangled polymers. Some available methods can attain uncrossability via inter-bond interactions instead of inter-bead interactions [5, 6]; however, given the very slow dynamics of the entangled polymers, the strategy of using uncrossability is difficult to practically implement. (Note that a

Y. Masubuchi (✉)
National Composite Center, Nagoya University, Nagoya, Japan
e-mail: mas@nuap.nagoya-u.ac.jp

Japan Association for Chemical Innovation, *Computer Simulation of Polymeric Materials*, DOI 10.1007/978-981-10-0815-3_6

naïve dissipative particle dynamics method cannot reproduce entangled polymer dynamics because the polymer chains cross each other owing to the soft-core nature of the inter-bead interactions [7].)

For further coarse graining, the modeling of entanglement has been challenging. One rigorous method involves determining the mobility matrix using the projection operator technique [8–10]. This approach has been realized but has not progressed owing to the fundamental difficulty of describing dynamics under fast flow, for which polymer rheology is of industrial importance. The other direction of theoretical development involves hypothetical modeling of entangled polymer dynamics. The use of the temporal network model [11, 12] is an intuitive approach based on the experimental finding that entangled polymers have a rubberlike plateau modulus. The rubberlike behavior can be modeled by temporal cross-links that have a lifetime comparable to the relaxation time of the material. However, the temporal network model alone cannot explain the dependence of the relaxation time on molecular shape.

To date, the most successful approach has been the tube model, in which the entangled polymer dynamics is represented by the motion of a single chain confined in a tube [13–15]. This model setup results in a marked suppression of the molecular mobility in directions perpendicular to the chain backbone. It is thus assumed that the transition between the unentangled and entangled dynamics corresponds to the switch between the isotropic mobility (Rouse model) to the anisotropic mobility (tube model). The chain motion along the backbone due to the anisotropic mobility is known as "reptation." Although the mechanism of this transition has not yet been elucidated, the tube model with the hypothetical anisotropy of the molecular motion has been remarkably successful in explaining the characteristic features of entangled polymer dynamics.

Owing to its simplicity and success in describing some basic features of entangled polymer dynamics, the tube model has been elaborated toward extensions that reproduce experimental data in a semiquantitative manner. In addition to reptation, several other aspects of chain dynamics are considered important. Contour length fluctuation (CLF) is the springlike motion of the chain inside the tube [16], whereas constraint release (CR) is the fluctuation of the tube in directions perpendicular to its backbone [17]. CR represents the entanglement/disentanglement of the surrounding chains with the test chain. The implementation of CLF has been largely established [18, 19]. Meanwhile, CR remains unsettled owing to its multichain nature. Many tube models in which the implementation and approximation for CR are elaborated in different ways have been proposed [17, 20–22]. Some of the theories that implement CLF and CR have successfully predicted the linear viscoelasticity of polymers with a molecular weight distribution [2]. However, the nonlinear viscoelasticity of such systems has not yet been solved via the tube picture owing to the difficulty of implementation for excess CR dynamics induced under fast flow (i.e., so-called convective constraint release (CCR) [23]).

To overcome the fundamental difficulties relating to the single-chain description employed in the tube picture, computer codes have been developed to solve the multichain reptation dynamics numerically (specifically for CR and CCR) [24–31].

PASTA and NAPLES are such codes. These codes are slow with respect to the original tube model owing to the multichain calculations. However, they are capable of calculating the rheology and dynamics of a complicated mixture of linear and branch polymers even for fast flows. Such calculation has not been realized in tube theories, and the calculation remains sufficiently faster than that of the coarse-grained molecular simulations in COGNAC. The efficiency of the calculation is due to the fact that the degree of freedom below the entanglement molecular weight is completely neglected (just as in the tube model). This model setup is incompatible with several problems in which molecular interactions at the atomic scale are important; for example, crystallization, glass transition, and interfacial phenomena cannot be considered.

6.2 Model

6.2.1 Slip-Link Model

PASTA and NAPLES are based on the slip-link description of the entangled polymers (see Fig. 6.1). In the slip-link models, an ensemble of reptating chains is considered. The chain consists of several segments for which the molecular weight is similar to the entanglement molecular weight. Each segment is constrained via a slip-link. The slip-link does not hinder the sliding motion of the chain along its backbone but suppresses the chain dynamics in directions perpendicular to the backbone. The slip-link is destroyed when one of the chains slides off. In contrast, a new slip-link is created if a certain amount of the Kuhn segment protrudes from the slip-link at the chain end. CR and CCR are implemented through the coupling of slip-links. Specifically, if one of the slip-links is removed as a result of the end

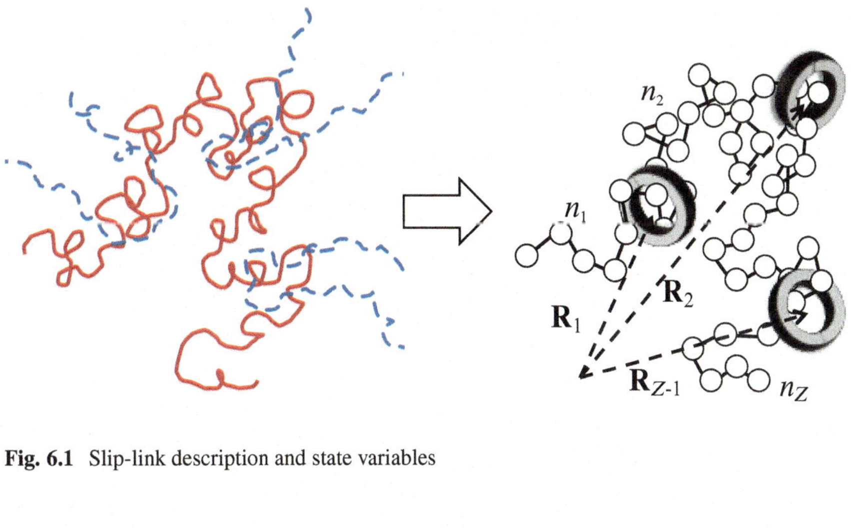

Fig. 6.1 Slip-link description and state variables

sliding, another slip-link on another chain is removed as well, reflecting the binary assumption of entanglement. With the creation of a new slip-link, another slip-link is created at a chosen segment. Note that the coupling between slip-links depends on the model assumptions, and its application with respect to PASTA and NAPLES is explained below.

The state variables of the system are (1) the slip-link position $\{\mathbf{R}\}$, (2) the number of Kuhn segments between entanglements $\{n\}$, and (3) the number of slip-links on each chain $\{Z\}$. The problem is how to describe their time developments. (Indeed, the difference between PASTA and NAPLES is the difference in the set of equations for the time development of these state variables.) In this description, the units of length, time, and energy are the average length between slip-links along the chain, the Rouse relaxation time of the subchain between entanglements and the thermal energy $k_B T$. The chemistry details of the polymer chain are embedded in the unit length and unit time.

6.2.2 Slip-Link Model for PASTA

The model for PASTA is called the Doi–Takimoto (DT) model [27]. The slip-link position $\{\mathbf{R}\}$ is developed affinely according to the flow:

$$\frac{d\mathbf{R}_i}{dt} - \boldsymbol{\kappa} \cdot \mathbf{R}_i = 0 \tag{6.1}$$

where κ is the velocity gradient tensor. The number of Kuhn segments between slip-links $\{n\}$ is assumed to be proportional to the distance between the slip-links. This means that the tension is not a function of the curvilinear position along the chain but an average value. The length of the end subchain changes with respect to the curvilinear diffusion of the chain along its backbone. (Note that the position of the chain end is traced in addition to the slip-link position.) The curvilinear diffusion corresponds to the reptation. The diffusion constant of the curvilinear diffusion is given by

$$D = \frac{a^2}{3\pi^2 \tau_0^{\mathrm{DT}} Z_0}, \tag{6.2}$$

where a and Z_0 are, respectively, the unit length (i.e., the average length between slip-links) and the average slip-link number per chain, both under equilibrium. τ_0^{DT} is the unit time of the DT model. In addition to the curvilinear diffusion of the chain, the chain length changes in time according to

$$\frac{dL}{dt} = -\frac{1}{\tau_{\mathrm{R}}}\left(L - L_{\mathrm{eq}}\right) + \dot{L}_{\mathrm{affine}} + \sqrt{\frac{2a^2}{3\tau_0^{\mathrm{DT}} Z_0 \Delta t}}\,g, \tag{6.3}$$

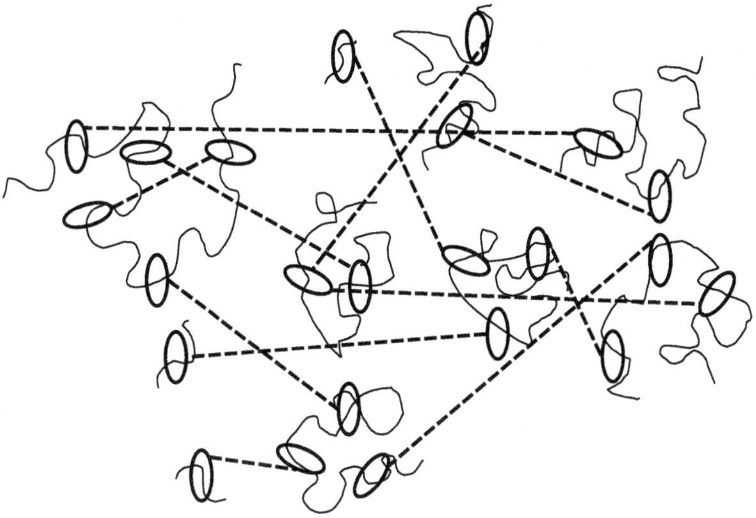

Fig. 6.2 Pairing between slip-links in the DT model. *Solid curves* are the polymers, and circles are the slip-links. *Broken lines* are the pairing between slip-links

where L is the curvilinear chain length, L_{eq} is its equilibrium value and $\dot{L}_{affine}$ is the contribution of the deformation/flow. g is a Gaussian random number with zero mean and variance at unity. Δt is the time step for numerical integration. τ_R is the Rouse relaxation time of the chain as a whole and given by $\tau_R = \tau_0^{DT} Z_0^2$. The springlike motion induced by the first term on the right-hand side corresponds to CLF.

The time development of $\{Z\}$ is induced by the reptation (curvilinear diffusion) and CLF (springlike motion) of the chain. When the chain end slides off from the adjoined slip-link, the subjected slip-link is removed. Meanwhile, when the chain end protrudes from the adjoined slip-link beyond unit length a, a new slip-link is created on the end segment. Upon slip-link creation, the direction of the end segment is randomized. According to the binary assumption of entanglement, each slip-link is coupled to another slip-link in a pair as shown in Fig. 6.2. When a slip-link is removed (as a result of sliding of the chain end), the coupled slip-link is also removed. Meanwhile, when a slip-link is created at a chain end, another slip-link is created at a randomly chosen subchain to be coupled.

The DT model can simulate the dynamics of star-branched polymers as well as linear polymers. In this case, the branch point position is fixed in space, and the reptation and CLF of the branching arm are prohibited beyond the branch point. Note that the diffusion of the star polymer is assumed to be zero. It is also noted that the calculated rheology is completely insensitive to the number of branching arms from the branch point unless the coupling between slip-links depends on the connectivity and the curvilinear distance of the slip-link from the branch point. Nevertheless, the idea of asymmetric reptation for star polymers has been introduced for the tube model earlier [32], and it has been experimentally confirmed that the rheology of the star polymer is almost insensitive to the number of arms given that the number of arms is chosen between 4 and 10 [33].

6.2.3 Slip-Link Model for NAPLES

The model for NAPLES is called the primitive chain network (PCN) model [26]. In this model, the coupling between slip-links is considered in a three-dimensional network as shown in Fig. 6.3. Owing to the real-space nature of the PCN model, the force balance around the slip-link, diffusion of the branch point and spatial correlation between chains can be considered.

In the PCN model, the time development of $\{\mathbf{R}\}$ and $\{n\}$ is described by kinetic equations that take account of the drag force, balance of tension around slip-links, osmotic force due to the density fluctuation, and thermal agitation:

$$\frac{d\mathbf{R}_i}{dt} - \boldsymbol{\kappa} \cdot \mathbf{R}_i = \frac{n_0}{2\tau_0^{\mathrm{PCN}}} \sum_j^4 \frac{\mathbf{r}_j}{n_j} - \frac{n_0 a^2}{3kT\tau_0^{\mathrm{PCN}}} \nabla \mu^{\mathrm{PCN}}(\mathbf{R}) + \sqrt{\frac{a^2}{2\tau_0^{\mathrm{PCN}} \Delta t}} \mathbf{W}_i \tag{6.4}$$

$$2\left(\frac{n_i}{r_i} + \frac{n_{i-1}}{r_{i-1}}\right)^{-1} \frac{dn_i}{dt} = \frac{n_0}{2\tau_0^{\mathrm{PCN}}} \left(\frac{r_i}{n_i} - \frac{r_{i-1}}{n_{i-1}}\right) - \frac{n_0 a^2}{12kT\tau_0^{\mathrm{PCN}}} \nabla \mu^{\mathrm{PCN}} + \sqrt{\frac{a^2 n_0^2}{6\tau_0^{\mathrm{PCN}} \Delta t}} W \tag{6.5}$$

where $\mathbf{r}_j$ is the subchain vector (the bond vector between adjacent slip-links), n_j is the number of Kuhn segments on the subchain, n_0 is the average value of n_j at

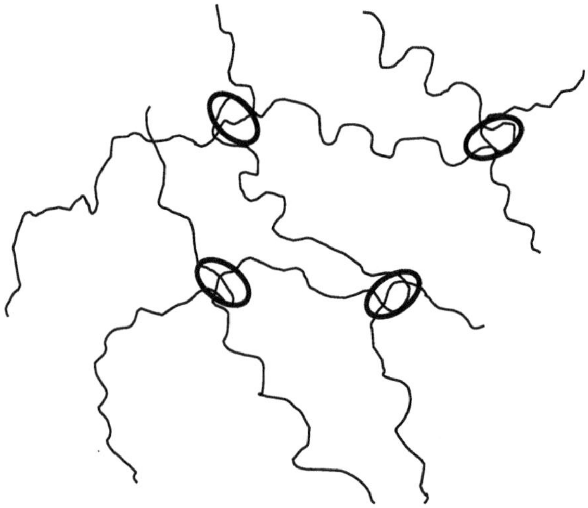

Fig. 6.3 PCN model

equilibrium, $\tau_0^{\mathrm{PCN}} = \zeta n_0 a^2/6kT$ is the unit time (see Sect. 6.3.4), ζ is the friction coefficient of the Kuhn segment, $\mathbf{W}$ is a random vector with length 1, and W is a random number taking a value of $+1$ or -1. μ^{PCN} is the chemical potential that reduces the density fluctuation induced by the artificial attractive interaction from the bundling of the chains [34]. μ^{PCN} is obtained from the free energy density written as [35]

$$\frac{F^{\mathrm{PCN}}(\mathbf{R})}{kT} = \begin{cases} \varepsilon\left(\dfrac{\phi_\alpha(\mathbf{R}) + \phi_\beta(\mathbf{R})}{\langle\phi_\alpha + \phi_\beta\rangle} - 1\right)^2 + \chi\phi_\alpha(\mathbf{R})\,\phi_\beta(\mathbf{R}) & \text{for } \phi_\alpha(\mathbf{R}) + \phi_\beta(\mathbf{R}) > \langle\phi_\alpha + \phi_\beta\rangle \\ 0 & \text{for } \phi_\alpha(\mathbf{R}) + \phi_\beta(\mathbf{R}) \le \langle\phi_\alpha + \phi_\beta\rangle \end{cases} \tag{6.6}$$

where ε is a numerical prefactor fixed at 0.5 empirically. $\phi_\alpha(\mathbf{R})$ and $\phi_\beta(\mathbf{R})$ are the number densities of the subchain at position $\mathbf{R}$ for chemistry α and β, respectively. $\langle\phi_\alpha + \phi_\beta\rangle$ denotes the average for the system as a whole. χ is the interaction parameter. Owing to the free energy description, the PCN model has been examined for entangled polymer blends and copolymers [35, 36]. However, such systems are not discussed further in this chapter.

The time development of $\{Z\}$ is induced by the creation/destruction of slip-links around chain ends. To trigger the creation/destruction events, the number of Kuhn segments at the end subchain, n, is monitored. If n/n_0 becomes smaller than the critical value fixed at 0.5, then the adjacent slip-link is removed and the partner chain is released from the constraint. Meanwhile, if n/n_0 becomes larger than a certain maximum value (fixed at 1.5), a new slip-link is created on the end subchain. Upon the creation of a new slip-link, another subchain is hooked for the coupling. This partner subchain is randomly chosen from the surroundings within a certain distance (fixed at unit length a) from the chain end. For the case of branch polymers, an additional mechanism that realizes relaxation of the backbone chain is introduced [37–39].

6.2.4 *Stress Tensor*

For PASTA and NAPLES, the stress tensor is written as

$$\sigma_{xy} = 3\upsilon kT\frac{n_0}{a^2}\left\langle\frac{r_x r_y}{n}\right\rangle, \tag{6.7}$$

where $\mathbf{r}$ is the subchain vector (with components r_x, r_y, and r_z), a is the average subchain length under equilibrium, n is the number of Kuhn segments on the subchain, n_0 is the average of n under equilibrium and υ is the number density of subchain. $\langle \dots \rangle$ indicates the ensemble average. Note that PASTA does not explicitly consider n, which is assumed to be proportional to the subchain length. Equation 6.7 demonstrates that the effects of intermolecular and intramolecular interactions from specific chemistries are embedded in the parameters and do not appear explicitly. Nevertheless, Eq. 6.7 is based on the stress-optical law that has been experimentally established [15].

6.3 Model Parameters

6.3.1 Overview

As discussed in Sect. 6.2.1, for slip-link models, the parameters are the unit of length, unit of time, and unit of energy. These units are chosen as follows. The unit of length is the average subchain length at equilibrium a. The unit of time is the Rouse relaxation time of the subchain containing n_0 Kuhn segments. The unit of energy is k_BT. For practical ease, instead of the parameters mentioned above, another set of parameters—the unit of molecular weight, unit of modulus, and unit of time—is used.

6.3.2 Molecular Weight

For PASTA and NAPLES, the average number of entanglements per chain (under equilibrium) Z_0 represents the molecular weight M of the polymer chain. The relationship between Z_0 and M is written as

$$Z_0 = \frac{M}{M_K n_0}, \tag{6.8}$$

where n_0 is the average number of Kuhn segments on the subchain under equilibrium and M_K is the molecular weight of the Kuhn segment. For the case of solutions (entangled solutions), n_0 depends on the polymer concentration. Because the value of n_0 depends on the model settings [40], the values of n_0 and Z_0 are discriminated for PASTA and NAPLES. For this reason, instead of n_0 and Z_0, Z_0^{PASTA}, Z_0^{NAPLES}, n_0^{PASTA}, and n_0^{NAPLES} will be used as necessary. The unit of molecular weight is defined as $M_0 = M_K n_0$ (such that $M_0^{\mathrm{PASTA}} = M_K n_0^{\mathrm{PASTA}}$ and $M_0^{\mathrm{NAPLES}} = M_K n_0^{\mathrm{NAPLES}}$ for PASTA and NAPLES, respectively).

For convenience, Z_0 is obtained from the plateau modulus G_N via the equation

$$\frac{4}{5} Z_0^{\mathrm{PASTA}} = \frac{2}{3} Z_0^{\mathrm{NAPLES}} = \frac{M}{M_e^{\mathrm{F}}} = \frac{M}{1.25 M_e^{\mathrm{G}}} = \frac{G_N}{\rho RT} M, \tag{6.9}$$

where M_e^F and M_e^G are the entanglement molecular weight according to the definitions proposed by Ferry and Graessley, respectively [41]. ρ is the density and R is the gas constant. The value of M_e^G can be found in POLYMERDATABASE included in the OCTA system. From GOURMET, we open the user definable format (UDF) file

"C:¥OCTA8¥ENGINES¥POLYMERDATABASE¥polymerdata.udf"

and set the view as Table mode. Tabulated M_e^G values for some polymers can then be seen in

PolymerDatabase.GeneralPolymers[].Properties[].Me

(see Fig. 6.4 for an example).

In the data table of polymerdata.udf, for some polymers, the value of M_e^G is indicated as −1.0E99. This value means that the M_e^G value is not available. For such polymers (and other such polymers not in the database), M_e^G can be determined from the packing length theory proposed by Fetters [42]. According to this theory, M_e^G can be obtained from the packing length p using the relations

$$p = \frac{M}{\langle R^2 \rangle \rho N_A}, \tag{6.10}$$

$$M_e^G = 218 \rho p^3, \tag{6.11}$$

where $\langle R^2 \rangle$ is the average square end-to-end distance of the polymer with molecular weight of M, ρ is the density, and N_A is Avogadro's number. The value $M/\langle R^2 \rangle$ can be obtained from atomistic molecular simulation via COGNAC. Another option using COGNAC is primitive path analysis [43]. In this approach, the geometrical entanglement network is extracted from a given snapshot of the atomistic simulation. Structural analysis of the obtained network gives the value of M_e^G. Details can be found in the literature [43].

6.3.3 Unit Modulus

The unit modulus is needed for the rheology calculation. The modulus can be obtained from the unit length and unit energy. Meanwhile, for convenience, the unit modulus can also be obtained from the plateau modulus found in the literature (or obtained experimentally). The unit modulus for PASTA, G_0^{PASTA}, and that for NAPLES, G_0^{NAPLES}, can be obtained as [44]

$$G_0^{\mathrm{PASTA}} = \frac{15}{4} G_N = 3 \frac{\rho R T}{M_e^G}, \tag{6.12}$$

$$G_0^{\mathrm{NAPLES}} = \frac{3}{2} G_N = \frac{6}{5} \frac{\rho R T}{M_e^G}. \tag{6.13}$$

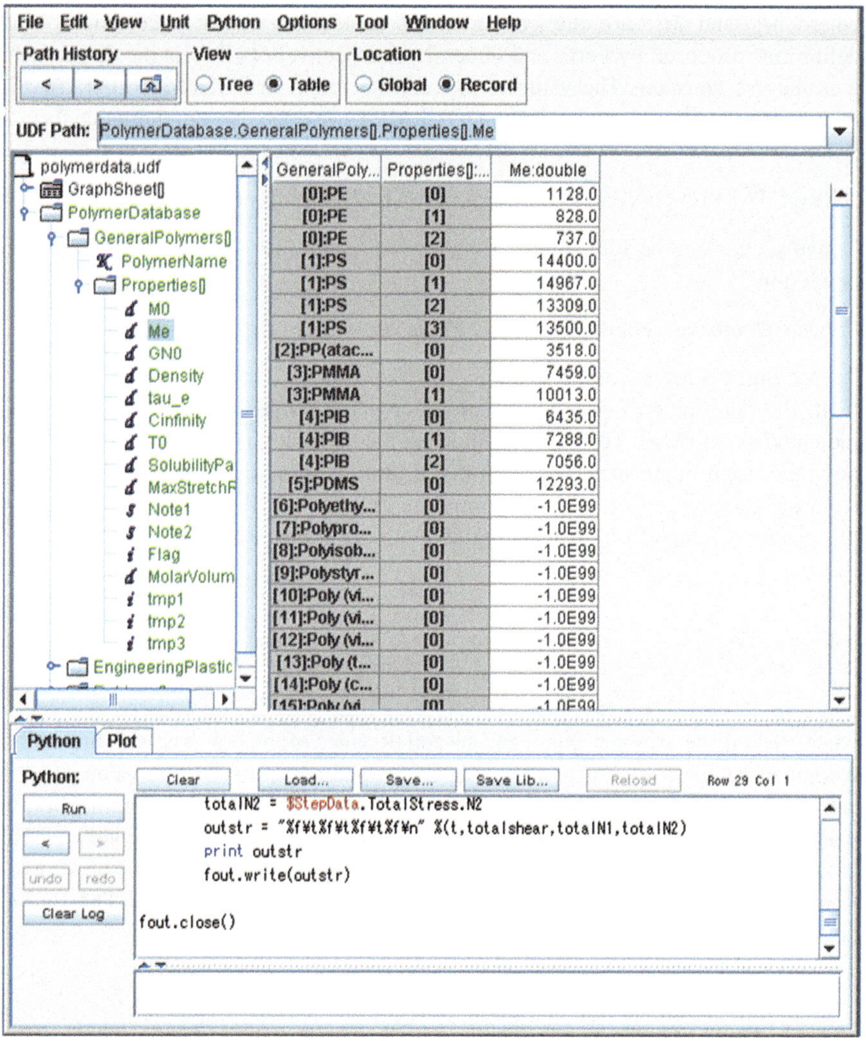

Fig. 6.4 Tabulated M_e^G values in polymerdata.udf shown via GOURMET

The plateau modulus G_N can be found in POLYMERDATABASE included in the OCTA system. From GOURMET, we open the UDF file

"C: ¥OCTA8¥ENGINES¥POLYMERDATABASE¥polymerdata.udf"

and set the view as Table mode. Tabulated G_N values for some polymers can then be seen in

PolymerDatabase.GeneralPolymers[ ].Properties[ ].GN0.

For the polymers for which G_N is not available, the packing length theory may be used as mentioned in the previous section to obtain M_e^G, and Eqs. 6.12 and 6.13 can then be used. The density ρ is available at

PolymerDatabase.GeneralPolymers[].Properties[].Density.

Note that M_e^G is insensitive to temperature whereas G_N depends on temperature as shown in Eqs. 6.12 and 6.13.

6.3.4 Unit Time

The unit times for PASTA and NAPLES, τ_0^{DT} and τ_0^{PCN}, were defined for the kinetic equation in Sect. 6.2. The unit time roughly corresponds to the Rouse relaxation time of a polymer chain with molecular weight of M_0^{PASTA} or M_0^{NAPLES}. However, owing to the numerical prefactor, τ_0^{NAPLES} is several times larger than τ_0^{DT}, even though $M_0^{PASTA} > M_0^{NAPLES}$. Nevertheless, τ_0^{DT} and τ_0^{PCN} can be determined from the chemistry and temperature via molecular dynamics simulations in COGNAC.

For practical convenience, τ_0^{DT} and τ_0^{PCN} are determined by fitting rheology data to those obtained experimentally.

1. Choose a target polymer and the flow conditions to be simulated.
2. Seek linear viscoelastic data for the target polymer or similar material. The chemistry should be the same, whereas other molecular characteristics such as the molecular weight, molecular weight distribution, and branching are not necessarily the same. The unit values are independent of these structural parameters. Rather, data for a monodisperse linear polymer are preferable. The temperature difference can be accommodated. The polymer concentration (in the case of a solution) can also be accommodated, but the dataset for the target concentration is better.
3. Determine the Z_0 value for the polymer examined in the experiment (see Sect. 6.3.2).
4. Construct the input UDF file with the determined Z_0 and conduct the simulation (via PASTA or NAPLES) to obtain the linear viscoelasticity (see Sect. 6.4).
5. Compare the calculated linear viscoelasticity with the experimental data. Note that the calculated modulus is normalized with respect to G_0 (see Sect. 6.3.3). At this stage, if the values of the modulus are appreciably different from each other, the estimate of Z_0 should be reconsidered. If not, the unit time can be obtained via the fitting (for the frequency in the case of dynamic viscoelasticity).
6. Accommodate the unit modulus and the unit time to the temperature if the target temperature is different from that for the fitted experimental data. For the unit time, the WLF equation [1]

$$\frac{\tau_0(T_1)}{\tau_0(T_2)} = \frac{a_T(T_1)}{a_T(T_2)} \tag{6.14}$$

can be used. Here, τ_0 (T) is the unit time and $a_T(T)$ is the shift factor for the frequency. The WLF equation for $a_T(T)$ is written as

$$\log a_T = \frac{-17.37\left(T - T_{\mathrm{g}}\right)}{51.65 + T - T_{\mathrm{g}}}, \tag{6.15}$$

where T_{g} is the glass transition temperature. Unfortunately, T_{g} is not available in POLYMERDATABASE in OCTA, but it can be found in the literature.

6.4 Example 1: Calculation of Linear Viscoelasticity and Determination of Unit Time

6.4.1 Overview

In this section, operations for PASTA and NAPLES are explained to determine the unit time via linear viscoelasticity. Unlike the case of an experiment, in simulations, the linear viscoelasticity can be obtained from the fluctuation of stress under equilibrium via the Green–Kubo formula. The obtained linear response function is then converted into dynamic viscoelasticity to be compared with experimental data for the determination of the unit time. In this section, the dataset for a linear monodisperse polystyrene melt ($M = 83$ k, $T = 167$ °C) obtained by Watanabe et al. [45] is used. Once the unit time is determined, simulations for nonlinear viscoelasticity and/or for different shapes of polymers (with the same chemistry) can be conducted as explained in Sect. 6.5.

6.4.2 Input UDF File for PASTA

From GOURMET, we open the template file for PASTA available as

"C: ¥OCTA8¥ENGINES¥PASTA¥PASTA¥sample¥pasta_test_in.udf".

The contents are as shown in Fig. 6.5.

We start from the settings of the sample to be simulated (see the Sample folder (in Tree View)). The values stored in boxes are the molecular weight (with respect to Z) in *Sample.Linear[0]*, the maximum stretch ratio in *Sample.MaxStretchRatio*, and the number of molecules in *Sample.NumChains*. The default values are already set in the file. For the case of a mixture of polymers with different molecular weights, we add another polymer by adding *Sample.Linear[1]* and so on. For the case of a star-branched polymer, we open the *Sample.Arm* folder and add the arm chain. (The detail of the operations for such modifications of the UDF file can be found in the GOURMET manual.)

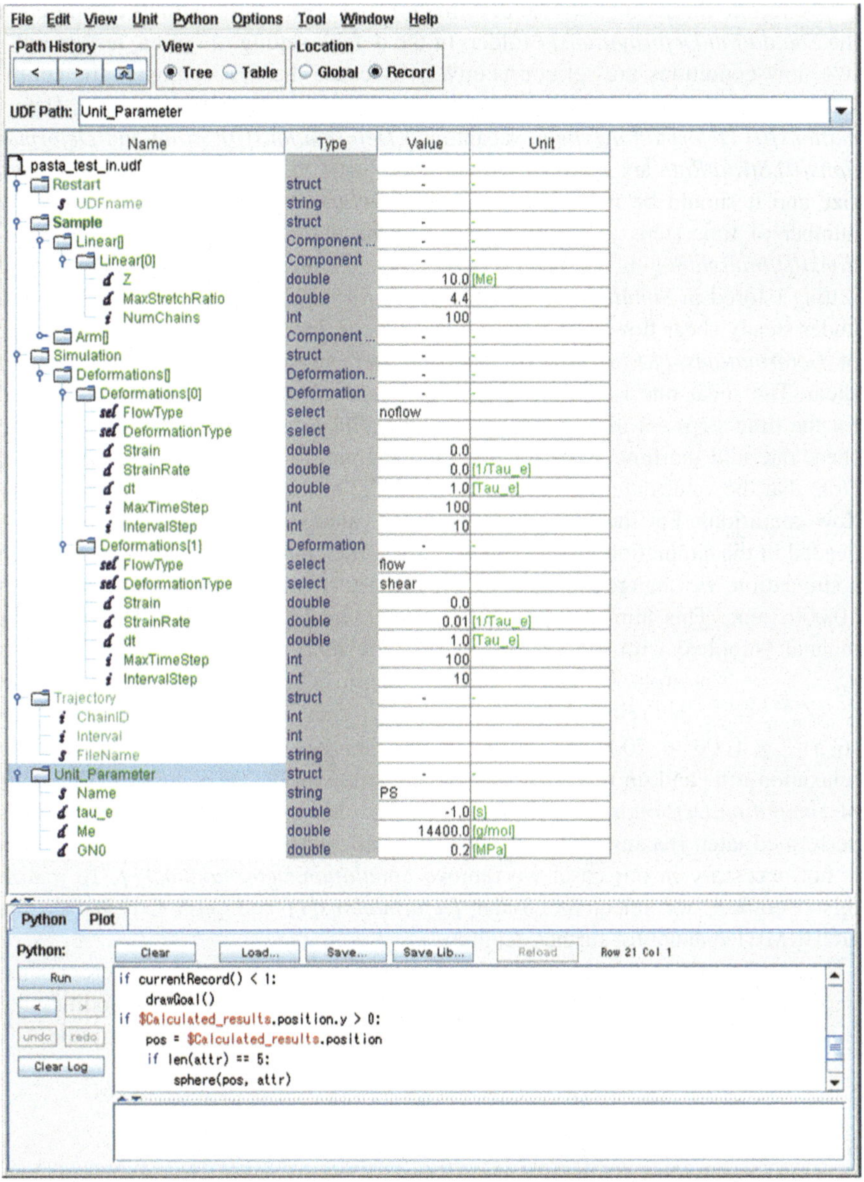

Fig. 6.5 "pasta_test_in.udf"

We set the value of Z for our target sample (i.e., the polystyrene melt with $M = 83{,}000$ as mentioned above) using Eq. 6.9. According to "polymerdata.udf" (see Sect. 6.3.2 for details), the M_e^G value of polystyrene melts is 14,400. According to this M_e^G value, $Z_0^{PASTA} = 83000/14400 \approx 6$. We put this Z_0^{PASTA} value in the *Sample.Linear[0]* box into the "pasta_test_in.udf" file.

We turn to the settings for the deformation and flow. These settings are stored in the *Simulation.Deformations[]* folder. In the default setting of "pasta_test_in.udf", two flow conditions are set consecutively. Because the first setting is the equilibration under no flow, *Deformations[0].FlowType* is "noflow". In this case, *Deformations[0].DeformationType* is vacant, and *Deformations[0].Strain* and *Deformations[0].StrainRate* are zero. *Deformations[0].dt* is the numerical integration step size and it should be no larger than unity. *Deformations[0].MaxTimeStep* is the number of time steps to be simulated, and the default setting is 100. *Deformations[0].IntervalStep* is the interval of the time step for the data output. The second setting (stored in *Simulation.Deformations.Deformations[1]*) is for the simulation under steady shear flow. The application of shear flow is achieved via the settings of *Deformations[1].FlowType* of flow and *Deformations[1].DeformationType* of shear. The shear rate is put into *Deformations[1].StrainRate*. The flow is applied for the time steps set in *Deformations[1].MaxTimeStep*. In the default setting, the shear rate and the time steps for flow application are 0.01 and 100, respectively. Note that the values are normalized with respect to the unit time. Now, we set our flow conditions. For the linear viscoelasticity calculation, the stress fluctuation is needed in the simulation under no flow for a sufficiently long time. To realize such a simulation, we change the value of *Simulation.Deformations[0].MaxTimeStep* to 10,000 steps. This number of time steps is estimated as follows. In the PASTA manual (supplied with the OCTA system), the relaxation time for a chain with $Z_0^{\mathrm{PASTA}} = 5$ is around 1000 steps. Because the relaxation time is proportional to $(Z_0^{\mathrm{PASTA}})^{3.4}$, the relaxation time for $Z_0^{\mathrm{PASTA}} = 6$ is roughly estimated as $(6/5)^{3.4} \times 1000 \approx 2000$. The value of *MaxTimeStep* should be several times the relaxation time and, in this particular case, a value of 10,000 is chosen. The value of *Simulation.Deformations[0].IntervalStep* should be set at unity for the analysis performed later. Because the shear flow part following the equilibrium calculation is not necessary in our case, we remove *Simulation.Deformations[1]*. To realize this operation, we select the folder *Deformations[1]* and push Ctrl-D (see the GOURMET manual for further details).

Figure 6.6 shows the modified UDF file. We save the file under a different name to preserve the original "pasta_test_in.udf". In this particular case, the modified file is saved as "pasta_Z06_in.udf", and the file can be found in the sample zip file.

6.4.3 Running PASTA

This section explains how to run PASTA code via the command line (see the PASTA manual for the graphical-user-interface-based method). The input file used hereafter is "¥pasta_Z06_in.udf", which was constructed in earlier sections.

We launch GourmetTerm (see the GOURMET manual for details) and start PASTA from GourmetTerm as shown in Fig. 6.7. In this specific case, the output UDF file is "pasta_Z06_out.udf", and the working directory is assumed to be C:¥OCTA_Textbook¥2_4_2.

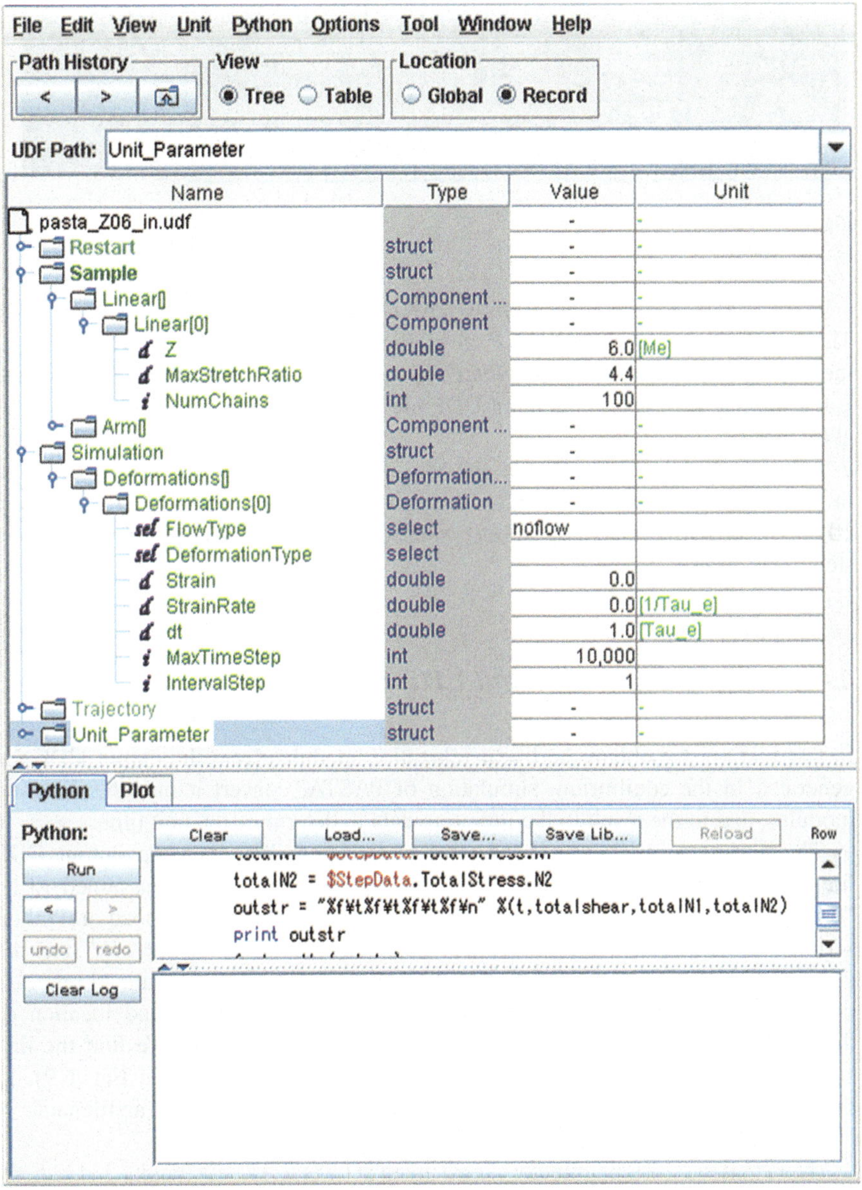

Fig. 6.6 "pasta_Z06_in.udf"

Fig. 6.7 Running PASTA via GourmetTerm

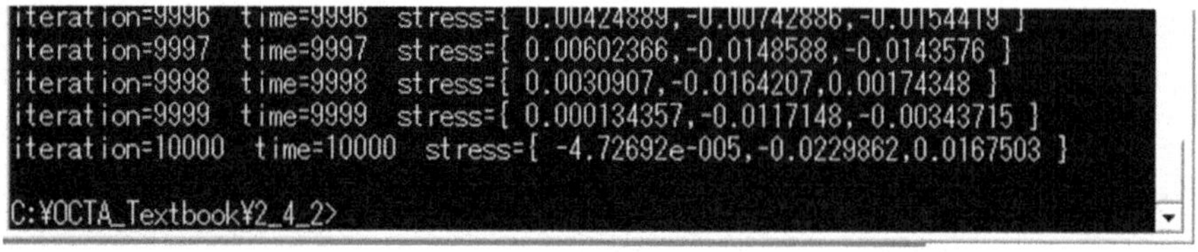

Fig. 6.8 Dumped table from PASTA

If PASTA is executed successfully, PASTA dumps a table as shown in Fig. 6.8. The leftmost column "iteration=" gives the number of steps that will be incremented up to the value *Simulation.Deformations[0].MaxTimeStep*. Because the value was set at 10,000 in the input UDF file in the previous section, the code stops at this value.

In addition to the dumped table, the output file "pasta_Z06_out.udf" is generated in the working directory. We open this output file from GOURMET. There are 10,000 records, and the values of stress tensor components are stored at each time step.

6.4.4 Analysis of the Output UDF File for PASTA

In this section, we obtain the linear relaxation modulus from the output UDF file generated in the equilibrium simulation of PASTA, convert it into the dynamic modulus, and fit the result to the literature data to determine the unit time.

The first step is to extract the stress fluctuation. We open the output UDF file via GOURMET (where, in this section, the output UDF is assumed to be "pasta_Z06_out.udf" owing to the previous section). We then load a Python script, "¥OCTA8¥ENGINES¥PASTA¥PASTA¥python¥pasta_stress.py", into the Python window in GOURMET. (To load the script, we push the "Load…" button in the Python window.) In the loaded script, the name and location of the result file containing the stress fluctuation should be given. We find the line starting with "filename=", and change the value accordingly (see Fig. 6.9). In this specific case, the file is to be "stress.txt" and we set the value as filename = "C:¥¥OCTA_Textbook¥¥2_4_2¥¥stress.txt".

(Note that the working directory is assumed at C:¥OCTA_Testbook¥2_4_2 here, and in the file path, the directory (folder) is given not by ¥ but by ¥¥ according to the Python rule.) The Python script in which the filename has already been modified is provided for convenience (i.e., "pasta_stress_rev.py").

After the setting of "pasta_stress.py" (or the loading of "pasta_stress_rev.py"), we run the Python script using the Run button located on the left side of the Python window. If successful, the stress.txt file is generated in the working directory. If not, error messages will appear in the window below the Python window.

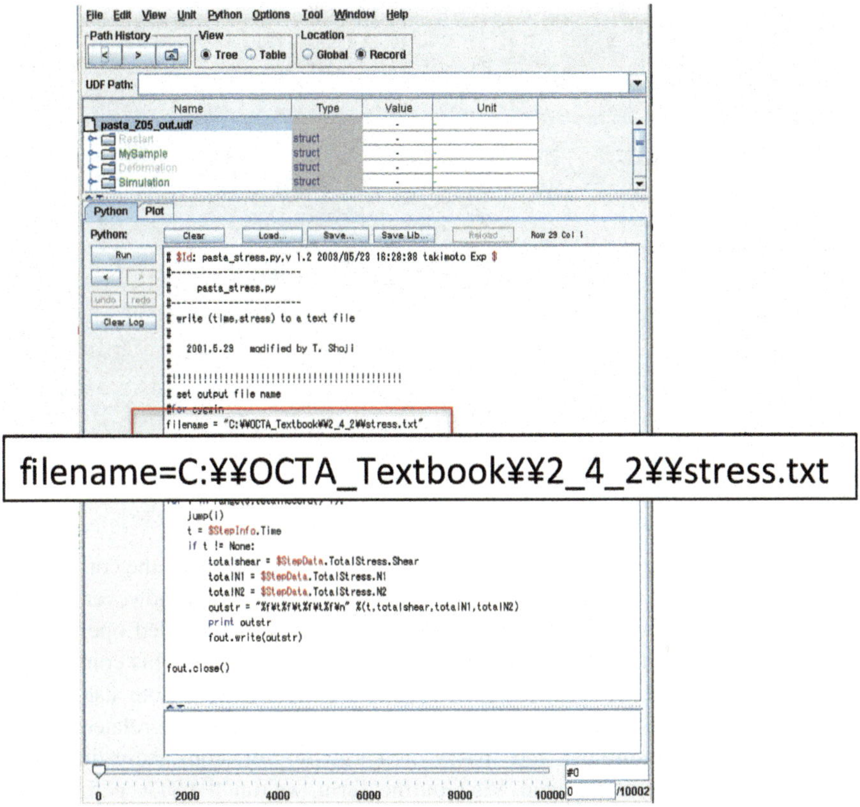

Fig. 6.9 "pasta_stress.py" on GOURMET

We now obtain the linear relaxation functions. The linear relaxation modulus $G(t)$ can be obtained from the stress fluctuation recorded in the stress.txt file. According to the Green–Kubo formula, $G(t)$ can be obtained from the stress fluctuation $\sigma(t)$:

$$G(t) = \frac{V}{kT}\left\langle \sigma\left(t' + t\right)\sigma\left(t'\right)\right\rangle_{t'}. \tag{6.16}$$

Once $G(t)$ is obtained, the other rheological functions can be obtained according to Boltzmann's relation. The linear viscosity growth curve $\eta(t)$ and the dynamic modulus $G^*(\omega)$ are written as

$$\eta(t) = \int_0^t G\left(t'\right)dt', \tag{6.17}$$

$$G^*(\omega) = i\omega\int_0^\infty G(t)e^{i\omega t}dt. \tag{6.18}$$

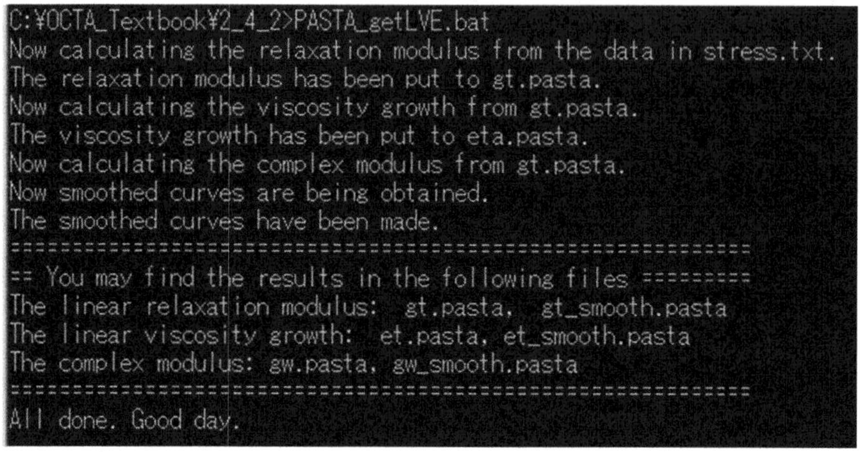

Fig. 6.10 Running "PASTA_getLVE.bat" via GourmetTerm

Each data processing step mentioned above can be carried out using the corresponding code embedded in the OCTA system. In this specific case, however, for the reader's convenience, a batch command is provided for the bundled operation as "PASTA_getLVE.bat" (enclosed in the sample zip file). Note that this command is a Windows batch file that properly works on GourmetTerm only. Note also that the file should be in the same folder as the data file stress.txt and the related Python script files "pasta_stsplit.py" and "pasta_cormarge.py". We copy these files when the working directory is changed. Via GourmetTerm, we run "PASTA_getLVE.bat" in the working directory for the data conversion. Messages appear as shown in Fig. 6.10.

As dumped, the files named "gt.pasta", "et.pasta", and "gw.pasta" store $G(t)$, $\eta(t)$, and $G^*(\omega)$, respectively. "gt_smooth.pasta", "eta_smooth.pasta", and "gw_smooth.pasta" store the smoothed functions. These files are in standard ASCII text format and can be managed easily.

Finally, we compare the obtained $G^*(\omega)$ with literature data to determine the unit time. The obtained moduli ($G(t)$ and $G^*(\omega)$) redeem the dimension according to the unit modulus $G_0^{\text{PASTA}} = 0.6$ MPa that can be determined according to the discussion in Sect. 6.3.3. $G^*(\omega)$ curves are then compared with literature data on a double-logarithmic plot. The unit time can be determined to attain the superposition of the crossing point of G' and G''. If the vertical positions of the crossing points are appreciably different from each other, the unit modulus should be reconsidered. (Note that the modification of the unit modulus affects Z_0^{PASTA} through a change in unit molecular weight, and thus recalculation with the modified parameter is necessary.) Figure 6.11 shows the example in which the unit time is set at $\tau_0^{\text{DT}} = 0.003$ s. Note that PASTA does not take account of the Rouse modes, and G'' thus decreases with increasing frequency in the high-frequency regime.

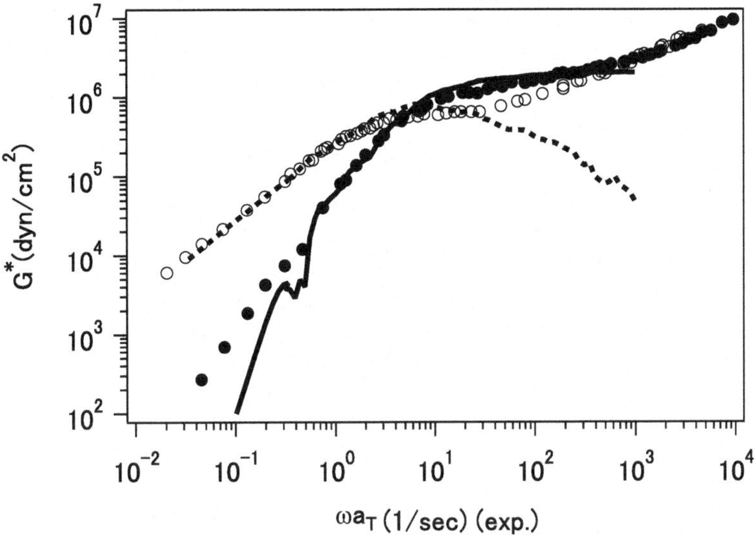

Fig. 6.11 Dynamic viscoelasticity of the polystyrene melt predicted by PASTA (*line*) compared with the literature data (*symbols*)

The obtained values of the unit modulus and time are insensitive to the molecular weight, molecular weight distribution, long-chain branching, deformation, and flow. Meanwhile, the unit modulus and time depend on the temperature according to a known relationship (written as Eqs. 6.12 and 6.14). Nevertheless, once the unit modulus and time are determined for a specific chemistry (with a given polymer concentration), the simulation for various kinds can be performed.

6.4.5 *Input UDF File for NAPLES*

Reflecting the difference of the model, the data structure of the input UDF file for NAPLES is different from that for PASTA. We open the sample UDF file for NAPLES "C:¥OCTA8¥ENGINES¥Naples_for_OCTA¥sample_input_udf¥ input step shear.udf" from GOURMET. The contents are shown in Fig. 6.12.

We set the value of Z_0^{NAPLES} for the sample in *Component[0].linear.Z.* According to Eq. 6.9 and the value of M_e^{G} written in "polymerdata.udf" $\left(M_e^{\mathrm{G}} = 14,400\right)$, for polystyrene with the molecular weight of 83 k, $Z_0^{\mathrm{NAPLES}} = 1.5 \times 83000/\,(1.25 \times 14400) \cong 7$.

We then change the flow condition set in the box *Simulation_Conditions. flow_type* from step_shear to equilibrium. (We click the box and choose "equilibrium" from the menu.) The simulation time step is set in *Simulation_Conditions.*

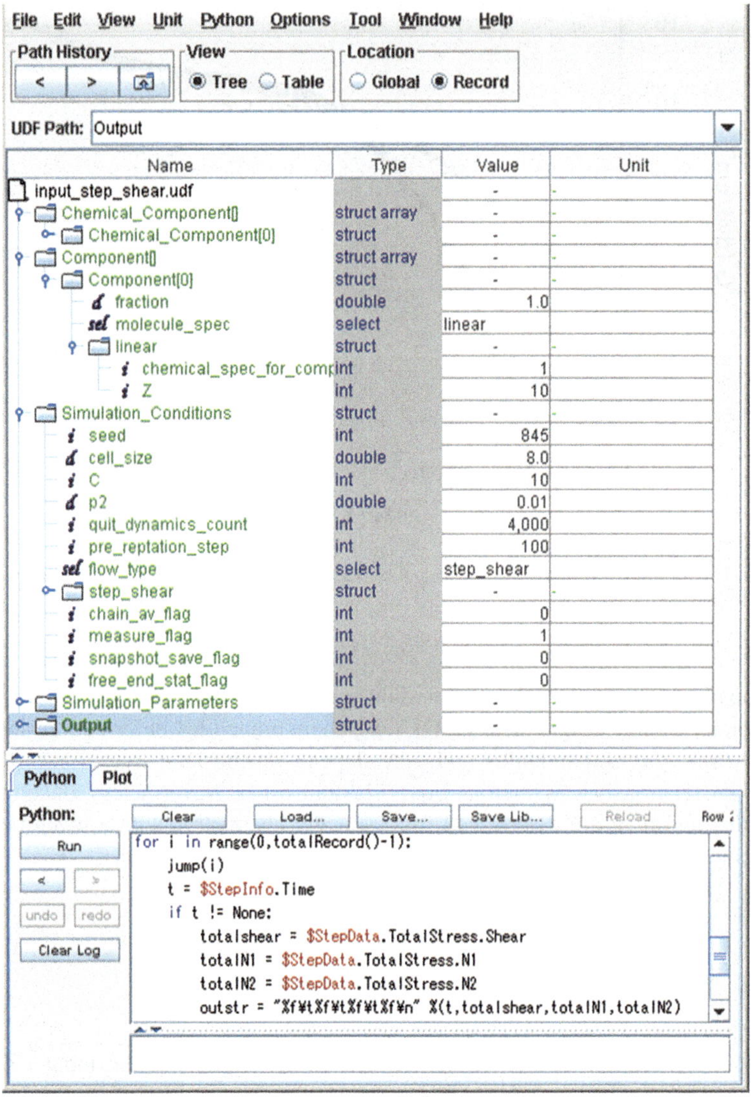

Fig. 6.12 "input_step_shear.udf"

quit_dynamics_count as 1000. The value *Simulation_Conditions.pre_reptation_step* gives the number of steps for equilibration. For this specific case, the value is set as zero.

At this stage, it is best to save the file by renaming it as "naples_Z07_in.udf". The contents of the file are shown in Fig. 6.13. For the reader's convenience, the file is available as "naples_Z07_in.udf" in the sample zip file.

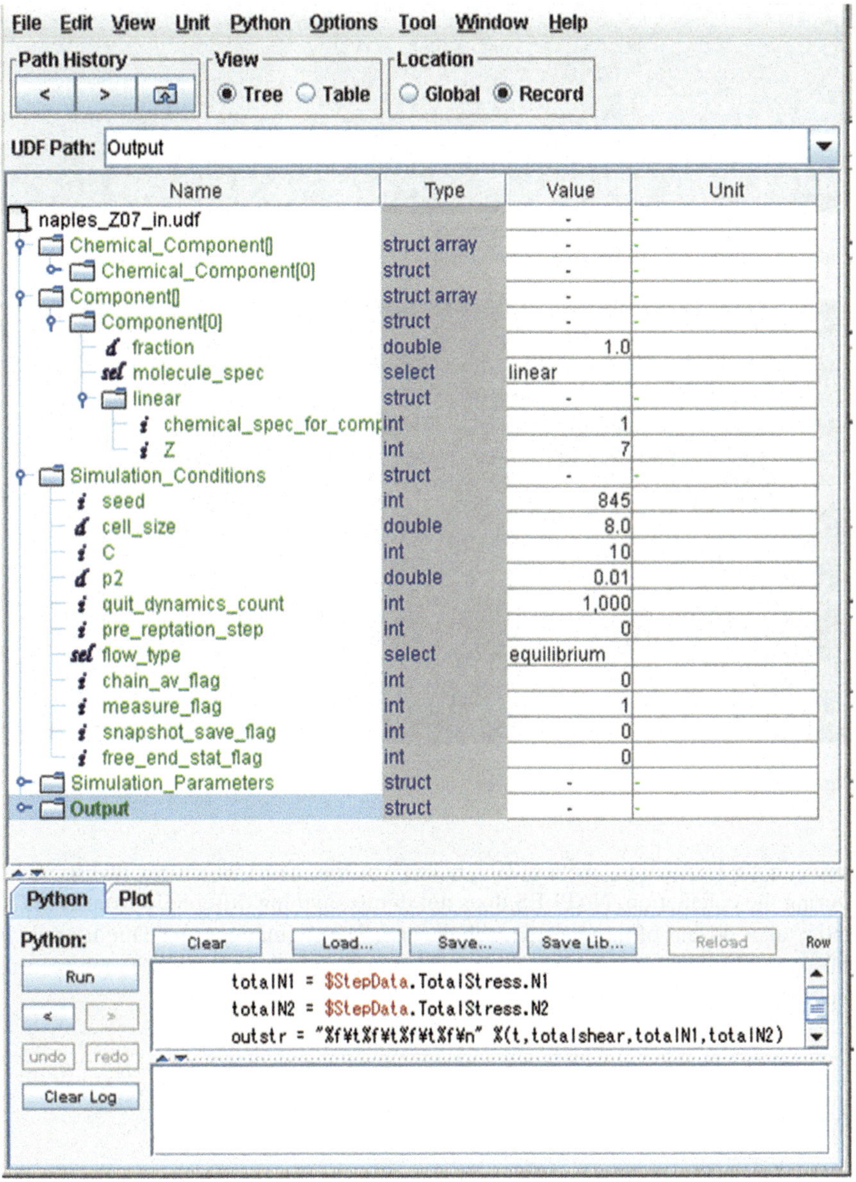

Fig. 6.13 "naples_Z07_in.udf"

6.4.6 Running NAPLES

We launch GourmetTerm and move to the directory in which the input UDF file is available (where in this example, the working directory is assumed to be C:¥OCTA_Textbook¥2_4_2). We run NAPLES using "naplesOCTA.bat" with a

```
C:¥OCTA_Textbook¥2_4_2>naplesOCTA.bat naples_Z07_in.udf
----------
Location Error:Chemical_Component[0].maximum_stretch_for_chemical
naples input file created from naples_Z07_in.udf.
naplesOCTA is started.  args : []
 :
naplesOCTA is completed.
```

Fig. 6.14 Running NAPLES via GourmetTerm

```
C:¥OCTA_Textbook¥2_4_2>NAPLES_getLVE.bat
Now calculating the relaxation modulus from the data in stress.txt.
The relaxation modulus has been put to gt.naples.
Now calculating the viscosity growth from gt.naples.
The viscosity growth has been put to eta.naples.
Now calculating the complex modulus from gt.naples.
Now smoothed curves are being obtained.
The smoothed curves have been made.
==============================================================
== You may find the results in the following files =========
The linear relaxation modulus:  gt.naples,  gt_smooth.naples
The linear viscosity growth:  et.naples, et_smooth.naples
The complex modulus: gw.naples, gw_smooth.naples
==============================================================
All done. Good day.
```

Fig. 6.15 Running "NAPLES_getLVE.bat" via GourmetTerm

given input UDF file ("naples_Z07_in.udf" in this case) as shown in Fig. 6.14. During the calculation, NAPLES does not dump anything differently from PASTA, and it takes dozens of minutes to reach the end. The simulation can be monitored via the "measure.npls" file (which stores stress) and movelog.npls file (which records messages).

6.4.7 Analysis of the Output File for NAPLES

NAPLES code does not generate a UDF file for output but creates its own result files. In particular, the stress fluctuation is stored in "measure.npls". From this file, with the calculations mentioned in Sect. 6.4.4, $G(t)$, $\eta(t)$ and $G^*(\omega)$ can be obtained. For this purpose, the batch command "NAPLES_getLVE.bat" is provided in the sample zip file. We run this command from GourmetTerm as demonstrated in Fig. 6.15. Note that "NAPLES_getLVE.bat" should be executed in the folder in which the "measure.npls" file exists.

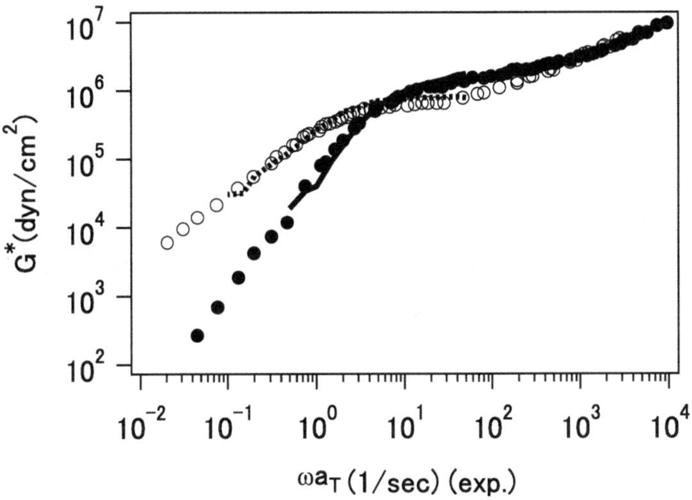

Fig. 6.16 Dynamic viscoelasticity of the polystyrene melt predicted by NAPLES (*line*) in comparison with the literature data (*symbols*)

The result files "gt.npls", "et.npls", and "gw.npls" store $G(t)$, $\eta(t)$, and $G^*(\omega)$, respectively. "gt_smooth.npls", "eta_smooth.npls", and "gw_smooth.npls" store the smoothed functions.

We obtain the unit time by fitting $G^*(\omega)$ with the literature data. With the unit modulus determined as $G_0^{\mathrm{NAPLES}} = 0.3$ MPa (according to Sect. 6.3.3), the modulus can be compared with the experimental data. We plot the simulation result with the experimental data on a double-logarithmic plot and tune the horizontal shift factor to attain the superposition of the crossing point of G' and G'' (see Fig. 6.16). This fitting gives $\tau_0^{\mathrm{PCN}} = 0.02\mathrm{s}$ for this specific case.

6.5 Example 2: Calculation of Nonlinear Viscoelasticity Under Fast Flow

6.5.1 Overview

In this section, on the basis of the unit time obtained in Sect. 6.4, the nonlinear viscoelasticity is calculated. Hereafter, the operation will be explained for NAPLES only. However, the operation for PASTA is essentially the same apart from differences in the parameter values and the structure of the UDF file. The target in the experiment is the transient shear viscosity for the monodisperse, linear polystyrene melt reported by Schweizer et al. [46]. The molecular weight is 200 k and the temperature is 175 °C.

6.5.2 Input UDF file

We open the template file "¥OCTA8¥ENGINES¥Naples_for_OCTA¥sample_input_udf¥input_step_shear.udf" via GOURMET. As Z_0^{NAPLES} for the polystyrene melt with $M = 200$ k is 17 (according to Eq. 6.9), we set this value in *Component[0].linear.Z.* The flow conditions described in *Simulation_Conditions* are set according to the target experiment. The flow type (*Simulation_Conditions.flow_type*) is changed to start_up_shear, and the flow rate should be accommodated on the basis of the value of τ_0^{PCN} that was determined in the previous section. Because the dataset (obtained by Watanabe et al.) for the determination of τ_0^{PCN} was obtained at $T = 167$ °C, the value of τ_0^{PCN} should be modified for the temperature of the target experiment, $T = 175$ °C, on the basis of the time–temperature superposition (see Eq. 6.15). Assuming that the glass transition temperature of polystyrene is 86.5 °C, τ_0^{PCN} ($T = 175$ °C) is obtained as 7×10^{-3} s. This τ_0^{PCN} value allows us to convert the experimental shear rate to the nondimensional shear rate for the simulation; in the experiment, the shear rate was varied as 30, 10, 3, 1, and 0.3 s^{-1}. These values correspond to nondimensional values of 0.21, 0.07, 0.021, 0.0071, and 0.0021, respectively, in the simulation. Here, we input the value 0.21 in *Simulation_Conditions.start_up_shear.strain_rate* to simulate the experiment for a shear rate of 30 s^{-1}. *Simulation_Conditions.quit_dynamics_count* stores the value of the time step up to which the code will run. Because the experiment was conducted for roughly 30 s, the time step is $30/\tau_0^{\mathrm{PCN}} \approx 4300$. We set this value in the corresponding box. The equilibration time given in *Simulation_Conditions.pre_reptation_step* should be appropriately set on the basis of the longest relaxation time. According to the manual for NAPLES, an estimate of the longest relaxation time for the chain with $Z = 10$ is approximately 50 steps. This value suggests the relaxation time for the chain with $Z = 20$ is 500 owing to the power law dependence of the relaxation time on the molecular weight with an exponent of 3.5. Note that NAPLES can attain the calculation from a given equilibrated configuration for the reduction of the equilibration. See the manual for details. Figure 6.17 shows the modified input UDF file for which the filename was changed to "naples_ss021_in.udf". For the reader's convenience, the file is available in the sample zip file.

6.5.3 NAPLES Simulation

Using the input UDF file created in the previous section, NAPLES simulation is carried out. Before the simulation, we protect the result files (such as "measure.npls") that were generated in the previous calculations by relocating the files from the working directory (assumed to be C:¥OCTA_Textbook¥2_4_2 in this tutorial) to another folder. Otherwise, the new simulation will affect the files and the expected results may not appear. If the working folder is clear, we launch GourmetTerm and run "naplesOCTA.bat" as shown in Fig. 6.14. Note that the input UDF file should be "naples_ss021_in.udf" this time.

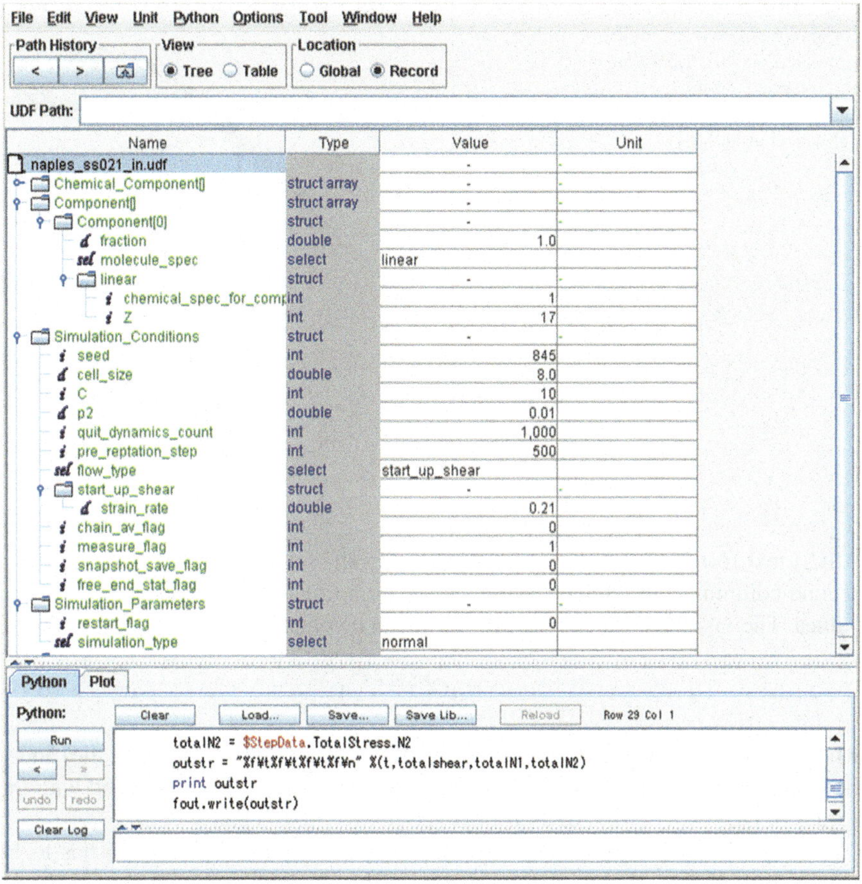

Fig. 6.17 "naples_ss021_in.udf"

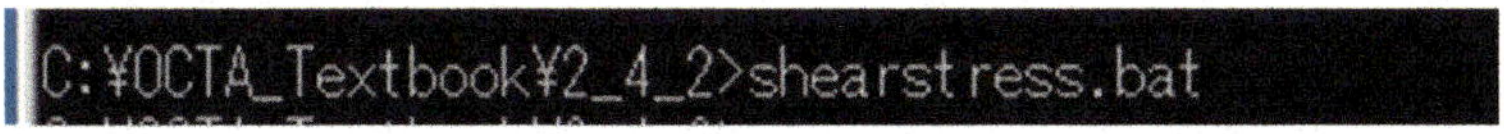

Fig. 6.18 Running "shearstress.bat"

6.5.4 Data Processing

The shear stress can be obtained from the generated "measure.npls" file. For this purpose, the batch command "shearstress.bat" is provided in the sample zip file. We run this command in the working directory (from GourmetTerm) as shown in Fig. 6.18, in which the file "shearstress.bat" exists in the working directory.

The "shearstress.bat" command generates the output file "shearstress.npls" in which the time development of the shear stress is stored. The file is in standard

Fig. 6.19 Viscosity growth curves for polystyrene melt simulated via NAPLES (*line*) and obtained from experiments (*symbols*) by Schweizer et al. [46]

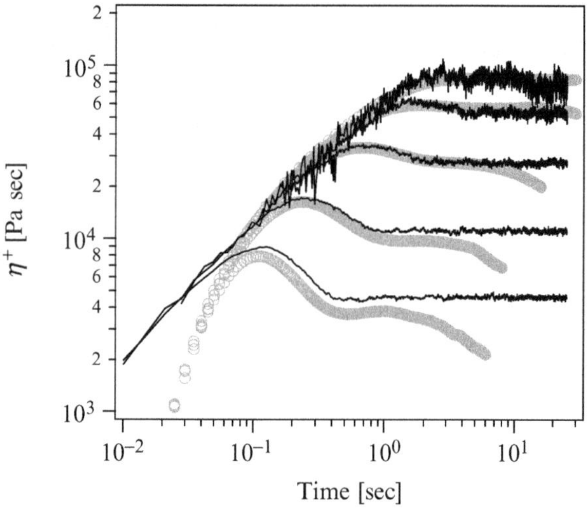

ASCII text format and it contains the time step and the shear stress in the first and second columns, respectively. These values are normalized with respect to the unit values. The viscosity with a real unit (Pa·s) can be obtained as

$$\eta = \tilde{\sigma}_{xy} G_0^{\mathrm{NAPLES}} / \dot{\gamma}, \tag{6.19}$$

where $\tilde{\sigma}_{xy}$ is the normalized shear stress (stored in the "shearstress.npls" file) and $\dot{\gamma}$ is the shear rate in the experiment. In this specific case, η can be obtained with the values of $G_0^{\mathrm{NAPLES}} = 0.3$ MPa (as obtained in Sect. 6.4.7) and $\dot{\gamma} = 30\ \mathrm{s}^{-1}$.

The series of operations explained from Sects. 6.5.2, 6.5.3 to 6.5.4 has to be repeated for various shear rates as necessary. The results from such simulations are shown in Fig. 6.19, in which the experimental data are also shown for comparison.

6.6 Download Instruction

Sample files are available from the OCTA website. See the instruction on http://octa.jp/download_bookdata.html. Password for the sample(zip file) of this chapter is "CvLbSxFY".

References

1. J.D. Ferry, *Viscoelastic Properties of Polymers*, 3rd edn. (Wiley, New Jersey, 1980)
2. Y. Masubuchi, Annu. Rev. Chem. Biomol. Eng. **5**, 11 (2014)
3. V.A. Harmandaris, V.G. Mavrantzas, D.N. Theodorou, M. Kröger, J. Ramírez, H.C. Öttinger, D. Vlassopoulos, Macromolecules **36**, 1376 (2003)

4. K. Kremer, G.S. Grest, I. Carmesin, Phys. Rev. Lett. **61**, 566 (1988)
5. J.T. Padding, W.J. Briels, J. Chem. Phys. **117**, 925 (2002)
6. G. Pan, C.W. Manke, Int. J. Mod. Phys. B **17**, 231 (2003)
7. G. Pan, C.W. Manke, J. Rheol. **46**, 1221 (2002)
8. G. Ronca, J. Chem. Phys. **79**, 1031 (1983)
9. W. Hess, Macromolecules **21**, 2620 (1988)
10. K.S. Schweizer, M. Fuchs, G. Szamel, M. Guenza, H. Tang, Macromol. Theor. Simul. **6**, 1037 (1997)
11. M.S. Green, A.V. Tobolsky, J. Chem. Phys. **14**, 80 (1946)
12. M. Yamamoto, J. Phys. Soc. Japan **11**, 413 (1956)
13. S. Edwards, Proc. Phys. Soc. **92**, 9 (1967)
14. P.G. de Gennes, J. Chem. Phys. **55**, 572 (1971)
15. M. Doi, S.F. Edwards, *The Theory of Polymer Dynamics* (Oxford University Press, Oxford, 1986)
16. M. Doi, J. Polym. Sci. Polym. Phys. Ed. **21**, 667 (1983)
17. W. Graessley, Adv. Polym. Sci. **47**, 67 (1982)
18. S. Milner, T. McLeish, Phys. Rev. Lett. **81**, 725 (1998)
19. A.E. Likhtman, T.C.B. McLeish, Macromolecules **35**, 6332 (2002)
20. C. Tsenoglou, ACS Polym. Prepr. **28**, 185 (1987)
21. J. des Cloizeaux, Europhys. Lett. **5**, 437 (1988)
22. G. Marrucci, J. Polym. Sci. Polym. Phys. Ed. **23**, 159 (1985)
23. G. Ianniruberto, G. Marrucci, J. Nonnewton. Fluid Mech. **65**, 241 (1996)
24. C.C. Hua, J.D. Schieber, J. Chem. Phys. **109**, 10018 (1998)
25. S. Shanbhag, R.G. Larson, J. Takimoto, M. Doi, Phys. Rev. Lett. **87**, 195502 (2001)
26. Y. Masubuchi, J.-I. Takimoto, K. Koyama, G. Ianniruberto, G. Marrucci, F. Greco, J. Chem. Phys. **115**, 4387 (2001)
27. M. Doi, J.-I. Takimoto, Philos. Trans. R. Soc. A Math. Phys. Eng. Sci. **361**, 641 (2003)
28. J.D. Schieber, J. Neergaard, S. Gupta, J. Rheol. **47**, 213 (2003)
29. A.E. Likhtman, Macromolecules **38**, 6128 (2005)
30. T. Uneyama, Y. Masubuchi, J. Chem. Phys. **137**, 154902 (2012)
31. M. Langeloth, Y. Masubuchi, M.C. Böhm, F. Müller-Plathe, J. Chem. Phys. **138**, 104907 (2013)
32. M. Doi, N.Y. Kuzuu, J. Polym. Sci. Polym. Lett. Ed. **18**, 775 (1980)
33. G.C. Berry, T.G. Fox, Adv. Polym. Sci. **5/3**, 261 (1968)
34. T. Uneyama, K. Horio, J. Polym. Sci. Part B Polym. Phys. **49**, 966 (2011)
35. Y. Masubuchi, G. Ianniruberto, F. Greco, G. Marrucci, Model. Simul. Mater. Sci. Eng. **12**, S91 (2004)
36. Y. Masubuchi, G. Ianniruberto, F. Greco, G. Marrucci, J. Non Cryst. Solids **352**, 5001 (2006)
37. Y. Masubuchi, G. Ianniruberto, F. Greco, G. Marrucci, Rheol. Acta **46**, 297 (2006)
38. Y. Masubuchi, Y. Matsumiya, H. Watanabe, S. Shiromoto, M. Tsutsubuchi, Y. Togawa, Rheol. Acta **51**, 1 (2011)
39. Y. Masubuchi, Y. Matsumiya, H. Watanabe, G. Marrucci, G. Ianniruberto, Macromolecules **47**, 3511 (2014)
40. Y. Masubuchi, G. Ianniruberto, F. Greco, G. Marrucci, J. Chem. Phys. **119**, 6925 (2003)
41. R.G. Larson, T. Sridhar, L.G. Leal, G.H. McKinley, A.E. Likhtman, T.C.B. McLeish, J. Rheol. **47**, 809 (2003)
42. L. Fetters, D. Lohse, W. Graessley, J. Polym. Sci. Part B-Polymer Phys. **37**, 1023 (1999)
43. R. Everaers, S.K. Sukumaran, G.S. Grest, C. Svaneborg, A. Sivasubramanian, K. Kremer, Science **303**, 823 (2004)
44. Y. Masubuchi, H. Watanabe, G. Ianniruberto, F. Greco, G. Marrucci, Macromolecules **41**, 8275 (2008)
45. H. Watanabe, T. Sakamoto, T. Kotaka, Macromolecules **18**, 1008 (1985)
46. T. Schweizer, Rheol. Acta **41**, 337 (2002)

Chapter 7
MUFFIN: Multiphase Simulator

Taku Ozawa

7.1 Introduction

MUFFIN (MUltiFarious FIeld simulator for a Non-equilibrium system) is a general simulation engine of the OCTA system based on the continuum model for multiphase structures of soft matter that uses the finite difference method (FDM) and finite element method (FEM).

MUFFIN treats various problems relating to soft matter, such as the phase separation of polymeric materials, the deformation of droplets under shear flow, the dynamics of electrolyte solutions, the chemical reactions and flow in a microchannel, the elastic model of composite materials, the swelling and shrinkage of gels, the growth of spherulites, and light transmittance. It is possible to evaluate an average property such as the average elastic modulus of a polymeric material with a phase-separated structure. Morphology data obtained using other simulation engines such as SUSHI and COGNAC (especially, dissipative particle dynamics) can be imported.

When performing various types of simulation, each type of field data has solvers and we can arrange an arbitrary combination of fields and order of solvers. This functionality is referred to as the "Dynamics Manager." Using this functionality, we can freely set the input data such as in the case of building the simulation algorithm.

Simulators (packages/modules) are included in MUFFIN version 5.1 as described below. User definable format (UDF) definitions of all simulators are unified, and FDM simulators are united as an executable module named "muffin5".

T. Ozawa (✉)
JSOL Corporation, Tokyo, Japan
e-mail: ozawa.taku@jsol.co.jp

Japan Association for Chemical Innovation, *Computer Simulation of Polymeric Materials*, DOI 10.1007/978-981-10-0815-3_7

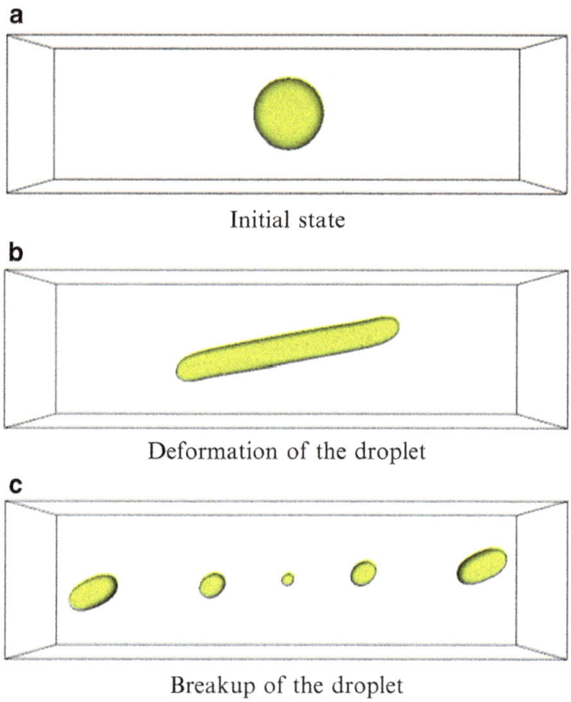

Fig. 7.1 Deformation and breakup of a droplet under shear flow

7.1.1 Multi-fluid Phase Dynamics Simulator

(A) FDM simulator; module name: "muffin5"
(B) FEM simulator; module name: "muffin5e_phaseseparation"

The multi-fluid phase dynamics simulator targets the dynamics of multiphase flow, e.g., phase separation under an external flow and near a wall, deformation of droplets in a flow, and evaporation process of a multicomponent thin film.

Figure 7.1 shows the deformation and breakup process of a droplet under shear flow. The shape of the deformed droplet and the criteria of breakup are related with the droplet size, viscosity of components, shear rate, and interfacial tension [1]. The simulator considers these effects and spontaneously creates interfaces. In Fig. 7.1c, small droplets form after the breakup.

Figure 7.2 shows the phase separation of a diblock copolymer in a thin film confined between walls [2, 3]. The block ratio of the diblock copolymer is 0.5:0.5 and this copolymer forms a lamellar structure. On the bottom wall, a chemical pre-pattern is set by introducing a partial surface potential that has an attractive (wettable) interaction with one component of the diblock copolymer. A straight lamellar structure forms along the edge of the simulation box in the late stage.

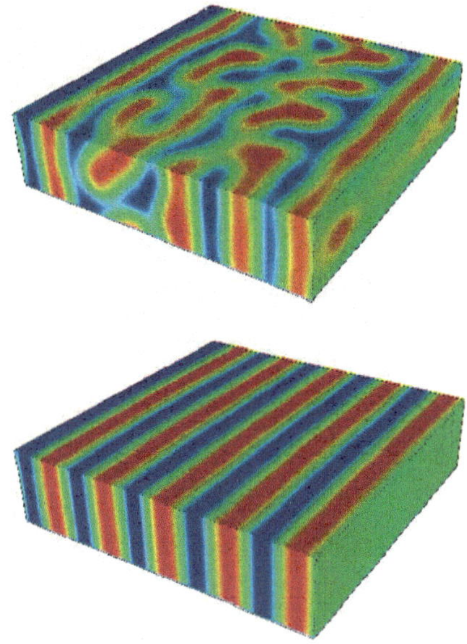

Fig. 7.2 Phase separation of a diblock copolymer system on a substrate with partial wettability: early stage (*top*) and late stage (*bottom*) of phase separation

7.1.2 *Electrolyte Fluid Dynamics Simulator*

(A) FDM simulator; module name: "muffin5"
(B) FEM simulator; module name: "muffin5e_electrolyte"

The electrolyte fluid dynamics simulator targets the dynamics of the electrolyte solution under an electric field (e.g., the dynamics of the diffusion layer near the charged plane and electrolyte flow around the charged particle).

Figure 7.3 shows the distribution of charge density around a particle with charge on its surface. The particle is surrounded by ions of opposite charge. By setting the pressure gradient as a boundary condition, flow is induced and the distribution of the ion density is deformed.

7.1.3 *Micro Electrochemical Fluidics Chip Simulator*

The micro electrochemical fluidics chip simulator targets the reaction, diffusion, and flow of fluid in micro-reactors and micro-TAS (total analysis system) using FEM. The module name is "muffin5e_memfluid".

Figure 7.4 shows the density contour of component C created by the reaction of components A and B. Components A and B are injected from the left boundary. Flow in the microchannel and the diffusion of each component are calculated. This example is of a simple rectangular region, but the simulator can treat a more complex channel structure because the FEM is used as the numerical method.

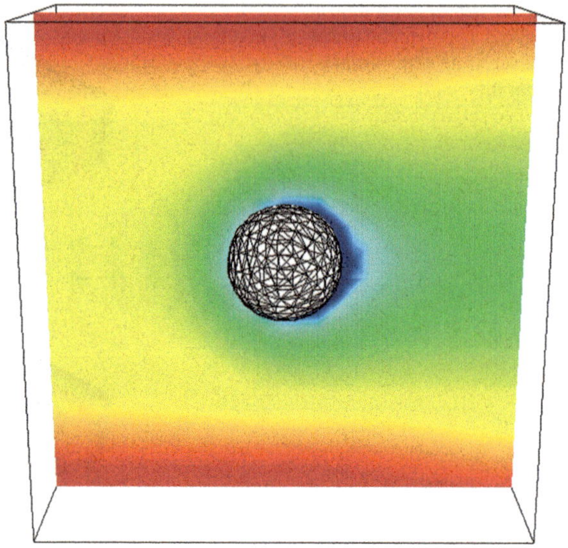

Fig. 7.3 Distribution of charge density around a charged particle in electrolyte flow

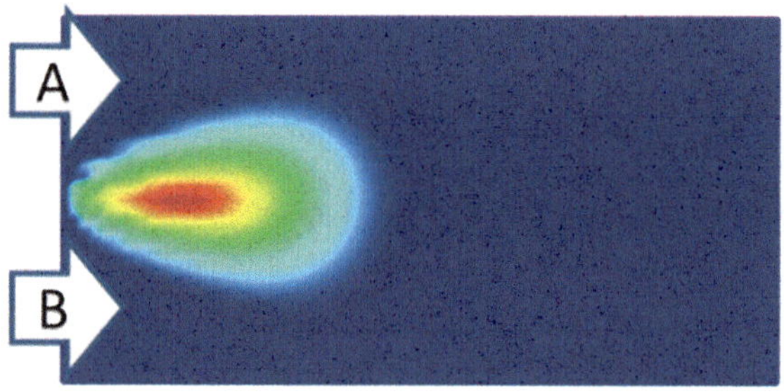

Fig. 7.4 Reaction of fluid in a microchannel. Component C is created by the reaction of components A and B

7.1.4 Multiphase Elasticity Simulator

The multiphase elasticity simulator targets the deformation of elastic material with a phase-separated structure using the FEM. The module name is "muffin5e_elastica".

Figure 7.5 shows the contour of free energy on interfaces between two components when the shear deformation is applied to an elastic material with a phase-separated structure. Phase-separated structures can be imported from simulation engines of OCTA as volume fraction distributions. The stress concentration inside the material and the average Young's, shear, and bulk moduli are evaluated.

7.1.5 Gel Dynamics Simulator

The gel dynamics simulator targets the large deformation, swelling, and shrinkage of a gel using the FEM. The module name is "muffin5e_geldyn".

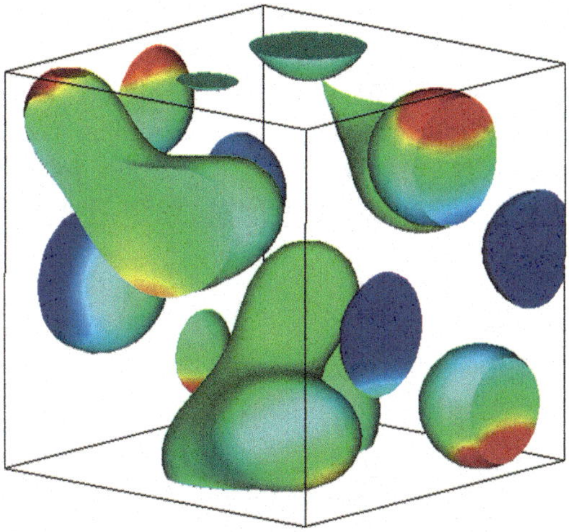

Fig. 7.5 Deformation of elastic material with a phase-separated structure. The distribution of free energy on interfaces is shown

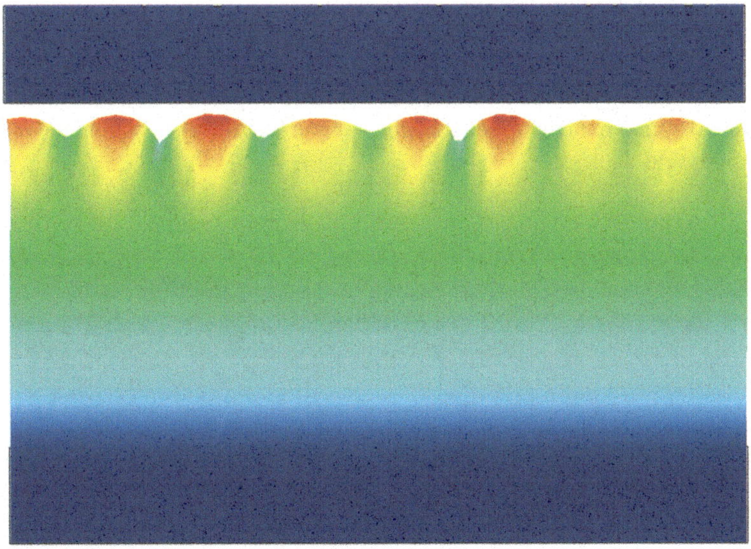

Fig. 7.6 Swelling dynamics of a polymer gel; initial structure (*top*) and swollen structure in the late stage (*bottom*) with the free energy contour

Figure 7.6 shows the swelling dynamics of a polymer gel in a two-dimensional model obtained using the collective diffusion model of gel networks and an explicit solver. The initial slab gel swells and a folding pattern forms on the surface in the late stage.

7.1.6 Light Transmittance Simulator

The light transmittance simulator targets the evaluation of the transparency of a polymeric material containing spherulites. The module name is "turban".

The simulator calculates the growth and filling of spherulites in a rectangular region. The spatial distribution of the dielectric tensor and the turbidity spectrum are then calculated.

7.1.7 Mesh Generator

MUFFIN has a two-dimensional/three-dimensional mesh generator named MILK. Generated mesh data are used in FEM simulators of MUFFIN. Module names are "milk2d5" and "milk3d5".

Figure 7.7 shows unstructured mesh data generated using MILK. Two spherical inclusions are set in a rectangular region. The region is meshed with tetrahedral elements. Figure 7.7 (right) shows the shear-deformed structure calculated using the multiphase elasticity simulator described above. Compared with Fig. 7.5, in which the internal structure is represented by the volume fraction distribution, the interface does not have thickness.

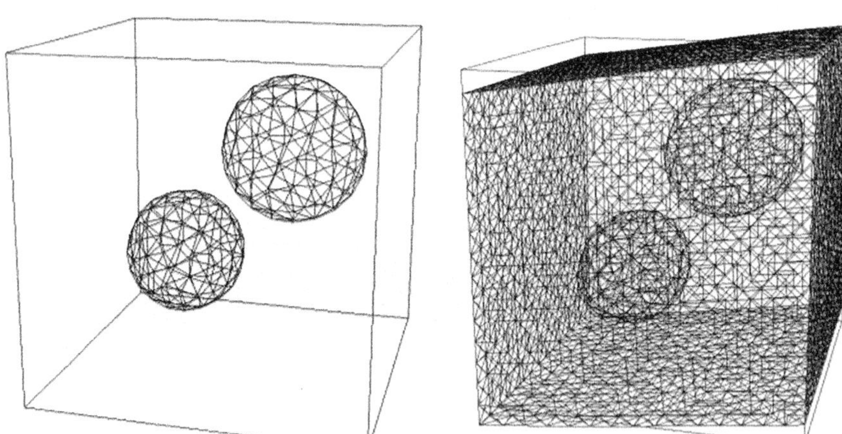

Fig. 7.7 Deformation of an elastic composite material. Mesh data are generated using MILK. *Left*: Surface mesh of two inclusions before deformation. *Right*: Shear-deformed structure

7.2 Theoretical Background

Basic theories of two simulators are described for reference in the tutorials presented in Sect. 7.3. First, the multi-fluid phase dynamics simulator employing the FDM is used to calculate the phase separation of polymer melts. Next, the multiphase elasticity simulator employing the FEM is used to calculate the deformation of an elastic material with a phase-separated structure after solidification. More detailed information is provided in the user's manual [4].

7.2.1 Multi-fluid Phase Dynamics Simulator

The multi-fluid phase dynamics simulator targets the phase separation of polymer melts or polymer solution. It is possible to calculate the (diffusion and fluid) dynamics of a multicomponent system considering the interactions of different components. Although this simulator has many functions that treat a complex system, as a simple tutorial, the basic theory that describes the phase separation of multicomponent polymer melts is explained below.

The time evolution equation of each component is

$$\frac{\partial \psi_\alpha}{\partial t} = -\nabla \cdot (\boldsymbol{v}\psi_\alpha) + \nabla \cdot (L_\alpha \nabla \mu_\alpha) \tag{7.1}$$

where ψ_α is the volume fraction of the component α, $\boldsymbol{v}$ is the velocity field of the fluid, L_α is the transport (diffusion) coefficient of the component α, and μ_α is the chemical potential of the component α.

To consider the hydrodynamic effect, the flow field is described using the Navier–Stokes equation. Because a low-velocity field is assumed, some terms are ignored and it follows that

$$0 = -\nabla p + \nabla \cdot \eta \left(\nabla \boldsymbol{v} + (\nabla \boldsymbol{v})^{\mathrm{T}} \right) + \boldsymbol{K} \tag{7.2}$$

where p is the pressure field, η is the viscosity coefficient, and $\boldsymbol{K}$ is the body force that works as the driving force of the fluid. Using the incompressibility condition, $\nabla \cdot \boldsymbol{v} = 0$, the Poisson's equation for the pressure field is obtained.

The chemical potential μ_α and body force $\boldsymbol{K}$ are calculated using the free energy based on the Flory–Huggins–deGennes theory for an M-component system:

$$F = \frac{k_B T}{b^3} \int dV \left[\sum_{\alpha=0}^{M-1} \frac{\psi_\alpha}{N_\alpha} \ln \psi_\alpha + \sum_{\alpha<\alpha'}^{M-1} \chi_{\alpha\alpha'} \psi_\alpha \psi_{\alpha'} \right.$$
$$\left. + \frac{1}{2} \sum_{\alpha<\alpha'}^{M-1} C_{\alpha\alpha'} \{\nabla (\psi_\alpha - \psi_{\alpha'})\}^2 \right], \tag{7.3}$$

where $\chi_{\alpha\alpha'}$ is the parameter of the interaction between components α and α', and b is the monomer (segment) size. $C_{\alpha\alpha'}$ is given as $C_{\alpha\alpha'} = b^2/\left(18\psi_\alpha\psi_{\alpha'}\right)$ by a random phase approximation (RPA).

The chemical potential and body force are obtained as

$$\mu_\alpha = \frac{\delta F}{\delta \psi_\alpha}, \quad \boldsymbol{K} = -\sum_{\alpha=1}^{M-1} \psi_\alpha \nabla \mu_\alpha. \tag{7.4}$$

In this simulator, other free energy models such as the Ohta–Kawasaki model for the diblock copolymer system (Fig. 7.2) are available. If another free energy model is selected in the input file, Eq. 7.3 is changed.

7.2.2 Multiphase Elasticity Simulator

It is possible to simulate the deformation of an elastic material with a phase-separated structure using the multiphase elasticity simulator. The phase-separated structure obtained using the multi-fluid phase dynamics simulator of Muffin, SUSHI, or COGNAC (especially, dissipative particle dynamics) affects the field data of the FEM. The volume fraction distribution is converted to the elastic modulus and the thermal expansion distribution. Isotropic and anisotropic elastic properties can be treated.

The basic theory for elastic material with a phase-separated structure is described below. The elastic modulus of each component is isotropic and a fixed displacement boundary condition is considered.

When considering a very small deformation, the strain tensor e_{ij} is described using the displacement vector $\boldsymbol{u}$:

$$e_{ij} = \frac{1}{2}\left(\nabla \boldsymbol{u} + (\nabla \boldsymbol{u})^{\mathrm{T}}\right) = \frac{1}{2}\left(\frac{\partial u_i}{\partial x_j} + \frac{\partial u_j}{\partial x_i}\right). \tag{7.5}$$

The free energy of linear elastic material is a scalar value. The strain tensor is divided into the components of compression (expansion) and shear, and the square of each component is used as a term in the free energy; the two terms are independent scalars of the second degree. Finally, the free energy is described as [5]

$$F = \int dV \left\{ G\left(e_{ij} - \frac{1}{3}\delta_{ij}e_{ll}\right)^2 + \frac{1}{2}K{e_{ll}}^2 \right\}, \tag{7.6}$$

where G is the local shear modulus, K is the local bulk modulus, δ_{ij} is Kronecker's delta, and the term $\left(e_{ij} - \frac{1}{3}\delta_{ij}e_{ll}\right)$ is the pure shear component of the strain tensor. The summation convention is applied.

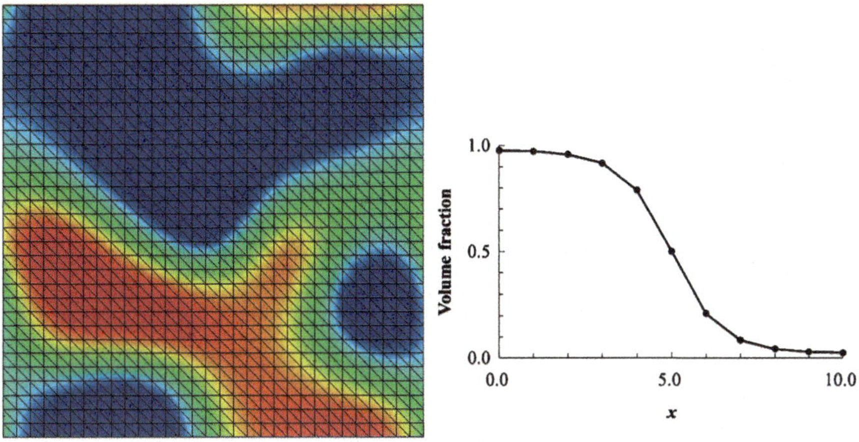

Fig. 7.8 Examples of the volume fraction distribution of a phase-separated structure. On both the left and right, only one component is shown. *Left*: Volume fraction contour and mesh structure of MUFFIN (FEM). *Right*: One-dimensional profile of the volume fraction at the interface calculated using SUSHI. Each dot represents a mesh node point

The local elastic modulus is calculated using the volume fraction distribution ψ_α for the phase-separated system:

$$K = \sum_{\alpha=0}^{M-1} \psi_\alpha K_\alpha, \quad G = \sum_{\alpha=0}^{M-1} \psi_\alpha G_\alpha . \tag{7.7}$$

The interface has finite thickness in the phase-separated structure calculated using engines of OCTA. In the simulation output data, the volume fraction changes gradually on a few mesh elements (Fig. 7.8). Considering Eq. 7.7, the converted elastic modulus also changes gradually in the interfacial region. This means it is possible to evaluate the effect of the thickness of the interface on a mechanical property.

The distribution of the displacement vector (for a deformed structure) is obtained under the condition that the free energy F is a minimum. The stress tensor distribution is calculated as $\sigma_{ij} = \partial F / \partial e_{ij}$.

To evaluate the average modulus of the system, the simulator calculates underlined terms in the equation

$$\underline{F_{\text{total}}} = G_{\text{average}} \underline{\int_{\text{total region}} dV \left(e_{ij} - \frac{1}{3}\delta_{ij} e_{ll} \right)^2} + K_{\text{average}} \underline{\int_{\text{total region}} dV \left(\frac{1}{2} e_{ll}{}^2 \right)} . \tag{7.8}$$

By performing two types of deformation (e.g., elongation and shear deformation), the average moduli G_{average} and K_{average} can be calculated by solving simultaneous linear equations.

7.3 Tutorial

7.3.1 Multi-fluid Phase Dynamics Simulator

Here, we use and modify a sample file that is included in the following folder.

```
OCTA8\ENGINE\MUFFIN5\sample\muffin5\
11.flow+diffusion(polymer)(p,v,psi)\
11.B.PhaseSeparationWithFlow
```

We copy this folder to another location and save the input file "EX11B_in.udf" under another name, "EX11B_3D_in.udf". The original file is for a two-dimensional system in which the dynamics of phase separation of polymers with the hydrodynamic effect are calculated. This file is modified for a three-dimensional system as described below.

We open the above UDF file and change the viewing style from "Tree" to "Table". First, the mesh parameters described in *parameter.mesh_parameter.axes[].values[]* are set. Here, we treat the $32 \times 32 \times 32$ system under a periodic boundary condition. In such a case, we set the cubic system to have an edge length of 31 (*values*[*1*]) and a division number of 31 (*values*[2], which means $\Delta x = 1$) as shown in Fig. 7.9.

We change the parameters regarding the calculation time as follows:

```
parameter.common_physical_parameter.DT.value[] = 1.0e−3
parameter.common_physical_parameter.FINAL_STEP.value[] = 50000
parameter.common_physical_parameter.
       INTERVAL_OF_MONITORING.value[] = 100
parameter.common_physical_parameter.
       INTERVAL_OF_UDF_OUTPUT.value[] = 10000
```

We change the physical parameters as shown in Fig. 7.10. The length of each polymer N is set to 10 and the parameter χ is set to 0.5. As other parameters, the

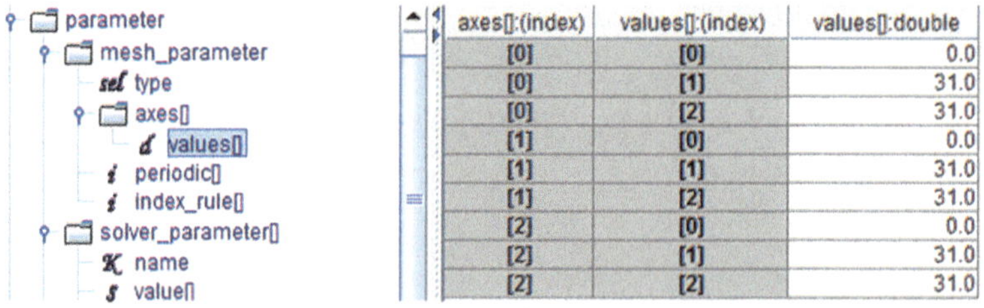

Fig. 7.9 Multi-fluid phase dynamics simulator; mesh parameter

physical_parameter[]:(index)	value[]:(index)	value[]:string
[0]:NUMBER_OF_COMPONENTS	[0]	2
[1]:POLYMERIZATION_INDEX_N	[0]	10
[1]:POLYMERIZATION_INDEX_N	[1]	10
[2]:AVERAGED_VOLUME_FRACTION	[0]	0.5
[2]:AVERAGED_VOLUME_FRACTION	[1]	0.5
[3]:DEVIATION_FROM_AVERAGED_VOLUME_FRACTION	[0]	0.01
[4]:SEED_OF_RANDOM_NUMBER	[0]	715
[5]:DIFFUSION_COEFFICIENT	[0]	1.0
[5]:DIFFUSION_COEFFICIENT	[1]	1.0
[6]:CHI_01	[0]	0.5
[7]:CA	[0]	1.0
[8]:VISCOSITY	[0]	1.0
[8]:VISCOSITY	[1]	1.0

Fig. 7.10 Multi-fluid phase dynamics simulator; physical parameters

boundary_condition[]:(index)	name_of_region:select	name_of_field:select	name_of_condition:select
[0]	YZ_BOUNDARY_PLANE_XM_AND_XP	K_Field	PERIODIC
[1]	ZX_BOUNDARY_PLANE_YM_AND_YP	K_Field	PERIODIC
[2]	XY_BOUNDARY_PLANE_ZM_AND_ZP	K_Field	PERIODIC
[3]	YZ_BOUNDARY_PLANE_XM_AND_XP	VolumeFraction	PERIODIC
[4]	ZX_BOUNDARY_PLANE_YM_AND_YP	VolumeFraction	PERIODIC
[5]	XY_BOUNDARY_PLANE_ZM_AND_ZP	VolumeFraction	PERIODIC
[6]	YZ_BOUNDARY_PLANE_XM_AND_XP	ChemicalPotential	PERIODIC
[7]	ZX_BOUNDARY_PLANE_YM_AND_YP	ChemicalPotential	PERIODIC
[8]	XY_BOUNDARY_PLANE_ZM_AND_ZP	ChemicalPotential	PERIODIC
[9]	YZ_BOUNDARY_PLANE_XM_AND_XP	ElectricPotential	PERIODIC
[10]	ZX_BOUNDARY_PLANE_YM_AND_YP	ElectricPotential	PERIODIC
[11]	XY_BOUNDARY_PLANE_ZM_AND_ZP	ElectricPotential	PERIODIC
[12]	YZ_BOUNDARY_PLANE_XM_AND_XP	Pressure	PERIODIC
[13]	ZX_BOUNDARY_PLANE_YM_AND_YP	Pressure	PERIODIC
[14]	XY_BOUNDARY_PLANE_ZM_AND_ZP	Pressure	PERIODIC
[15]	YZ_BOUNDARY_PLANE_XM_AND_XP	Velocity	PERIODIC
[16]	ZX_BOUNDARY_PLANE_YM_AND_YP	Velocity	PERIODIC
[17]	XY_BOUNDARY_PLANE_ZM_AND_ZP	Velocity	PERIODIC

Fig. 7.11 Multi-fluid phase dynamics simulator; boundary condition

volume fraction of each component is set to 0.5 and all diffusion and viscosity coefficients are set to 1.0. The capillary number, which is defined through the process of nondimensionalization, is set to 1.0.

For each field, the periodic boundary condition is set in all directions as shown in Fig. 7.11.

Figure 7.12 shows the settings of the dynamics manger, which are given in *dynamics_manager.procedures_table[]*. Using the set of fields and functions (solvers), the diffusion (phase separation) of polymers is calculated considering the hydrodynamic effect.

After saving the file, we start the calculation using this file as input. Here we use "GourmetTerm" to run this simulator. On a Windows system, we run "StartGourmetTerm" included in "OCTA8" from the Start menu (Fig. 7.13).

Where prompted, we enter the "cd" command to move to the directory where we created the input file above and run the following command:

```
muffin5 -I EX11B_3D_in.udf -O EX11B_3D_out.udf
```

procedures_table[]:(index)	command_li...	field:select	name_of_func:string
[0]:TEST_MODEL_H	[0]	ChemicalPotential	FLORY_HUGGINS
[0]:TEST_MODEL_H	[1]	K_Field	GRADIENT_CHEMICAL_POTENTIAL
[0]:TEST_MODEL_H	[2]	VolumeFraction	SOLVE_EQUATION_OF_CONTINUITY_WITH_FLOW
[0]:TEST_MODEL_H	[3]	Velocity	SOLVE_STOKES_EQUATION_AND_PRESSURE
[1]:SINGLE_PHASE	[0]	VolumeFraction	CONSTANT_VOLUME_FRACTION_WITH_NOISE
[1]:SINGLE_PHASE	[1]	ChemicalPotential	FLORY_HUGGINS
[1]:SINGLE_PHASE	[2]	K_Field	GRADIENT_CHEMICAL_POTENTIAL

Fig. 7.12 Multi-fluid phase dynamics simulator; dynamics manager

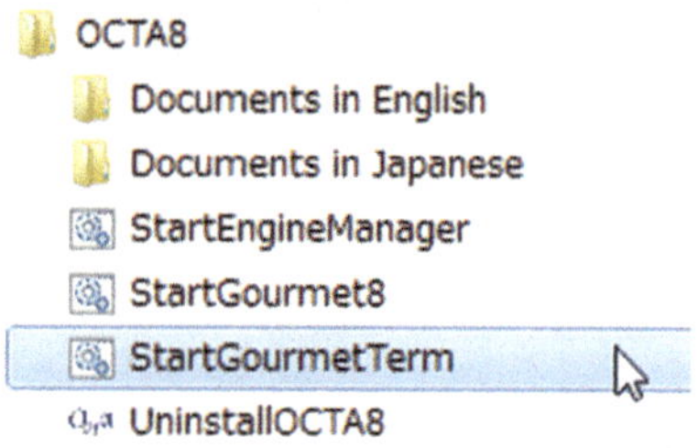

Fig. 7.13 Start menu of windows

```
**********************************************************************
*                                                                    *
*                 Program has finished successfully !                *
*                   M U F F I N  version  5.1                        *
*                                                                    *
**********************************************************************
```

Fig. 7.14 Message issued by the multi-fluid phase dynamics simulator when the calculation finishes successfully

Here, "muffin5" is the name of the simulator. Input and output file names are described after "-I" and "-O", respectively.

If the calculation finishes successfully, we see the message shown in Fig. 7.14.

We open the output file "EX11B_3D_out.udf". We then move to the final record and right-click on the .udf name and select **show_ field...** as shown in Fig. 7.15 (left). In the dialog "show_field", we set *field_name* to "VolumeFraction" and *all_boundary* to "on", as shown in Fig. 7.15 (center). Figure 7.16 shows the volume fraction distribution contour of component 0. By changing the parameters of this dialog, various types of visualization are possible. For example, Fig. 7.16 (right) shows the isosurface of the volume fraction 0.5 (component 0). For this visualization, we set *field_name* to "VolumeFraction", *cut_plane* to "off" and *isosurface* to [0.5] as shown in Fig. 7.15 (right).

7.3.2 Multiphase Elasticity Simulator

We use and modify a sample file included in the following folder:

```
OCTA8\ENGINE\MUFFIN5\sample\muffin5ebeta\Elastica\EX04
```

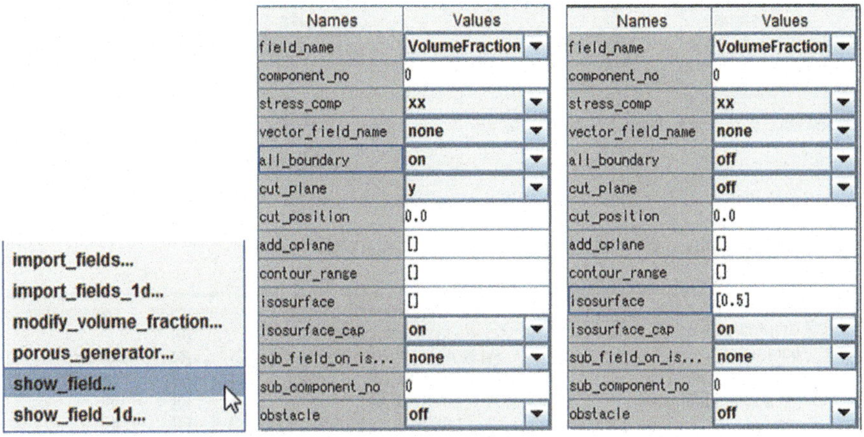

Fig. 7.15 Multi-fluid phase dynamics simulator; visualization: action menu (*left*), parameters for the volume fraction contour (*center*), parameters for the isosurface of the volume fraction (*right*)

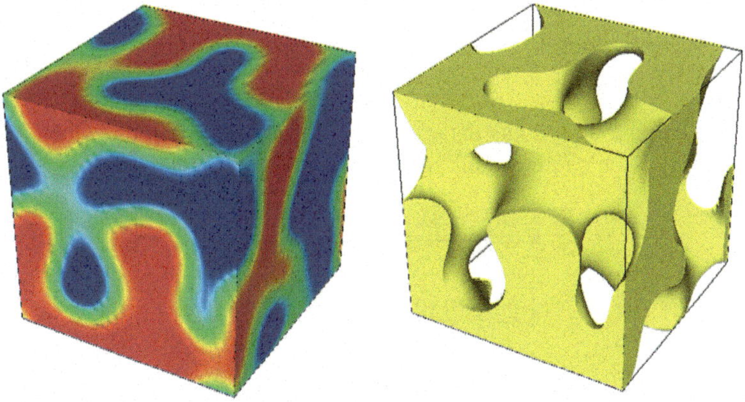

Fig. 7.16 Multi-fluid phase dynamics simulator, volume fraction contour of component 0 (*left*) and isosurface of the volume fraction 0.5 (*right*)

We copy this folder to another location and open the input file "EX04_in.udf" and right-click on the .udf name and select **import_fields...**. In the dialog that appears, we right-click on the text box of *import_udf_filepath* and select the output file of the multi-fluid phase dynamics simulator that was obtained above. We set *import_record_no* to 5 and *save_as* to "EX04_2_in.udf" (Fig. 7.17).

We open the created file and check the region size and division number described in *parameter.mesh_parameter.axes[].values[]*. These values are automatically changed to match the region size of the imported data. Meanwhile, we do not apply the periodic boundary condition in this simulation.

We check the physical parameters. For two components, isotropic mechanical properties are set as shown in Fig. 7.18.

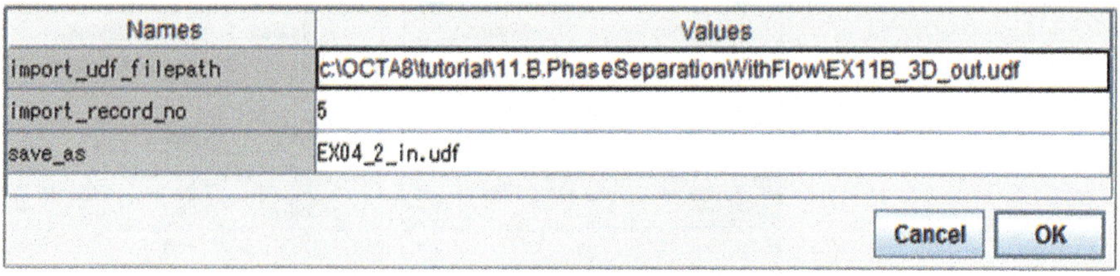

Fig. 7.17 Multiphase elasticity simulator, import of morphology data

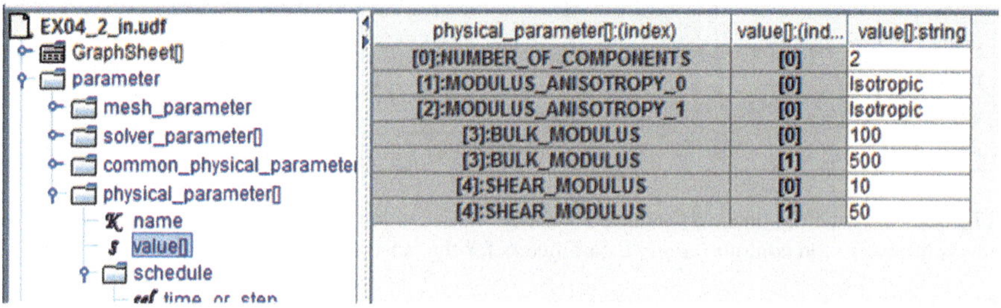

Fig. 7.18 Multiphase elasticity simulator, physical parameters

When setting the modulus data, we can refer to relationships among Young's modulus E, Poisson's ratio ν, bulk modulus K, and shear modulus G:

$$K = \frac{E}{3(1-2\nu)}, \quad G = \frac{E}{2(1+\nu)},$$
$$E = \frac{9KG}{3K+G}, \quad \nu = \frac{3K-2G}{6K+2G}. \tag{7.9}$$

We change the boundary condition as shown in Fig. 7.19. In the X direction, a small displacement on the YZ plane is introduced to apply the elongation.

After saving the file, we start the calculation using this file as input. In "GourmetTerm", we move to the directory where we put the input file above using the "cd" command and run the following command:

```
muffin5e_elastica -I EX04_2_in.udf -O EX04_2_out.udf
```

Here, "muffin5e_elastica" is the name of the simulator, and input and output file names are described after "-I" and "-O", respectively.

If the calculation finishes successfully, we see the message shown in Fig. 7.20.

We open the output file "EX04_2_out.udf" and move to record 1. Record 0 is the initial structure, and record 1 is the deformed structure. In *results[]*, the total free energy and the underlined terms in Eq. 7.8 are outputted (Fig. 7.21).

Next, we right-click on the .udf name and select **show_field...**. In the dialog that appears, we set *field_name* to "FreeEnergy" and *range* to [0, 0.01]. Fig. 7.22 shows the free energy distribution on the boundary of the region.

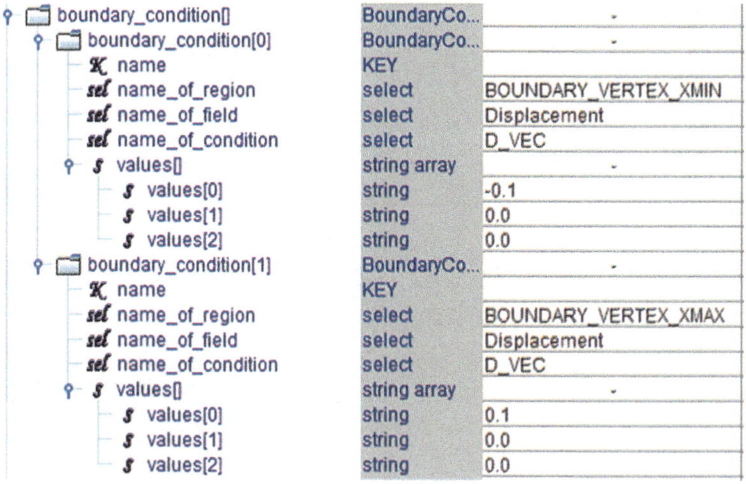

Fig. 7.19 Multiphase elasticity simulator, boundary condition for the elongation simulation (tree view)

```
**************************************************
The One Step Evolution Procedure has been Finished !
**************************************************
**********************************************************************
*                                                                    *
*                  Program has finished successfully !               *
*                   M U F F I N  version 5.0 E beta                  *
*                                                                    *
**********************************************************************
```

Fig. 7.20 Message issued by the multiphase elasticity simulator when the calculation finishes successfully

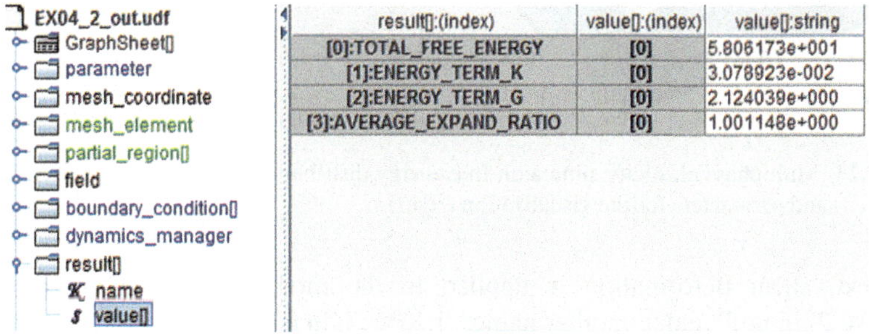

Fig. 7.21 Multiphase elasticity simulator, output of energy

According to the parameters of the dialog "**show_field**", various visualizations are available. Figure 7.23 (left) shows the isosurface of the volume fraction 0.5 (component 0) and the contour shows the distribution of free energy. Applying this visualization, the internal structure of the phase-separated material and the concentration of free energy can be evaluated.

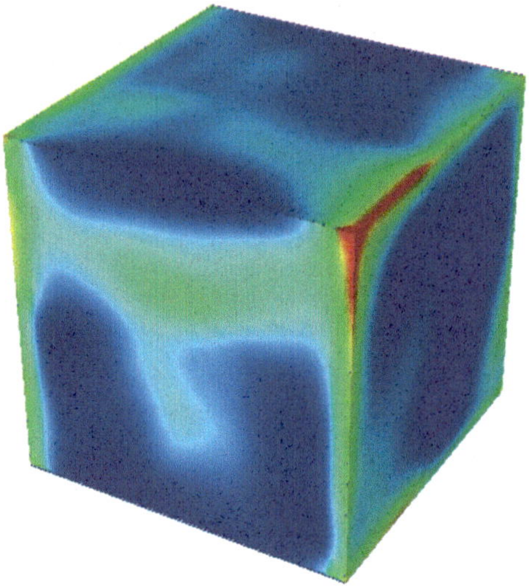

Fig. 7.22 Multiphase elasticity simulator, free energy distribution after deformation

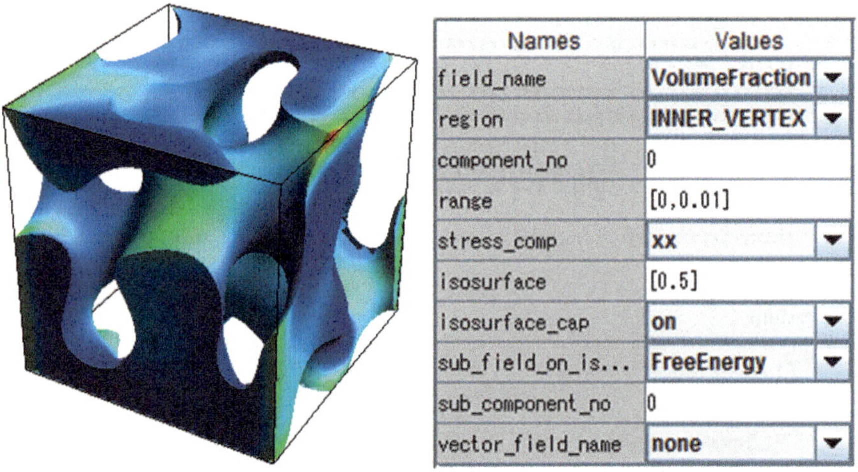

Fig. 7.23 Multiphase elasticity simulator, free energy distribution on the interface after deformation (*left*) and parameters for the visualization (*right*)

Next, shear deformation is applied to the model. We save the input file "EX04_2_in.udf" under another name, "EX04_3_in.udf", and change the boundary condition as shown in Fig. 7.24. In the Y direction, displacement on the YZ plane is introduced to apply shear deformation.

We run "muffin5e_elastica" as explained above (output file name is "EX04_3_out.udf"). After the calculation, we open the output file and check the data of *results[]* (Fig. 7.25) in record 1.

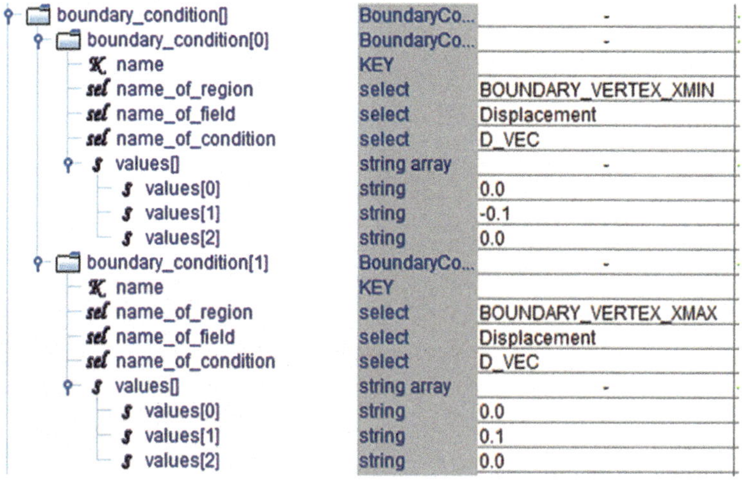

Fig. 7.24 Multiphase elasticity simulator, boundary condition for shear deformation

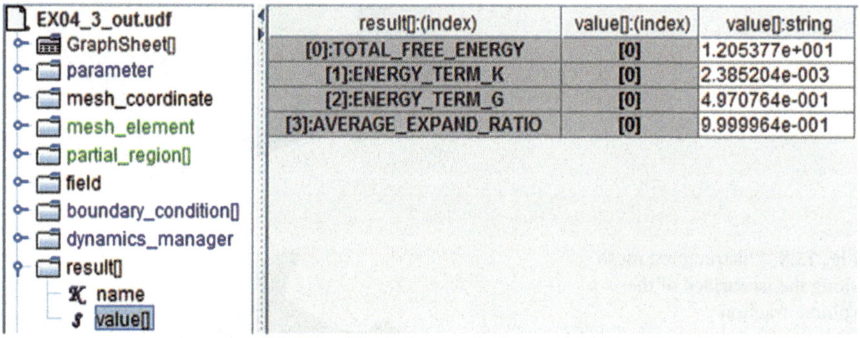

Fig. 7.25 Multiphase elasticity simulator, output of energy (shear deformation)

Total free energy and the underlined terms in Eq. 7.8 are outputted as shown in Fig. 7.25. Using these values obtained in the two types of simulation (for elongation and shear deformation), simultaneous linear equations for the average moduli can be defined. By solving these equations, the average moduli are obtained; the bulk modulus is about 318 and the shear modulus is about 23. Naturally, these values change depending on the phase-separated structure.

When solving the simultaneous linear equations using a Python script, numpy is useful and it is possible to use numpy in the Python window of GOURMET. In the sample script shown in Fig. 7.26, the equations $4x + 2y = 10$ and $x + y = 3$ are solved in a matrix calculation.

Fig. 7.26 Python script using numpy to calculate simultaneous linear equations

```
import numpy
from numpy import *
a = array([[4, 2], [1, 1]])
b = array([10, 3])
print numpy.linalg.solve(a, b)
```

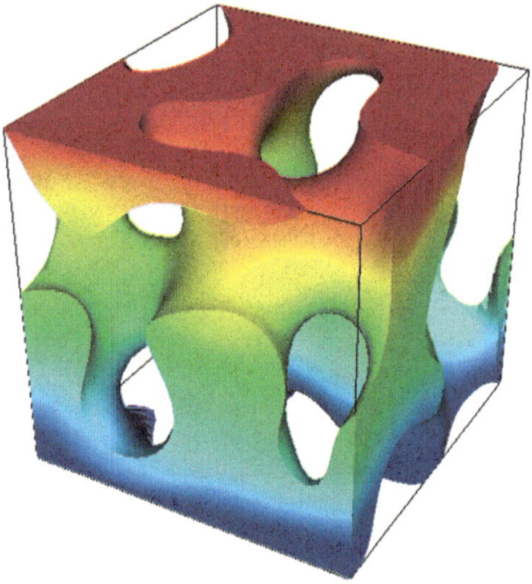

Fig. 7.27 Heat conduction simulation, temperature distribution on the isosurface of the volume fraction

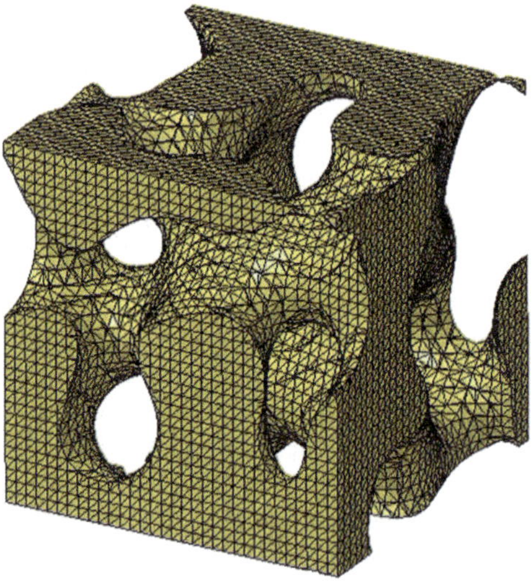

Fig. 7.28 Unstructured mesh along the isosurface of the volume fraction

Appendix

Using the multi-fluid phase dynamics simulator of MUFFIN, it is possible to simulate heat conduction employing the FDM and the phase-separated structure obtained in advance. The volume fraction distribution is converted to the thermal conductivity distribution. Figure 7.27 shows the temperature distribution on the interface. Average thermal conductivity can be evaluated using the temperature difference and heat flux.

To use the calculated phase-separated structure in external FEM software such as that used for nonlinear structural analysis or fluid dynamics, an unstructured mesh with a thin interface (mesh along the interface) is needed. Using J-OCTA (http://www.j-octa.com/), which is a commercial version of OCTA, it is possible to create and export triangular mesh data along the isosurface of the volume fraction or tetrahedral mesh data of a component. Figure 7.28 shows an unstructured triangular mesh along the interface of the phase-separated structure obtained using the multi-fluid phase dynamics simulator. This figure was visualized using the preprocessor of general FEM software LS-DYNA (http://www.lstc.com/) after importing mesh data created by J-OCTA.

References

1. T. Inamuro, R. Tomita, F. Ogino, Int. J. Mod. Phys. B **17**, 21 (2003)
2. K. Yoshimoto, T. Taniguchi, in *Proceedings of SPIE*, vol. 8680, 86801I, Mar 2013, http://proceedings.spiedigitallibrary.org/proceeding.aspx?articleid=1672894
3. R. Dessí, M. Pinna, A.V. Zvelindovsky, Macromolecules **46**, 1923 (2013)
4. Muffin User's Manual, http://octa.jp, Accessed 23 Nov 2015
5. L.D. Landau, E.M. Lifshitz, *Theory of Elasticity*, 3rd edn. (Butterworth-Heinemann, Burlington, 1986), p. 10

Chapter 8
KAPSEL: Colloidal Dispersion Simulator

Ryoichi Yamamoto and John Jairo Molina

8.1 What Is KAPSEL?

KAPSEL (Kyoto Advanced Particle Simulator for Electro-hydrodynamics) is a simulation software package designed to study the dynamics of colloidal particle dispersions. It aims to provide an efficient direct numerical calculation for the governing equations of these complicated systems, while maintaining a high degree of accuracy. This allows one to study, among other phenomena, the sedimentation and rheology of colloidal dispersions, as well as the electrophoresis of charged colloidal particles. The software and the underlying numerical methods were developed as part of a research project funded by the Japan Science and Technology Agency (PRESTO/CREST program) and carried out at Kyoto University.

When attempting to simulate the dynamics of colloidal dispersions, the main obstacle one encounters is the large separation of length and time scales governing the dynamics of the system, which can be orders of magnitude larger than the scales of the constituent atoms and molecules. In addition to describing the individual dynamics of the microscopic particles, with times scales on the order of $\sim 10^{-10}$ s, we must also consider the collective long wavelength motion of the fluid and ions, with relaxation times on the order of 10^{-3}–10^{3} s. Such slow relaxation phenomena occur at characteristic time scales that are several orders of magnitude larger than the maximum time accessible to conventional molecular dynamic simulations ($\sim 10^{6}$ s). Thus, such a microscopic approach is ill suited for studying the dynamics of colloidal dispersions, as it is not possible to perform a simulation that is long enough to capture all the relevant phenomena. Mesoscopic simulation methods, relying on coarse-grained models, have thus been developed to study colloidal particle dispersions.

R. Yamamoto (✉) • J.J. Molina
Graduate School of Engineering, Kyoto University, Kyoto, Japan
e-mail: ryoichi@cheme.kyoto-u.ac.jp

Japan Association for Chemical Innovation, *Computer Simulation of Polymeric Materials*, DOI 10.1007/978-981-10-0815-3_8

Simulation methods for colloidal particle dispersions that do not directly calculate the fluid motion, such as Brownian dynamics [1] and Stokesian dynamics [2], have been extensively used, given their relatively low computational costs. However, such methods are restricted to simple Newtonian host fluids, and extensions to complex fluids possessing internal degrees of freedom (such as polymeric fluids or electrolyte solutions) have proven to be extremely difficult. In addition, these approaches are only applicable in the low-Reynolds-number regime (Re $\ll$ 1), defined in terms of the size of the dispersed particles. This limits the field of study to quasi-steady problems ($t \gg \tau$, t is the time required to observe the given phenomena, and τ is the characteristic time required for the fluid motion to follow the particle motion) for relatively small particle sizes. After the introduction of the Stokesian dynamic method in the late 1980s, there was no notable progress for over a decade. This changed around the turn of the century, after which important contributions allowed the development of direct numerical simulation methods, which directly solve the coupled fluid and particle equations of motion, using a coarse-grained description for the fluid, thus allowing one to access the characteristic time scales of the dispersed particle dynamics.

There are two key aspects to consider when conducting a direct numerical simulation for colloidal particle dispersions; the first is how to model the motion of the host fluid, and the second is how to deal with the effects of particle motions on the fluid (the particle/fluid coupling). The motion of the host fluid can be described using the Navier–Stokes equation. This motion can be effectively simulated using (from lowest to highest degree of coarse graining) the lattice Boltzmann method and the multiparticle collision and dissipative particle dynamic methods, which replace the solvent with representative particles. To couple the fluid and particle dynamics, it is necessary to take into account the proper boundary conditions at the interface between the solvent (fluid) and particles (solid) when solving the fluid equations of motion. For host solvents described using the Navier–Stokes equation, it would seem appropriate to use a non-regular body-fitted mesh to solve the equations, as is done in the finite element method. However, the computational costs of this approach are prohibitive. Although the use of a regular fixed grid would appreciably reduce the computational cost, difficulties arise when enforcing the boundary conditions, due to the mismatch between the (typically cubic) grid and (spherical) particles. The fluid/particle dynamic method [3] was developed precisely to overcome this problem; it achieves this by considering the solid particles as a highly viscous fluid and by introducing an auxiliary field variable that specifies this fluid/particle domain. Various definitions for this auxiliary variable, which should depend only on the distance from the grid points to the particle surface, have been proposed [4]. The smoothed profile method (SPM), used in KAPSEL, adopts such an auxiliary field to distinguish between the domains inside (solid) and outside (fluid) of the particle, as shown in Fig. 8.1. The use of this continuous auxiliary field, with a finite-width interface, allows one to simulate the coupled fluid and particle dynamics with a high degree of accuracy in an efficient manner [5]. In addition, the method can be easily extended to consider complex host fluids, such as electrolyte solutions, within the same framework [6, 7]. The formal definition of the auxiliary field $\phi(x)$ used in KAPSEL is given by Eqs. 8.1–8.3:

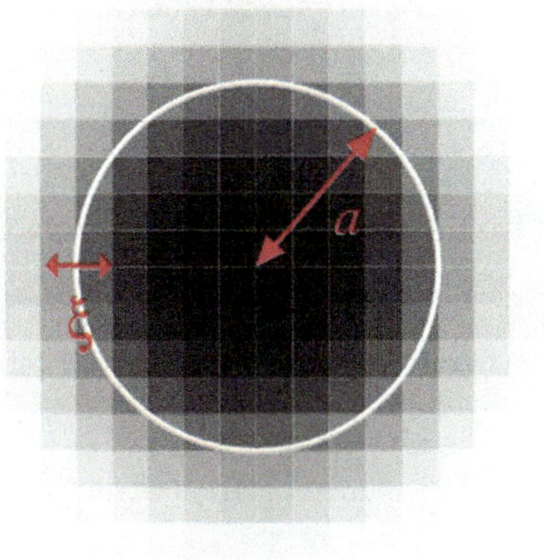

Fig. 8.1 In the SPM, the solid domain corresponding to particles of radius a at positions $\mathbf{R}_i(i = 1, \ldots, N)$ is obtained by introducing an interface of finite thickness ξ at the surface of the particles and by defining an auxiliary phase-field variable $0 \leq \phi_i \leq 1$ (where Δ is the grid spacing). This use of a finite-width interface improves the computational efficiency

$$\phi(\mathbf{x}) = \sum_{i=1}^{N} \phi_i(\mathbf{x}), \quad \phi_i(\mathbf{x}) = g(|\mathbf{x} - \mathbf{R}_i|) \tag{8.1}$$

$$g(x) = \frac{h\left[\left(a + \frac{\xi}{2}\right) - x\right]}{h\left[\left(a + \frac{\xi}{2}\right) - x\right] + h\left[x - \left(a + \frac{\xi}{2}\right)\right]} \tag{8.2}$$

$$h(x) = \begin{cases} \exp\left(-\Delta^2/x^2\right) & x \geq 0 \\ 0 & x < 0 \end{cases} \tag{8.3}$$

In this paper, we introduce the basic theory underlying the KAPSEL simulation package and provide detailed installation instructions, sample simulation procedures, and descriptions for advanced applications. The KAPSEL simulation code is an open source and available for free to anyone who agrees to our license agreement. The latest version of the code, along with various sample files, can be found on the official project website (KAPSEL-HP, http://www-tph.cheme.kyoto-u.ac.jp/kapsel/) [8]. We hope the current paper can be used in conjunction with the information in KAPSEL-HP as a reference for all work related to KAPSEL.

8.2 KAPSEL Installation and Basic Operations

8.2.1 OCTA Installation

We assume that OCTA8 is to be installed at /usr/local/OCTA8/ (under Linux) or C: \OCTA8\ (under Windows). All the following commands should be input at the command prompt (where we assume a bash shell).

```
> su        (Linux only)
> cd /usr/local/OCTA8/GOURMET/src
>./configure
> make
> make install
```

If we are using Cygwin (Windows), we need to add the following symbolic link:

```
> ln -s /cygdrive/c/OCTA8/ /usr/local/.
```

After completing the OCTA installation, we build libplatform as follows:

```
> su        (Linux only),
> cd /usr/local/OCTA8/GOURMET/src
>./configure
> make
> make install
```

8.2.2 KAPSEL Installation

At the KAPSEL-HP website, download the latest KAPSEL source code (kapsel3.*.zip), and unzip the file to the desired installation directory as follows:

```
> unzip kapsel3.X.zip
```

We see the following directory structure.

kapsel3.X/	KAPSEL source code
Doxygen/	Source code documentation
Examples/	Sample files
Tools/	Auxiliary tools
UDF/	Input and scripts

The software is built as follows;

```
> cd kapsel3.X
> make 'ENV=CYGWIN'      (Cygwin only)
> make 'ENV=GCC'         (Linux only)
```

An input file for a simple test case can be found in the "UDF" folder, and we can use this file to check that the installation was successful and that we are able to run KAPSEL:

```
> cd UDF
>../kapsel -Iinput.udf -Ooutput.udf -Ddefine.udf \
 -Rrestart.udf
```

If the KAPSEL installation was successful, the simulation should start, and we should see the following after approximately 30 s:

```
#Simulation has ended!
#Total Running Time (s):    24.77
#                   (m):     0.41
#                   (h):     0.01
#Average Step Time  (s):     0.02
#                   (m):     0.00
#                   (h):     0.00
```

8.2.3 Analysis with GOURMET

The test simulation performed above is for a sedimenting system of ten particles, five of which are heavier than the fluid and five lighter than the fluid, using a cubic grid of size $32 \times 32 \times 32$ to solve the fluid equations of motion. The output data are saved in the "output.udf" file, and by using GOURMET and/or Python, we can easily visualize these results. As an example, we now explain how to animate the particle trajectories using GOURMET.

- Start GOURMET, select "File" (item 1 in Fig. 8.2) → choose "Open" and read "output.udf."
- Load "particleshow.py" into the Python console at the bottom of the window (item 2 in Fig. 8.2).
- Click on the "Run" button on the right-hand side of the Python console (item 3 in Fig. 8.2). Next, click on the "play" button of the newly opened viewer window to visualize the animation of the particle trajectories.

Any Python script file can be executed in this manner within GOURMET, giving access to the simulation data in "output.udf" in order to perform the required data analysis.

8.2.4 Analysis Without GOURMET

Analyzing very large datasets using GOURMET can be time consuming. We therefore suggest two alternatives to GOURMET:

1. Python script (executed directly from the command line)
   In the "UDF" folder, we find a sample script "sk.py" that calculates the dynamic structure factor $S(k)$. If we have the "numpy" Python package installed, we can use this script to reproduce the results in Fig. 8.3 as follows (within Windows).

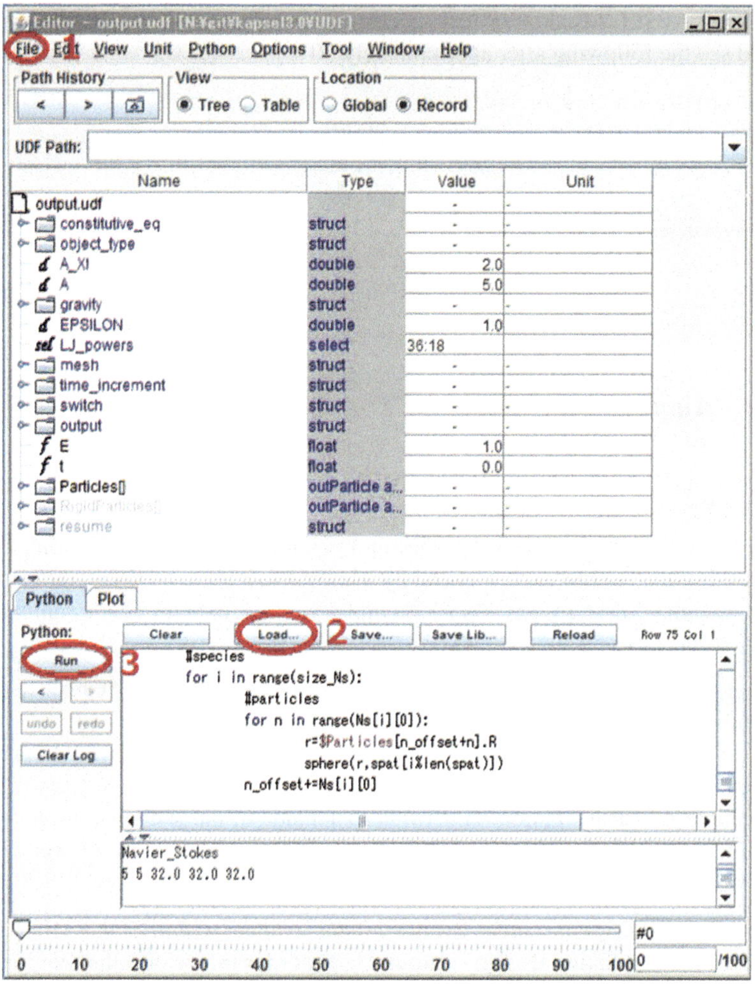

Fig. 8.2 Procedure of visualizing the simulation results (creating an animation) using Gourmet

From the start menu, we select "all programs" → "OCTA8" → "Start-GourmetTerm" and type the following commands at the prompt:

```
> python sk.py
> gnuplot
>> plot 'sk.dat' w line
```

For details regarding the Python interface within GOURMET, we refer to the "GOURMET PYTHON SCRIPT – reference" manual provided as part of the OCTA documentation.

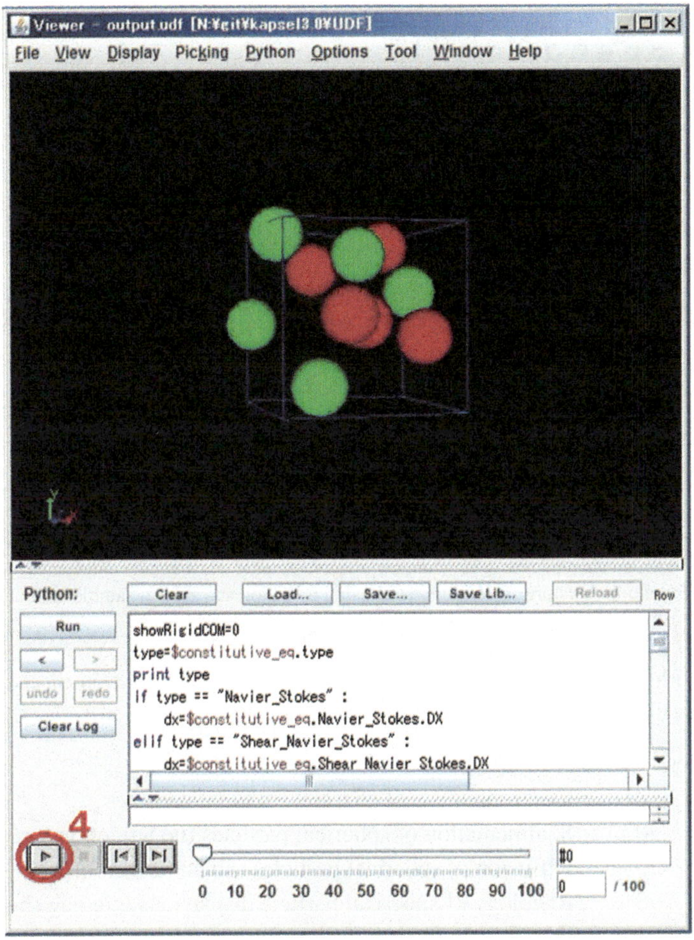

Fig. 8.2 (continued)

2. Fortran or C code

   OCTA provides a C/Fortran library that we can use to access the data stored in the KAPSEL input/output user definable format (UDF) files. We refer to the "PLATFORM INTERFACE LIBRARY – libplatform – reference" manual provided as part of the OCTA documentation for detailed information.

8.2.5 Sample Simulations

The following is a list of sample simulations that are distributed as part of the KAPSEL source code. The required input files can be found in the "Examples"

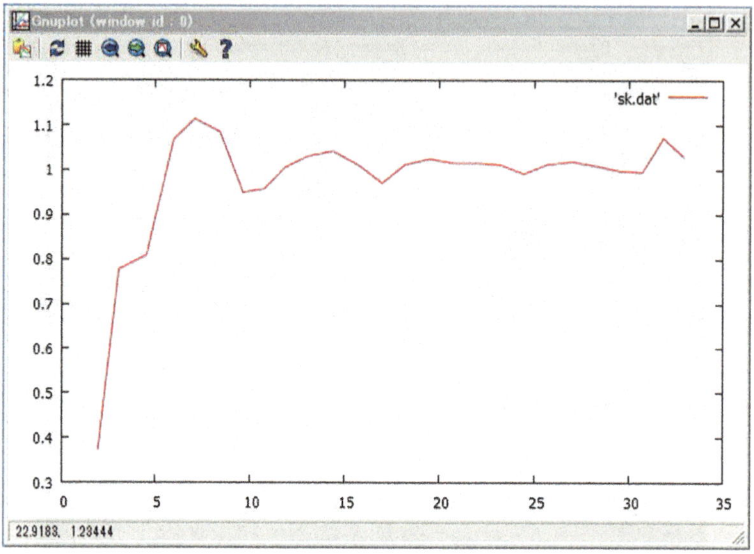

Fig. 8.3 Dynamic structure factor generated using a Python script for a sample simulation

folder within the main KAPSEL directory. A detailed description of each sample case is available at the KAPSEL-HP website [8].

Examples/

/01	Electrophoresis of charged colloidal particles
/02	Sedimentation of spherical particles (no attraction force)
/03	Diffusion of spherical particles (no attraction force)
/04	Rheology of spherical particle dispersions (zig-zag shear flow)
/05	Condensation of spherical particles (with attraction force)
/06	Thermal fluctuations and rotational dynamics of flexible bead-chain particles under shear flow
/07	Rheology of spherical particle dispersions (Lees–Edwards periodic boundary conditions)
/08	Sedimentation and rotational dynamics of non-spherical rigid particles under shear flow
/09	Dynamics of self-propelled particles
/10	Sedimentation of non-spherical rigid particles at high Reynolds number

8.3 Dynamics of Particle Dispersions

Particle dispersion refers to a system consisting of particles dispersed in a host fluid (solution), examples of which are ubiquitous in everyday life and include many food products, paints, pigments, cosmetics, and slurry. KAPSEL has been designed to carry out direct numerical simulations of particle dispersions undergoing diffusion, aggregation or coagulation [9], or sedimentation [10, 11]. In addition, it can be used to perform rheological calculations of particle dispersions under shear flow [12–14].

8.3.1 Basic Equations

Let us consider the motion of N spherical particles, of radius a, immersed in a Newtonian host fluid with viscosity η and density ρ. The host fluid is an incompressible Newtonian fluid, and as such, its time-dependent motion is determined by solving the Navier–Stokes equation to obtain the fluid velocity field $\mathbf{u}(\mathbf{x},t)$. The motions of the colloidal particles are obtained by solving the Langevin equation, which considers the random forces and torques caused by the thermal fluctuations, for the position $\mathbf{R}_i(t)$, velocity $\mathbf{V}_i(t)$, orientation $\mathbf{Q}_i(t)$, and angular velocity $\Omega_i(t)$ of the particles (where $i = 1,\cdots,N$ denotes the particle number). We use a fixed orthonormal grid to solve the coupled equations of motion for the fluid and particles in an efficient manner and introduce an additional field variable, the smooth profile function $\phi = \sum_{i=1}^{N} \phi_i$ ($\phi = 1$ in the solid phase and $\phi = 0$ in the fluid phase), to describe the interfacial region between the fluid and particle domains [5, 6]. In the following, we introduce the basic equations used to simulate these fluid/particle systems:

1. *Fluid equations of motion: the Navier–Stokes equation*
   The equations of motion for the fluid/particle velocity $\mathbf{u}(\mathbf{x},t)$ and pressure $p(\mathbf{x},t)$ fields are the Navier–Stokes equation:

$$\rho\,(\partial_t + \mathbf{u}\cdot\nabla)\,\mathbf{u} = -\nabla p + \eta\nabla^2\mathbf{u} + \rho\phi\mathbf{f}_p \tag{8.4}$$

   and the continuity equation:

$$\nabla\cdot\mathbf{u} = 0, \tag{8.5}$$

   where $\phi\mathbf{f}_p$ is a constraint force required to satisfy the boundary conditions over the surface of the particles. Within the SPM formalism, this constraint force also maintains the rigidity of the particles. Further details regarding the specific form of the constraint force can be found in the literature [5, 6]. Although we only consider the simple case of spherical particles in this paper, arbitrarily complex shapes, such as linear chains and helices, can be treated in the same manner.

2. *Particle equations of motion*: *the Langevin equation*
The time-dependent configuration of the i-th particle $\{\mathbf{R}_i(t), \mathbf{V}_i(t), \mathbf{Q}_i(t), \boldsymbol{\Omega}_i(t)\}$, with mass M_p and moment of inertia $\mathbf{I}_p$, is obtained using the set of equations:

$$\dot{\mathbf{R}}_i = \mathbf{V}_i, \quad \dot{\mathbf{Q}}_i = \mathbf{Q}_i \cdot \begin{pmatrix} 0 & \Omega_i^z & \Omega_i^y \\ \Omega_i^z & 0 & \Omega_i^x \\ \Omega_i^y & \Omega_i^x & 0 \end{pmatrix}, \tag{8.6}$$

$$M_p \dot{\mathbf{V}}_i = \mathbf{F}_i^H + \mathbf{F}_i^{\text{other}} + \mathbf{G}_i^V, \quad \mathbf{I}_p \cdot \dot{\Omega}_i = \mathbf{N}_i^H + \mathbf{N}_i^{\text{other}} + \mathbf{G}_i^{\Omega}, \tag{8.7}$$

where $\mathbf{F}_i{}^H$ and $\mathbf{N}_i{}^H$ respectively denote the hydrodynamic force and torque (arising from the fluid/particle interaction), which are determined by the constraint force $\phi\mathbf{f}_p$, such that the conservation of momentum is maintained. $\mathbf{F}_i{}^{other}$ and $\mathbf{N}_i{}^{other}$ respectively correspond to any other forces or torques (apart from the hydrodynamic forces due to the fluid), including direct interparticle interactions and external force fields such as the gravitational field. $\mathbf{G}_i{}^V$ and $\mathbf{G}_i^{\Omega}$ are random forces and torques, respectively, due to thermal fluctuations in the system, which we describe assuming Gaussian white noise with a zero mean:

$$\left\langle \mathbf{G}_i^V(t) \cdot \mathbf{G}_j^V(0) \right\rangle = 3k_B T \alpha^V \delta(t)\delta_{ij}, \quad \left\langle \mathbf{G}_i^{\Omega}(t) \cdot \mathbf{G}_j^{\Omega}(0) \right\rangle = 3k_B T \alpha^{\Omega} \delta(t)\delta_{ij}, \tag{8.8}$$

where $\langle \cdots \rangle$ denotes an ensemble average.

We should introduce the thermal fluctuations directly into the Navier–Stokes in Eq. 8.4; however, this would considerably increase the computational cost of the simulations, which is why, in KAPSEL, the thermal fluctuations are only treated within the Langevin dynamic framework. Therefore, we cannot expect to rigorously satisfy the fluctuation–dissipation theorem. This is not an insurmountable problem because the time scale over which this discrepancy appears is much smaller than the characteristic diffusion time scale of the particles. However, because the fluctuation–dissipation theorem cannot be used, we cannot guarantee that the temperature introduced in Eq. 8.8 corresponds to the actual physical temperature of the particle system. An additional fine-tuning operation is required to precisely control the temperature. In KAPSEL, we define the particle temperature T as the temperature obtained from the translational and rotational diffusion coefficients, D^V and D^{Ω}, respectively. This is achieved by introducing two additional parameters, α^V and α^{Ω}, which control the intensity of the fluctuating forces in Eq. 8.8. In general, setting $\alpha^V = \alpha^{\Omega} = 1$ should pose no difficulty, but if precise control of the temperature is required, the following procedure can be used to set the control parameters. A more detailed description on how to control the temperature can be found in previous publications [15, 16]:

- First, perform equilibrium simulations of a dilute dispersion (volume fraction $\varphi \ll 1$) at the desired temperature T, with $\alpha^V = \alpha^{\Omega} = 1$. Compute the translational and rotational diffusion coefficients, denoted as D^V_{sim} and D^{Ω}_{sim}, respectively.

- Second, compare the previously computed values of the diffusion coefficients with the expected theoretical values, D^V_{theoty} and $D^\Omega_{\mathrm{theoty}}$, for the given volume fraction φ. Fix the control parameters, α^V and α^Ω such that $T^\Omega = T^V = T$.
- Finally, conduct the desired simulation for an arbitrary volume fraction φ, with the values of α^V and α^Ω obtained above.

8.3.2 A Note on the Units

In KAPSEL, the basic simulation units are the grid spacing Δ and the density ρ and viscosity η of the host fluid. The corresponding time units are then defined by the viscous diffusion time $\tau_0 = \rho\Delta^2/\eta$. As an example, we consider a particle dispersion in water ($\eta = 10^{-3}$Pa·s, $\rho = 10^3$kg/m^2) with a characteristic length scale (grid spacing) of $\Delta = 1\mu$m. The unit of time for this simulation would be given by $\tau_0 = 10^{-6}$s. If the basic simulation units are defined in the input file (input.udf) as RHO $= A$, ETA $= B$ and DX $= C$, then the simulation time step Δt is automatically determined as $\Delta t = T_{\mathrm{dump}} = (A/B)/k^2_{\mathrm{max}}$, with k_{max} (proportional to C^{-1}) being the largest resolved wavevector in the solution scheme of the fluid equations of motion. For most simulations, this default value will be adequate; however, when an external force is added or a high shear rate is imposed, the simulation can become unstable and crash if $\Delta t = T_{\mathrm{dump}}$. In such cases, we should consider reducing the time step using a suitable multiplication factor, such that $\Delta t = T_{\mathrm{dump}} \times$ factor.

8.3.3 Particle Types

The current version of KAPSEL can handle the following four types of particles:

- Spherical particle
  - Particle interactions: 2n-n Lennard–Jones potential ($n = 6, 12, 18$).
  - Attraction forces can be turned *on*/*off*.
- Chain particle
  - Flexible bead-chain particles with a finitely extensible nonlinear elastic potential (FENE) [17, 18]
- Rigid particle
  - Arbitrarily shaped rigid particles composed of agglomerations of spherical beads
- Self-propelled particle
  - Spherical "squirmer" swimmers [19]

8.3.4 Input UDF File

To simulate particle dispersions in a Newtonian fluid, KAPSEL provides three options for the constitutive equation (*constitutive_eq*): Navier_Stokes, Shear_Navier_Stokes, and Shear_Navier_Stokes_Lees_Edwards.

constitutive_eq = Navier_Stokes

- *DX* grid spacing Δ (unit of length)
- *RHO* fluid density
- *ETA* fluid viscosity
- *kBT* particle temperature
- *alpha_v* additional control parameter for the translational temperature
- *alpha_o* additional control parameter for the rotational temperature

constitutive_eq = Shear_Navier_Stokes
(shear flow simulations using zigzag flow)

- *DC* steady shear *shear_rate* steady shear rate
- *AC* oscillatory shear *shear_rate* shear rate amplitude
  frequency shear rate frequency

constitutive_eq = Shear_Navier_Stokes_Lees_Edwards
(shear flow simulations using Lees–Edwards boundary conditions)

- *DC* steady shear *shear_rate* steady shear rate
  For the dispersed particles, we can choose one of three basic types: spherical, chain, or rigid particle.

object_type.type = spherical particle

- *Particle_number* number of particles
- *MASS_RATIO* ratio of particle density to fluid density
- *janus_** self-propelled particle parameters (not considered in this paper)

object_type.type = chain

- *Beads_number* number of spherical beads per chain
- *Chain_number* number of chains
- *MASS_RATIO* ratio of bead density to fluid density

object_type.type = rigid

- *Beads_number* number of spherical beads per rigid particle
- *Chain_number* number of rigid particles
- *MASS_RATIO* ratio of bead density to fluid density
- *Rigid_motion* free: free motion
  Fix: fixed velocity and angular velocity

Common parameters, such as the radius of the spherical particle (bead), are given below:

- *A_XI* thickness of the interfacial boundary layer ξ
- *A* particle radius
- *gravity.G* gravitational acceleration
- *gravity.direction* direction of gravity
- *EPSILON* magnitude of the Lennard–Jones potential energy
- *LJ_powers* exponent of the Lennard–Jones-type potential
- *mesh.NPX* simulation cell size in the x direction: $L_x = 2^{\mathrm{NPX}}$
- *mesh.NPY* simulation cell size in the y direction: $L_x = 2^{\mathrm{NPY}}$
- *mesh.NPZ* simulation cell size in the z direction: $L_x = 2^{\mathrm{NPZ}}$
- *time_increment.type* auto: $\Delta T = \mathrm{T}_{\mathrm{dump}} \times \text{factor}$

$$T_{\mathrm{dump}} = \rho / \left(\eta k_{\mathrm{max}}^2\right)$$

Manual: manually set the time step.

The following options specify the detailed simulation conditions:

- *switch.ROTATION*: Consider the rotational motion of the system (*on* by default)
- *switch.LJ_truncate on*: Ignore attraction force between particles
  Off: Include attraction force between particles
  None: No Lennard–Jones interactions between particles
- *switch.INIT_distribution* Controls the initial particle distribution
  uniform_random: generate a random distribution
  random_walk: generate a random distribution
  FCC: use face-centered cubic lattice points
  BCC: use body-centered cubic lattice points
  user_specify: manually specify in input.udf

If user_specify is chosen, the initial particle configuration (namely the positions and velocities) should be set in the *user_specify.Particles[].R* and *user_specify.Particles[].v* arrays. If the number of elements in the list is smaller than the number specified in Particle_number, we either increase it using Gourmet (Edit – > Add an array Element) or directly edit the UDF file to add the remaining entries. In the case of rigid particles, this option allows us to define any arbitrary shape as a collection of (possibly overlapping) spherical particles.

- *switch.FIX_CELL* ON: set the system drift velocity to zero ($\hat{\mathbf{u}}_{k=0} = 0$)
- *output.GTS* number of steps between output intervals
- *output.Num_snap* number of output data frames
  (the total number of time steps is GTS × Num_snap)
- *output.AVS on* if we want to output the data in AVS format
- *output.AVS.ON.Out_dir* directory name for the AVS-formatted data
  For example, if we wish to save files within the "data" folder, we ensure that both ./data and ./data/avs sub-folders exist within the current directory. KAPSEL will create an AVS-formatted field file at ./data/data.fld. The data files will be located at ./data/avs/data_X.dat (where X is the step number).

- *output.AVS.ON.FileType* AVS data file format (Binary/ASCII)
- *output.UDF on* if we want the output data in UDF format

8.4 Electrophoresis of Charged Colloidal Particles

When colloidal particles are dispersed in a solution with very high permittivity, such as water, functional groups at the surface of the particles become ionized, which leaves the particle with a net surface charge. The ions released into the solution, which are electrostatically attracted to the particle surface and undergo diffusion due to thermal fluctuations, form a cloud-like ionic atmosphere called an electric double layer around the colloidal particles. Under equilibrium conditions, the properties of such colloidal dispersions are well described by the Poisson–Boltzmann equation. However, for so-called electrokinetic phenomena, such as electrophoresis, the dynamics of the colloidal particles and their surrounding ionic distributions are determined by a nontrivial interplay between the hydrodynamic and electrostatic interactions. As a result of this competition, it is possible to find situations in which the ionic distribution is unable to immediately follow the motion of the colloids. This gives rise to a distortion of the electric double layer, into a nonspherical and nonequilibrium state. Such dynamic phenomena can only be quantitatively reproduced in computer simulations, and the number of previous studies that have achieved this is rather limited. Within the SPM framework used by KAPSEL, this complex multi-scale electro-hydrodynamic phenomenon is described employing direct numerical simulations to solve the coupled set of equations governing the dynamics of the fluid, the ionic distribution and the colloidal particles. The dynamics of the system are determined by the Navier–Stokes equation for the solvent velocity field, coupled to the advection–diffusion equation for the ionic distribution and the Newton–Euler equations for the colloids, with the electrostatic interactions given by the Poisson equation.

8.4.1 Basic Equations

We now describe the basic equations needed to simulate the electro-hydrodynamic problem, following references [6, 7, 20]. Let us consider a system of N spherical colloidal particles of radius a, dispersed in an electrolyte solution. The permittivity of the solvent is assumed to be spatially uniform. The surface of the colloidal particle is assumed to carry with it a uniform charge density, with a total charge, per particle, given by Ze. Within an ordinary continuum description, the particle surface charge distribution would be given by a delta function. Consequently, a suitable boundary-fitted mesh would need to be used within a finite element method, which substantially reduces the computational efficiency. In contrast, the SPM uses a continuous distribution for the surface charge density $eq(\mathbf{x})$. For example, using a first-order derivative of the smooth profile function $\phi(\mathbf{x})$, we define the surface charge as

$$eq(\mathbf{x}) = \frac{Ze\,|\nabla\phi(\mathbf{x})|}{4\pi a^2}. \tag{8.9}$$

In the same way that the smooth profile function $\phi(\mathbf{x})$ is defined to reduce to a delta function in the limit $\xi \to 0$, $q(\mathbf{x})$ is also reduced to a delta function in this limit:

i. *Ionic equations of motion: the advection–diffusion equation*

Let C_α be the density distribution of α-type ions (of valence Z_α), defined as

$$C_\alpha(\mathbf{x},t) = (1-\phi(\mathbf{x},t))\,C_\alpha^*(\mathbf{x},t) \tag{8.10}$$

over the entire computational domain. The domain inside the particle region, where ions do not exist, is effectively expressed using the smooth profile function $(1-\phi) \approx 0$. The density is defined in terms of C_α^*, an auxiliary field variable, which facilitates the numerical calculations because $C_\alpha^*(\mathbf{x},t)$ is chosen to be smooth and continuous throughout the entire computational domain. The total charge distribution, including the particle surface charge distribution, is then given by

$$\rho_e(\mathbf{x}) = e\sum_\alpha Z_\alpha C_\alpha(\mathbf{x}) + eq(\mathbf{x}). \tag{8.11}$$

The dynamics of the ions are determined by assuming that the auxiliary density distribution C_α^* follows an advection–diffusion equation:

$$\partial_t C_\alpha^* = -\nabla\cdot C_\alpha^*\mathbf{u} + \Gamma_\alpha\nabla\cdot\left(C_\alpha^*\nabla\mu_\alpha\right). \tag{8.12}$$

This last equation consists of two terms; the first expresses the advection of the ions due to the solvent velocity field $\mathbf{u}$, while the second expresses the diffusion of the ions due to a gradient in the chemical potential μ_α. Here, Γ_α is the Onsager transport coefficient for α-type ions, which is related to the friction and diffusion coefficients according to $f_\alpha = 1/\Gamma_\alpha$ and $D_\alpha = k_BT\;\Gamma_\alpha$, respectively. The friction coefficient is conveniently expressed in reduced units as $m_\alpha = 2\cdot k_BTf_\alpha/3\eta e^2$, which, for an aqueous KCl solution at 25 °C gives $m_{K+} \sim m_{Cl+} \sim 0.184$. Finally, the chemical potential of the ions is assumed to take the form

$$\mu_\alpha = k_BTlnC_\alpha^* + Z_\alpha e(\Psi - \mathbf{E}\cdot\mathbf{x}), \tag{8.13}$$

where E denotes the external electric field and $\Psi(\mathbf{x})$ is the electrostatic potential obtained by solving the Poisson equation $\epsilon\nabla^2\Psi = -\rho_e$.

ii. *Solvent equations of motion*: *the Navier–Stokes equation*
We assume that the solvent is an incompressible ($\nabla \cdot \mathbf{u} = 0$) Newtonian fluid. The solvent velocity field $\mathbf{u}$ is then determined by the Navier–Stokes equation

$$\rho\left(\partial_t + \mathbf{u}\cdot\nabla\right)\mathbf{u} = -\nabla p + \eta\nabla^2\mathbf{u} - \rho_e\left(\nabla\Psi - E\right) + \rho\phi\mathbf{f}_p. \tag{8.14}$$

We note that, in contrast to Eq. 8.4, for neutral colloidal dispersions, an extra forcing term due to the electrostatic interactions $-\rho_e\left(\nabla\Psi - E\right)$ now appears in the Navier–Stokes equation.

iii. *Particle equations of motion*
The time dependence of the position, velocity, angular velocity, and orientation of particle $\{\mathbf{R}_i(t), \mathbf{V}_i(t), \mathbf{Q}_i(t), \Omega_i(t)\}$, $\{\mathbf{R}_i(t), \mathbf{V}_i(t), \mathbf{Q}_i(t), \Omega_i(t)\}$,with mass M_p and moment of inertia $\mathbf{I}_p$, is determined by the Newton–Euler equations:

$$\dot{\mathbf{R}}_i = \mathbf{V}_i, \quad \dot{\mathbf{Q}}_i = \mathbf{Q}_i \cdot \begin{pmatrix} 0 & \Omega_i^z & \Omega_i^y \\ \Omega_i^z & 0 & \Omega_i^x \\ \Omega_i^y & \Omega_i^x & 0 \end{pmatrix}, \tag{8.15}$$

$$M_p\dot{\mathbf{V}}_i = \mathbf{F}_i^H + \mathbf{F}_i^{\text{other}}, \quad \mathbf{I}_p \cdot \dot{\Omega}_i = \mathbf{N}_i^H + \mathbf{N}_i^{\text{other}}. \tag{8.16}$$

The only difference with respect to the equations of motion for the neutral particles is the absence of any thermal fluctuating forces (Brownian motion).

8.4.2 Electric Double-Layer Properties

We determine the equilibrium distribution for a concentration of ions evolving in time according to Eq. 8.12 in the absence of any electric field (i.e., $E = 0$). For a uniform chemical potential, such that μ_α is constant over all space, the equilibrium ion distribution is given by

$$C_\alpha^*(\mathbf{r}) = \overline{C}_\alpha \exp\left(-\frac{Z_\alpha e\Psi(\mathbf{r})}{k_B T}\right). \tag{8.17}$$

This is the Boltzmann distribution under an electrostatic potential Ψ. Inserting the Boltzmann distribution into the Poisson equation gives the celebrated Poisson–Boltzmann equation, which lies at the heart of any quantitative analysis of the electric double-layer structure. In the case of a spherical colloidal particle immersed in a symmetric $\pm z$ electrolyte solution, this Poisson–Boltzmann equation takes the form

$$\nabla^2\Psi(\mathbf{r}) = \frac{2ze\overline{C}}{\epsilon}\sinh\left(\frac{ze\Psi(\mathbf{r})}{k_B T}\right). \tag{8.18}$$

The appropriate boundary conditions for a point infinitely far from the particle are $\Psi\,|_{r=\infty} = 0$ and $C^*\,\big|_{r=\infty} = \overline{C}$, while the (constant surface charge) boundary condition at the surface of the particle is given as

$$\nabla\Psi\,\Big|_{\text{surface}} = -\frac{\sigma e}{\epsilon}, \quad \sigma e = Ze/4\pi a^2 \tag{8.19}$$

The popular Debye–Hückel approximation assumes $ze\Psi/k_B T \ll 1$ and linearizes the exponential appearing in Eq. 8.18 to yield

$$\nabla^2\Psi\,(\mathbf{r}) = \frac{2z^2e^2\overline{C}}{k_B T\epsilon}\Psi = \kappa^2\Psi. \tag{8.20}$$

This expression for the electrostatic potential depends on a single parameter, with units of length, which defines the Debye screening length

$$\kappa^{-1} = \frac{1}{\sqrt{8\pi\lambda_B z^2\overline{C}}}. \tag{8.21}$$

We also introduce the Bjerrum length $\lambda_B = e^2/4\pi k_B T\epsilon$ of the solution, which is approximately 0.72 nm for water at 25 °C (and fixed according to $\Delta = 4\pi\lambda_B$ in KAPSEL). In the case of a general electrolyte solution, with Z_α the valence of α-type ions, the screening length takes the same general form

$$\kappa^{-1} = \frac{1}{\sqrt{4\pi\lambda_B\sum_\alpha Z_\alpha^2\overline{C}_\alpha}}. \tag{8.22}$$

Given the spherical symmetry of the system, we need only consider the $r = |\mathbf{r}|$ dependence when solving the Poisson–Boltzmann equation. Within the Debye–Hückel approximation (8.20), we are left with the simple second-order differential equation

$$\frac{d^2\Psi}{dr^2} + \frac{2}{r}\frac{d\Psi}{dr} = \kappa^2\Psi. \tag{8.23}$$

A general solution to this equation is given by the Yukawa-type potential

$$\Psi(r) = \Psi_0\frac{a}{r}exp\left[-\kappa\,(r-a)\right], \tag{8.24}$$

from which we see that the electrostatic interactions arising from the surface charge of the colloids are screened over a characteristic length scale of κ^{-1}. It is easy to see that this screening length increases with temperature and decreases with the ionic strength $\sum_\alpha Z_\alpha^2\overline{C}_\alpha/2$.

8.4.3 UDF Description

To simulate charged colloidal dispersions in electrolyte solutions, allowing for the study of electrophoretic phenomena under external fields, we simply select "Electrolyte" as the "constitutive_eq" option in the input UDF file. We note that the common simulation parameters are the same as those in the case of neutral particles, which have already been explained in Sect. 8.3.3 and will not be repeated here. We detail below the specific input options related to the simulation of charged dispersions.

constitutive_eq = Electrolyte

- *kBT* temperature of the system T (used to determine the ion distribution)
- *Dielectric_cst* dielectric constant of the solvent
- *Init_profile* initial ionic distribution (Uniform/Poisson_Boltzmann)
- *Add_salt* saltfree: only add counter-ions to the solution
  salt: add counter-ions plus a binary salt (with +/− ions),
- *Add_salt.saltfree.Valency_counterion* valence of the counter-ions
- *Add_salt.saltfree.Onsager_coeff_counterion* Onsager transport coefficient for the counter-ions (Defined in Eq. 8.12, it corresponds to the diffusion coefficient divided by $k_B T$)
- *Add_salt.salt.Valency_positive_ion* valence of the positive ions
- *Add_salt.salt.Valency_negative_ion* valence of the negative ions
- *Add_salt.salt.Onsager_coeff_positive_ion* Onsager transport coefficient for the positive ions
- *Add_salt.salt.Onsager_coeff_negative_ion* Onsager transport coefficient for the negative ions
- *Add_salt.salt.*Debye_length Debye screening length (Defined in Eq. 8.22, it represents the thickness of the electric double layer),
- *Electric_field* external electric field (on/off)
- *Electric_field.ON* DC: steady field, AC: alternating field
- *Electric_field.ON.DC.Ex* strength of the electric field in the x direction
- *Electric_field.ON.DC.Ey* strength of the electric field in the y direction
- *Electric_field.ON.DC.Ez* strength of the electric field in the z direction
- *Electric_field.ON.AC.Frequency* frequency of the alternating electric field

For charged dispersions, we can only choose spherical particles as particle type. We must specify the charge of the particles as follows.
object_type.type = spherical particle

- *Surface_charge* total particle surface charge

The calculation of the time step now takes into account the time scale determined by the ionic diffusion.

- *time_increment.type* auto: $\Delta T = T_{\text{dump}} \times$ factor
  (T_{dump} is the minimum of $T_{\text{dump}} = \rho/(\eta {k_{\text{max}}}^2)$ and $1/(k_B T \Gamma_\alpha k_{\text{max}}^2)$),
  Manual: manually set the time step.

References

1. D.L. Ermak, J.A. McCammon, J. Chem. Phys. **69**, 1352 (1978)
2. J.F. Brady, G. Bossis, Annu. Rev. Fluid Mech. **20**, 111 (1988)
3. H. Tanaka, T. Araki, Phys. Rev. Lett. **85**, 1338 (2000)
4. T. Kajishima, S. Takiguchi, H. Hamasaki, Y. Miyake, JSME Int. J. Ser. B **44**, 526 (2001)
5. Y. Nakayama, R. Yamamoto, Phys. Rev. E **71**, 036707 (2005)
6. Y. Nakayama, K. Kim, R. Yamamoto, Eur. Phys. J. E **26**, 361 (2008)
7. K. Kim, Y. Nakayama, R. Yamamoto, Phys. Rev. Lett. **96**, 208302 (2006)
8. KAPSEL Home Page. (Transport Phenomena Lab, Kyoto University, 2015), http://www-tph.cheme.kyoto-u.ac.jp/kapsel. Accessed 17 Nov 2015
9. Y. Matsuoka, T. Fukasawa, K. Higashitani, R. Yamamoto, Phys. Rev. E **86**, 051403 (2012)
10. A. Hamid, R. Yamamoto, J. Phys. Soc. Jpn. **82**, 024004 (2013)
11. A. Hamid, R. Yamamoto, Phys. Rev. E **87**, 022310 (2013)
12. T. Iwashita, R. Yamamoto, Phys. Rev. E **80**, 061402 (2009)
13. T. Iwashita, T. Kumagai, R. Yamamoto, Eur. Phys. J. E **32**, 357 (2010)
14. H. Kobayashi, R. Yamamoto, J. Chem. Phys. **134**, 064110 (2011)
15. T. Iwashita, Y. Nakayama, R. Yamamoto, J. Phys. Soc. Jpn. **77**, 074007 (2008)
16. T. Iwashita, R. Yamamoto, Phys. Rev. E **79**, 031401 (2009)
17. H. Kobayashi, R. Yamamoto, Phys. Rev. E **81**, 041807 (2010)
18. H. Kobayashi, R. Yamamoto, Phys. Rev. E **84**, 051404 (2011)
19. J.J. Molina, Y. Nakayama, R. Yamamoto, Soft Matter **9**, 4923 (2013)
20. H. Oshima, *Theory of Colloidal and Interfacial Electric Phenomena* (Academic, London, 2006)

Part III
Examples of the Application of OCTA

Chapter 9
Melt Viscoelasticity

Naoki Kobayashi

9.1 Introduction

In the field of plastics modeling, it is important to know the characteristics of polymer melts. Polymers cannot be handled as simple Newtonian fluids owing to the effect of entanglements between molecular chains; rather, they must be handled as viscoelastic materials having both viscosity and elasticity. A typical relaxation spectrum of a polymer melt is shown in Fig. 9.1 [1]. Step shears were applied to the polymer melt, after which the change in stress during the relaxation process was observed to obtain the spectrum. The behavior can be divided into different regions. The glass region and transition region represent relaxation over a short period of time, which depends on the molecular structure of the monomers mainly the rubber region represents relaxation that is attributable to entanglements; the flow region represents relaxation of the whole polymer chain; and the time that distinguishes the rubber region from the flow region is referred to as the maximum relaxation time (τ_m).

It is considered difficult to reproduce such rheological characteristics of polymer melts in molecular simulations given the time scale of relaxation and the spatial scale of polymer entanglements. In fact, it is not realistic to calculate a relaxation spectrum in rheology using a full atomistic-type molecular dynamics method because such calculation requires considerable time, even if we were to use the fastest-class large-scale parallel computer in the world. However, in the case in which a Kremer–Grest-type bead–spring model is used, even though a certain amount of time is required for calculation, there has been a report of performing the

N. Kobayashi (✉)
Mitsui Chemicals Inc., Chiba, Japan
e-mail: Naoki.Kobayashi@mitsuichemicals.com

Japan Association for Chemical Innovation, *Computer Simulation of Polymeric Materials*, DOI 10.1007/978-981-10-0815-3_9

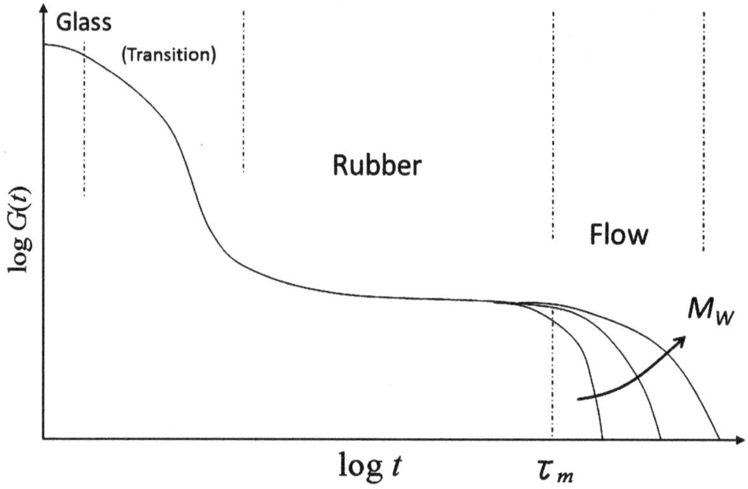

Fig. 9.1 Typical behavior of the relaxation modulus of a polymer

calculation [2]. This section introduces examples of calculating relaxation spectra of Kremer–Grest-type model melts having some extent of entanglement and of calculating various rheological properties.

9.2 Calculation Model

In the Kremer–Grest-type bead–spring model, the entanglement molecular weight has been reported to approximately range from 35 [3] to 54 [4] beads. The bead diameter is denoted σ and the bead weight *m*. Here, with *N* representing the degree of polymerization, we discuss the following cases: $N = 30$, where entanglements hardly exist on average; $N = 100$, where two to three entanglements exist on average; and $N = 200$, where four to six entanglements exist on average. The simulation temperature was set to $T = 1.0\ \varepsilon/k_B$ (where ε is the energy scale of the potential and k_B is Boltzmann's constant), and the remaining parameters were set to Kremer–Grest-type values [3].

9.3 Calculation of Stress Relaxation by COGNAC

COGNAC can calculate the stress relaxation using an autocorrelation function of stress obtained from the equilibrium calculation in the field of molecular dynamics according to a Green–Kubo formula. This function has been implemented since COGNAC version 8; because the autocorrelation function of stress is calculated concurrently with the simulation using an algorithm of Magatti [5], the correlation function can be run over several hours with a lower memory footprint.

9.4 Creation of the Initial Structure

Generating an initial structure is generally a sensitive operation, and considerably complicated procedures are often required. However, such procedures are omitted here, and instead, a function for generating an input UDF file of SILK [6] (the COGNAC default) making use of "potential_map.udf" is used to simply generate (as an initial structure) ten molecular chains randomly in a cubic cell with a density of $\rho = 0.85\, m/\sigma^{-3}$. A thermal equilibration calculation is performed until the energy of the molecular chains does not change (related output UDF files: "30x10out.udf", "100x10out.udf", "200x10out.udf").

9.5 Start of Simulation

The maximum relaxation time of a Kremer–Grest chain for which $N = 200$ has been reported to be approximately $1 \times 10^5 \tau$ (where $\tau = \sigma(m/\varepsilon)^{1/2}$) [2]. The settings are determined such that the calculation is performed over 1×10^8 steps with $dt = 0.01\tau$ and the results are written out every 10,000 steps. NVT_Kremer_Grest is used as the algorithm, and defaults are used for other potentials without change. In the settings, *Stress* in *Correlation_Function* of *Output_Flags* is set to ON as shown in Fig. 9.2 to calculate the autocorrelation function of stress (related input UDF files: "30x10gin.udf", "100x10gin.udf", "200x10gin.udf").

To reduce capacity, only the final record is output as an attached output UDF file.

9.6 Output and Plotting of Simulation Results

The obtained output UDF file is read by GOURMET to find that a tag named *Correlation_Functions* has been created; this tag is clicked on to find that calculation results are stored in *Stress.Correlation[]*. *Gt*() is obtained by plotting *Time[]* and *G_t[]* in the final record. In addition, an action command is prepared in the *Stress* tag of *Correlation_Function*, and $G(t)$ can be drawn automatically by placing the mouse cursor over the action command and right-clicking to run **Plot** (Fig. 9.3).

Various pieces of rheological information obtained from $G(t)$, such as the maximum relaxation time of the polymer chain, can be estimated by observing the spectrum (related output UDF files: "30x10gout.udf", "100x10gout.udf", "200x10gout.udf").

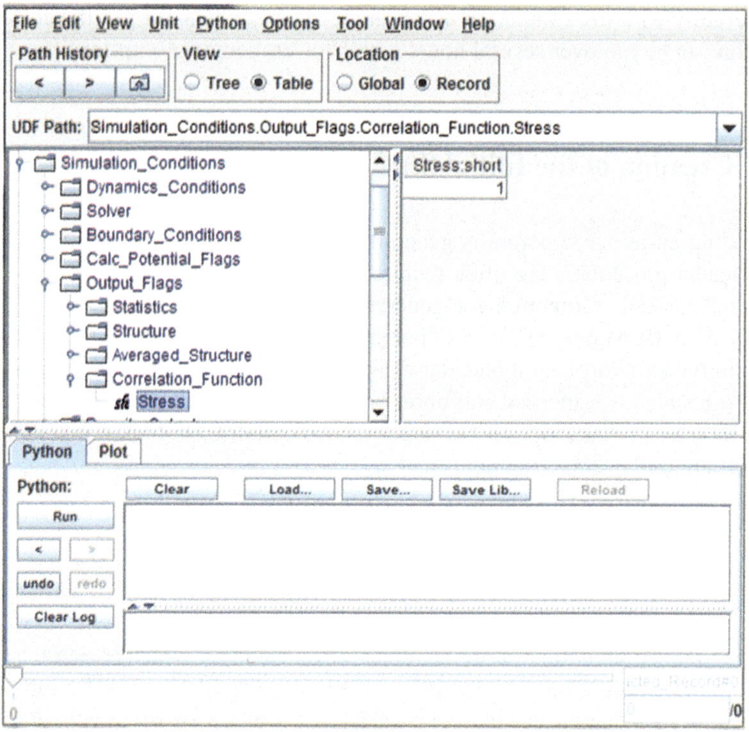

Fig. 9.2 Calculation of the correlation function in an input UDF file

9.7 Analysis of Simulation Results

The obtained relaxation spectrum for each molecular weight is shown in Fig. 9.4. The initial relaxation processes completely overlap regardless of whether entanglements exist because the molecule species are the same. The difference in the maximum relaxation time clearly appears to be based on the difference in the relaxation of the rubber region due to the existence of entanglements and the difference in molecular weight. Moreover, the molecular weight dependence of the maximum relaxation time agrees with the literature. The effect of entanglements was studied from the values of $G(t) \times t^{0.5}$ normalized by the time dependence of Rouse relaxation according to the method of Likhtman et al. [2] to obtain the results shown in Fig. 9.5, and the results almost completely agree with the literature.

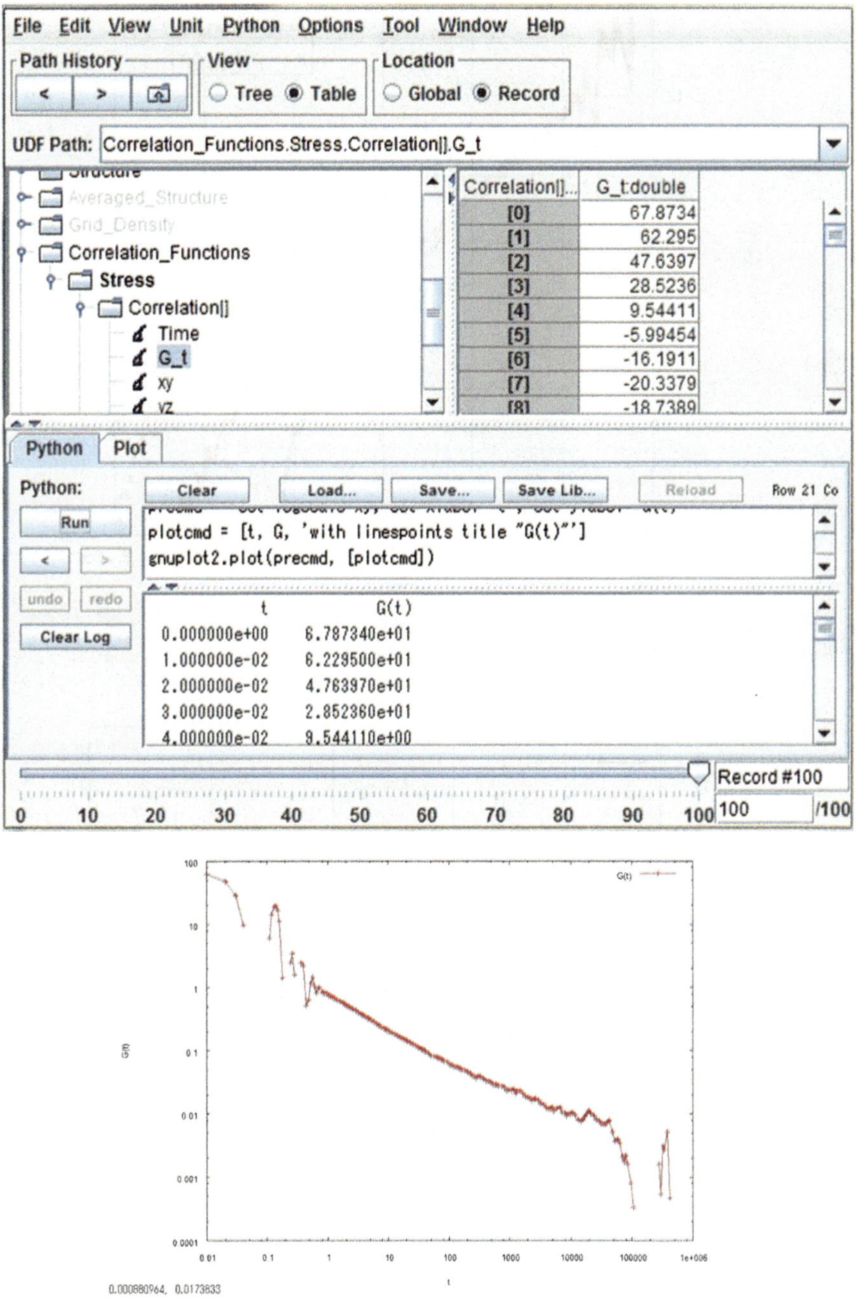

Fig. 9.3 Plot of the stress relaxation spectrum

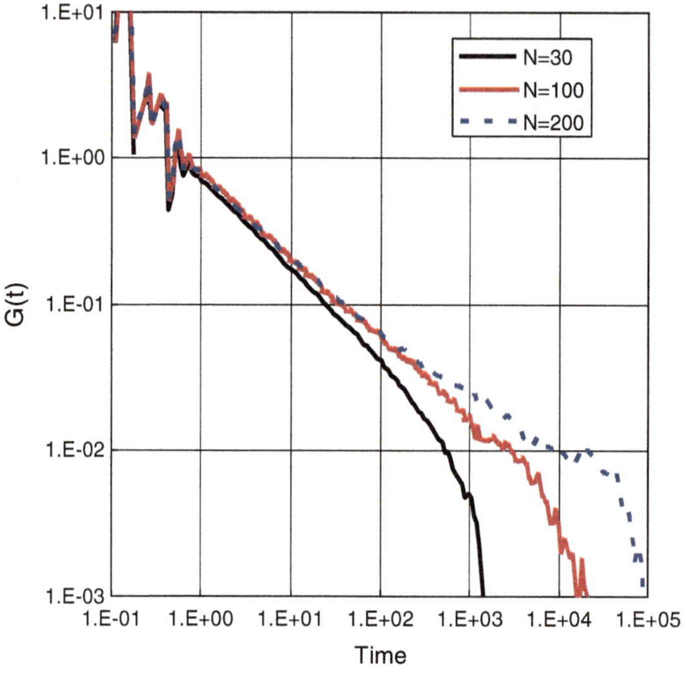

Fig. 9.4 Stress relaxation in the standard Kremer–Grest model

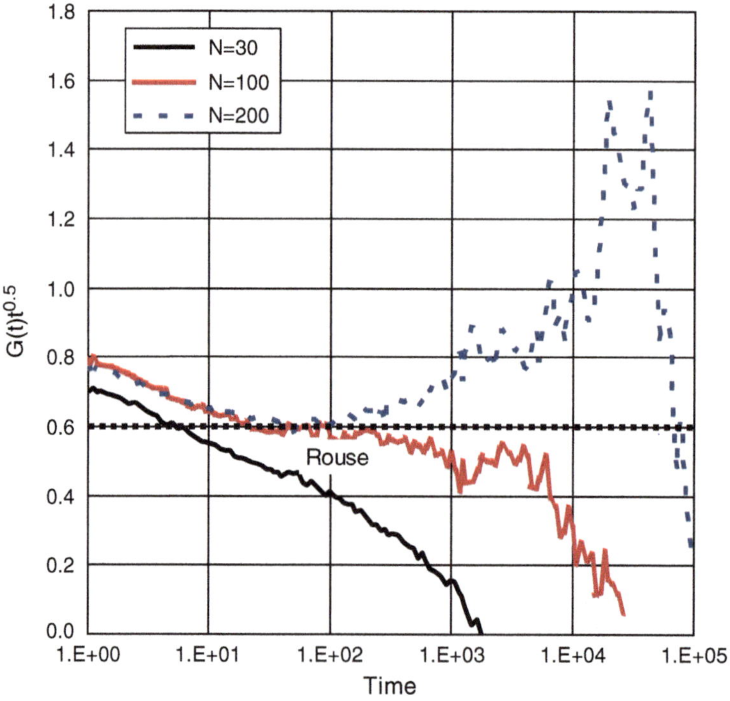

Fig. 9.5 Stress relaxation spectra normalized by Rouse relaxation $t^{-0.5}$

9.8 Concluding Remarks

When using the Kremer–Grest chain as an analogy of a real chain and applying it to actual polymer material development, the effects of the molecular weight distribution, long-chain branching, and cross-linking on the rheological characteristics can be directly estimated by simulation. In other words, by comparing the entanglement, relaxation time, and the like obtained from calculation with the rheological properties of real polymer materials, useful information on melt viscoelasticity can be obtained from relative values in such a comparison.

9.9 Download Instruction

Sample files are available from the OCTA website. See the instruction on http://octa.jp/download_bookdata.html. Password for the sample (zip file) of this chapter is “hsy4pmN7”.

References

1. C.W. Macosko, *Rheology: Principles, Measurements, and Applications* (Wiley, New York, 1994), p. 118
2. A.E. Likhtman, S.K. Sukumaran, J. Ramirez, Macromolecules **40**, 6748 (2007)
3. K. Kremer, G.S. Grest, J. Chem. Phys. **92**, 5057 (1990)
4. S.K. Sukumaran, G.S. Grest, K. Kremer, R. Everaers, J. Polym. Sci. B Polym. Phys. **43**, 917 (2005)
5. D. Magatti, F. Ferri, Appl. Optics **40**, 4011 (2001)
6. F. Sawa, *COGNAC User’s Manual* (OCTA Users Group, 2013), Chapter 6

Chapter 10
Crystallization of Polymers

Takashi Yamamoto

10.1 Introduction

Plastics and fibers are mostly made of crystalline polymers, where higher-order structures of coexisting crystalline and amorphous regions are known to dominate the materials' properties, such as the materials' density and mechanical properties. In composites of polymers and inorganic materials, local ordered structures at the interfaces play important roles. For example, recent investigations of semiconducting polymers have shown that crystallinity and crystal orientation near electrodes control the electric properties.

The crystal structure and organization of crystalline lamellae have long been the central subjects of research on crystalline polymers, and an enormous amount of knowledge has been accumulated since the discovery of polymer single crystals. Development of specific crystalline textures is a result of complicated irreversible processes taking place far from equilibrium, and therefore, thermodynamical and statistical mechanical analyses of the processes often face serious problems. In addition, the crystalline and amorphous regions are closely intertwined, making experimental investigations of the structure and properties difficult.

In recent years, molecular simulations of polymers have attracted increasing attention. However, simulations of crystalline polymers seem to lag far behind those of amorphous polymers because of the many difficulties that stem from the structural complexity and extreme sluggishness of polymer crystallization. As described before, the self-organization in crystalline polymers has great significance in many fields of scientific as well as having technological importance. Uncovering molecular processes using molecular simulations is thus expected to make great contributions to these fields [1].

T. Yamamoto (✉)
Graduate School of Science and Engineering, Yamaguchi University, Yamaguchi, Japan
e-mail: yamamoto@mms.sci.yamaguchi-u.ac.jp

Japan Association for Chemical Innovation, *Computer Simulation of Polymeric Materials*, DOI 10.1007/978-981-10-0815-3_10

Stereoregular polymers are mostly crystalline, and they crystallize into more or less regularly folded thin lamellae. These lamellar crystals further aggregate to form higher-order structures such as spherulites or fibers depending on the crystallization condition. The relevant spatial scales of polymer crystallization range from the atomistic scale (on the order of the nanometer) to the lamella scale (on the order of tens of nanometers) and to the macroscopic scale (on the order of microns). The mechanism of self-organization at each scale is dominated by distinct physics of individual interest. In this chapter, however, we restrict our attention to the smallest-scale problems, the molecular mechanisms of crystallization. Polymers are flexible, giant molecules, but remarkably crystallize into regularly chain-folded lamellae through the tugging of molecular tails in highly entangled states. To understand how these molecular processes emerge has long been the most fundamental subject of polymer crystallization. In this chapter, we explain some of the attempts to observe directly crystallizing polymers by conducting molecular simulations.

10.2 Molecular Models and Simulation Conditions

The crystal structure of a polymer depends sensitively on the molecular structure. This is indeed the very foundation of how we determine a molecular structure through crystal analysis. In principle, the molecular pathway of polymer crystallization should also depend on the molecular structure, and it might appear that we must always adopt full atomistic models. However, owing to the extremely slow process of crystallization, such direct approaches are often beyond execution even when using present-day computers. In searching for a proper way around the difficulty, the following points are important in selecting molecular models and simulation conditions:

1. We can accelerate simulation by adopting a model as simple as possible but having the necessary features of the polymer concerned. Such a selection of the minimal model is not unique and depends on researcher's interest. Like the case when taking usual coarse-grain approaches, various important chemical characteristics of the polymer molecule are lost in the minimal model. However, it is well acknowledged that polymer crystallization has rather universal features that are quite independent of the chemical structure [2]. For example, the crystal growth rate and lamella thickness are known to follow very simple universal rules. To understand these universal features of polymer crystallization, Lauritzen and Hoffman proposed long ago a phenomenological molecular model (LH model), which has been used with great success and has long been deemed the standard model of polymer crystallization [2]. The model is described with a limited number of parameters, such as the surface free energy and the enthalpy of fusion, which are reflections of the chain stiffness and the cohesive energy. It is now clear that proper minimal models are useful in unveiling the universal molecular mechanism of polymer crystallization.

2. We can accelerate crystallization itself by selecting proper crystallization conditions. We can then expect direct observation of crystallizing molecules even when using realistic molecular models. A typical example is crystallization through stretching or under shear flow. It is well known that polymer molecules crystallize rapidly when they undergo large elongational or experience shear flow. This feature provides favorable grounds for molecular simulations, and indeed several investigations taking advantage of this feature have been reported. It is also known that modification of the chain stiffness facilitates crystallization. Stiff chains should have higher melting temperatures and therefore provide greater supercooling for a given temperature of crystallization. Several studies of the crystallization of stiffened polymer molecules have been reported [3], but care needs to be taken in interpreting the data.

10.3 Results of Simulations

10.3.1 Chain-Folded Crystallization of a Single Molecule

In homogeneous solutions or melts in supersaturated or supercooled states, small clusters or embryos are created and annihilated by thermal fluctuations. The accidental emergence of clusters larger than a critical size, the size of the critical nucleus, gives rise to the continuous growth of clusters into stable clusters. Because the critical free energy determines the frequency of nucleation that dominates the overall rate of crystallization, a real molecular image of the critical nucleus has been a subject of great interest. However, experimental observations of the nuclei are still difficult, especially for polymers, and molecular simulations therefore provide great opportunities.

When a random coiled molecule is placed in a proper supercooled state, the molecule collapses and transforms into a crystalline cluster through chain folding. Figure 10.1 shows a model of polyethylene made of 500 CH_2 (C_{500}) units that is collapsing and transforming into a chain-folded conformation within a few hundred picoseconds, which was reproduced by COGNAC at 300 K.

In this simulation, we adopted the same force field (i.e., the Dreiding force field) as that used in the original paper of Kavassalis and Sundararajan [4]. Selection of the Dreiding force field and relevant parameters in COGNAC are shown in Fig. 10.2.

Care must be taken in the setting of parameters for the nonbonded interactions, where the Dreiding force field demands the inclusion of interactions among one–four atoms separated by three C–C bonds (Fig. 10.3).

The early simulation described above was successful in reproducing the collapse and crystallization of a single polyethylene molecule. Many similar works have been reported on polyethylene [5] and polyethylene derivatives with various chemical modifications. However, it should be noted that the crystallization rate or whether the molecules crystallize at all within a very short time on the order of a nanosecond

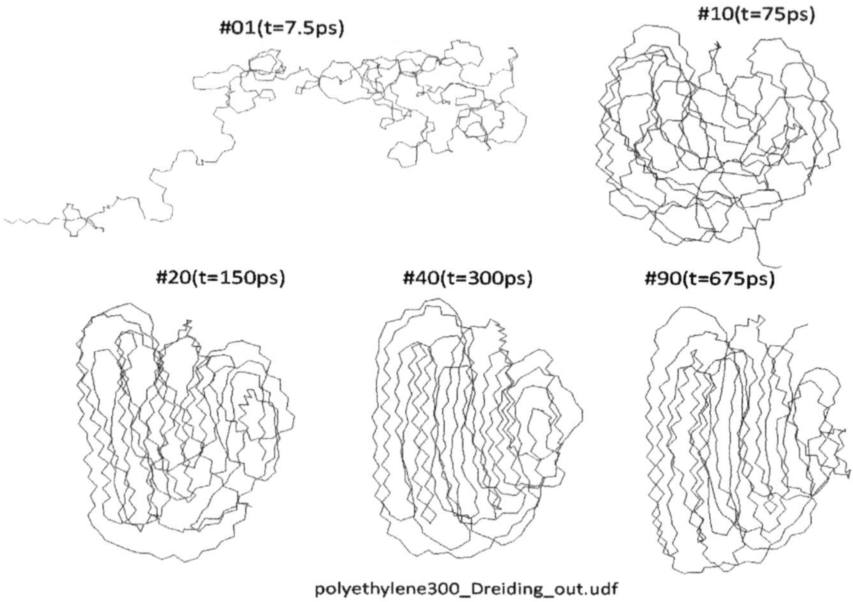

Fig. 10.1 Collapse and crystallization of a single chain of polyethylene at 300 K simulated using the Dreiding force field (polyethylene300_Dreiding_in.udf, polyethylene300_Dreiding_out.udf)

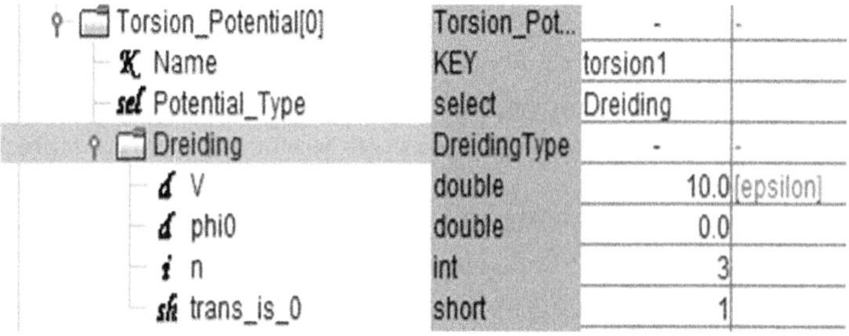

Fig. 10.2 Torsion energy parameters for the Dreiding force field taken from ref. [5]

is sensitive to the force field chosen. With the adoption of a more realistic force field that is known to reproduce polyethylene properties better than the Dreiding force field, the crystallization was shown to be much more sluggish [6]. For example, Fig. 10.4 shows a result of our MD simulation of a single polyethylene of the same length C_{500} conducted with the force field by given by Rigby and Roe.

As clearly seen in the figure, we only observe the collapsing of the molecule without any sign of crystallization. This is evidently owing to the difference in the torsion potentials between the Dreiding and Rigby–Roe force fields; the gauche energy for the Dreiding force field is very high, making the polymer model

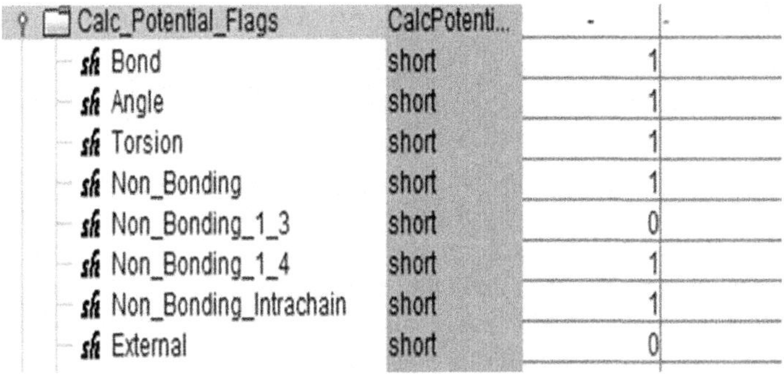

Fig. 10.3 Selection of nonbonding_1_4 for the Dreiding force field

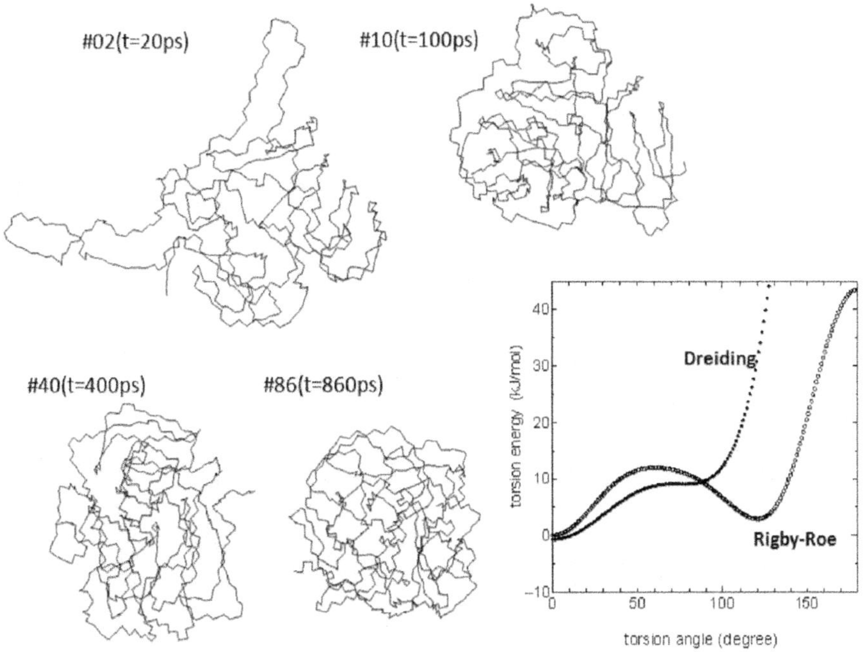

Fig. 10.4 Collapse and crystallization of a single chain of polyethylene at 300 K simulated using the Rigby–Roe force field (polyethylene300_RigbyRoe_in.udf, polyethylene300_RigbyRoe_out.udf) and a comparison of the torsion potentials for the Dreiding and Rigby–Roe force fields

rather stiff. Polymer stiffness generally facilitates crystallization as described in the previous section. It seems that the early success in simulating polyethylene crystallization was enabled by the somewhat unintended choice of the stiff polymer model.

10.3.2 Crystallization from a Highly Stretched Melt

In almost every example of polymer processing, such as the production of fibers or thin films, polymer molecules crystallize under large deformation or flow, where the initial polymer molecules in the melt are highly deformed or stretched. Owing to the competition between the conformational relaxation back to the original random coil and the ordering into the stretched crystalline state, the crystallization becomes much more complicated than that in the quiescent state.

Nevertheless, as described before, the highly accelerated crystallization from the stretched or sheared melt is favorable for molecular simulation. Several MD simulations of polymer crystallization under uniaxial stretching and shear flow in n-alkanes, polyethylene, and isotactic polypropylene have hitherto been reported [6–10].

Figure 10.5 shows the initial elongation of polyethylene melt along the Z-direction.

A system consisting of 21 chains of polyethylene (C_{513}) was initially generated and then elongated along the Z-direction by the rate $\dot{Z}/Z = 0.0012\ [1/tau] = 6.0 \times 10^{-4}\ [1/ps]$ at 360 K, near the melting temperature, under a constant volume condition (Poisson ratio of 0.5) (Fig. 10.6), where the force field adopted was again that of Rigby and Roe.

The highly stretched melt prepared above was then maintained at various temperatures below the melting point under a constant stress condition, where we

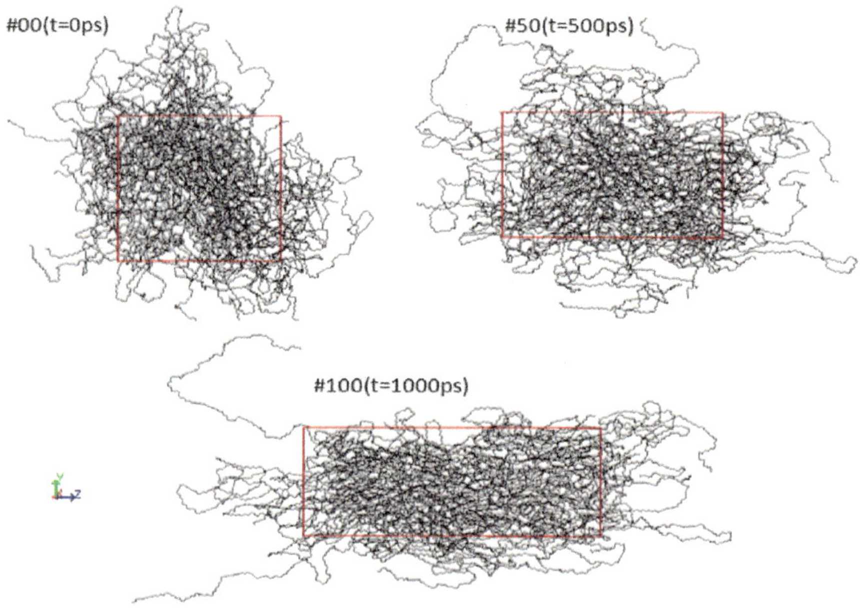

Fig. 10.5 Molecular trajectory during the uniaxial stretching of polyethylene melt at 360 K (alkane_C513_Initial_Draw_in.udf, alkane_C513_Initial_Draw_out.udf)

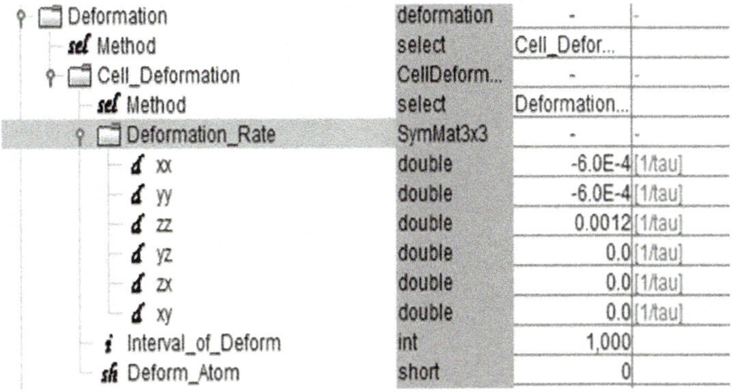

Deformation	deformation	-	-
Method	select	Cell_Defor...	
Cell_Deformation	CellDeform...	-	-
Method	select	Deformation...	
Deformation_Rate	SymMat3x3	-	-
xx	double	-6.0E-4	[1/tau]
yy	double	-6.0E-4	[1/tau]
zz	double	0.0012	[1/tau]
yz	double	0.0	[1/tau]
zx	double	0.0	[1/tau]
xy	double	0.0	[1/tau]
Interval_of_Deform	int	1,000	
Deform_Atom	short	0	

Fig. 10.6 Parameter setting for uniaxial stretching with a constant volume condition; the rate of elongation along the Z-axis direction is twice the compression along the transverse directions of the X- and Y-axes

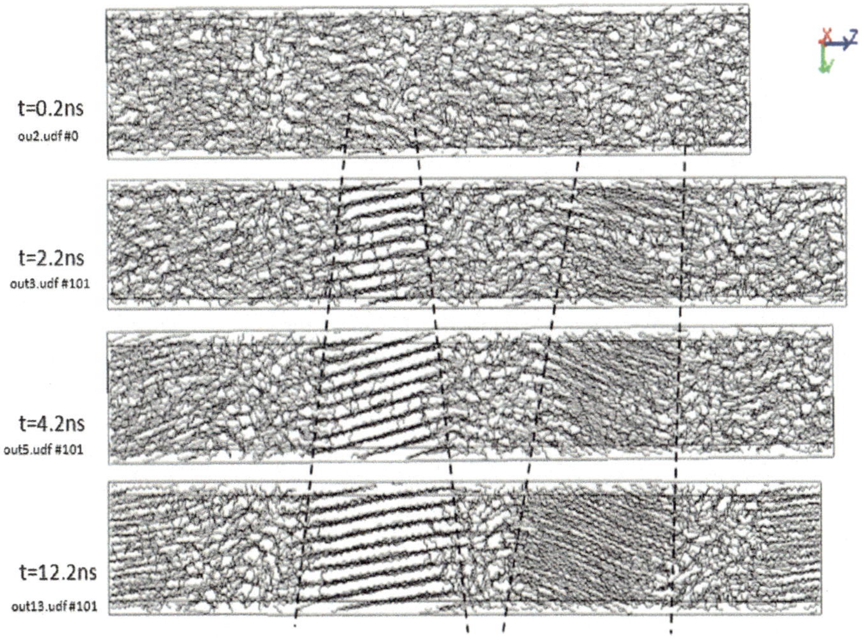

Fig. 10.7 Isothermal crystallization of polyethylene at 360 K from highly stretched melt (alkane_C513_Xtallz_PR-RectCell_T6_in.udf, alkane_C513_Xtallz_PR-RectCell_T6_out2.udf ~ out13.udf)

could observe rapid crystallization into fiber structures. Typical snapshots of the crystallizing polyethylene molecules during the simulation (Fig. 10.7) clearly show that well-ordered stacked lamellae formed within a very short time of about 10 ns.

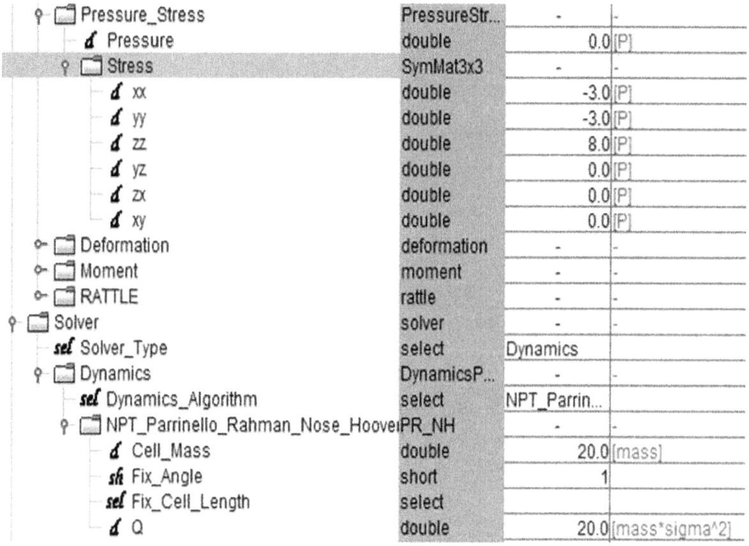

Fig. 10.8 Parameter setting for the crystallization under a constant elongational stress condition using the Parrinello–Rahman method

It was found that a characteristic of fiber formation was the formation of very thin lamellae followed by the gradual thickening of the lamellae. Detailed analyses of the molecular processes involved in the early stages of fiber formation have been reported in several papers [6–9].

At the end of this section, we add a remark about the stress condition adopted in the simulation. In this example, we applied highly anisotropic stress employing the Parrinello–Rahman–Nose–Hoover method (Fig. 10.8) to avoid sample contraction.

Release of the stress to the atmospheric pressure level, which is the usual practice in previously published papers [6–8], results in the highly stretched melt rapidly shrinking into a more or less relaxed state and much slower crystallization.

10.3.3 Crystal Growth of the Chain-Folded Lamellae

Polymer crystallization in quiescent melt is a most fundamental and well-investigated problem. The crystallization starts from primary nucleation in either a homogeneous or heterogeneous manner, but the final structures of the polymer materials are governed by the process of crystal growth.

On their way to crystallization, the random coil molecules first attach partially to the growth front and then gradually pull their tails to transform into the chain-folded conformations. Lauritzen and Hoffman proposed a phenomenological model that has long served as the standard model of polymer crystallization more than

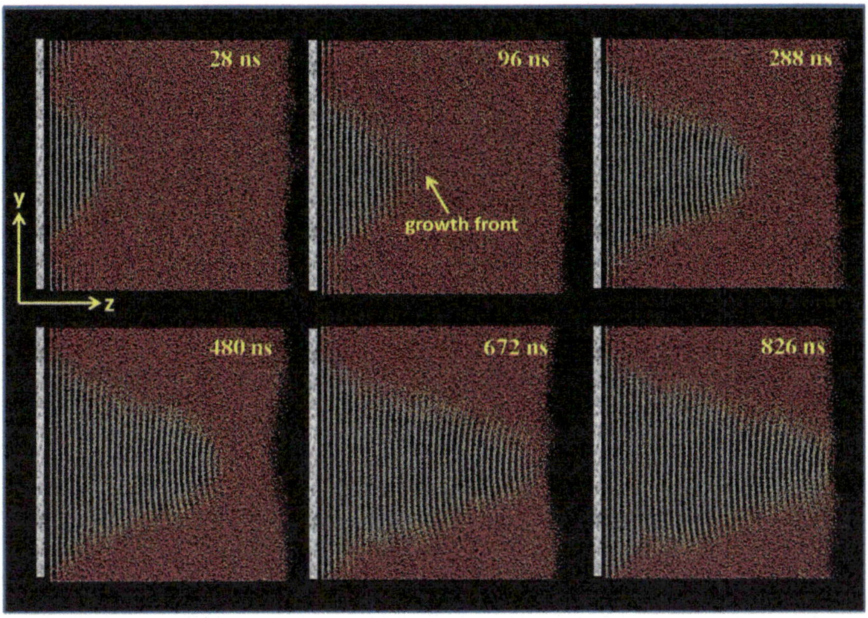

Fig. 10.9 Images of a typical crystallization process for polyethylene at given successive times. The chain-folded lamella grows along the Z-axis with a marked tapered growth front

half a century. However, the model construction was heavily dependent on the molecular processes assumed, and the very bases of the standard model have often been subjects of fierce debate. Because the complicated molecular processes of crystallization are difficult to uncover in experiments, molecular simulations are expected to be the only tools capable of "seeing" the crystallizing molecules directly.

The crystal growth of polymers is a very slow molecular process taking place on a mesoscopic scale. The molecular simulation must therefore deal with large systems for very long simulation times. Figure 10.9 shows typical snapshots of a model system of about 200 chains of C_{500} crystallizing on a substrate obtained from a microsecond of simulation with a parallel code for a minimal model of polyethylene [11], where the crystalline stems run along the Y-axis and the crystal grows along the Z-axis.

Through slow and complex molecular processes, the molecules were found to crystallize into chain-folded lamellae. The lamellae showed a marked tapered growth front, which clearly indicated the presence of pronounced lamella thickening at the growth front. According to recent molecular simulations, the molecular image of polymer crystallization and the structure of chain-folded crystals have greatly advanced from those that were conceived for many years.

10.4 Conclusions and Comments

Molecular simulation studies of polymer crystallization are advancing rapidly. However, there remain innumerable subjects of research for future simulations. Besides understanding the fundamental molecular process of crystallization, crystallization in nanospace and that under a flow field are subjects of great technological importance. Crystallization in molecularly constrained systems such as gels and rubbers and crystallization in realistic and in more and more complex polymers are undoubtedly important and challenging subjects awaiting further investigation [10, 12–14].

10.5 Download Instruction

Sample files are available from the OCTA website. See the instruction on http://octa.jp/download_bookdata.html. Password for the sample (zip file) of this chapter is "FyH4y8ch."

References

1. T. Yamamoto, Polymer **50**, 1975 (2009)
2. J.D. Hoffman, R.L. Miller, Polymer **38**, 3151 (1997)
3. T. Miura, M. Mikami, Phys. Rev. E. **75**, 031804 (2007)
4. T.A. Kavassalis, P.R. Sandararajan, Macromolecules **26**, 4144 (1993)
5. S. Fujiwara, T. Sato, J. Chem. Phys. **107**, 613 (1997)
6. M.S. Lavin, N. Waheed, G.C. Rutledge, Polymer **44**, 1771 (2003)
7. A. Koyama, T. Yamamoto, K. Fukao, Y. Miyamoto, Phys. Rev. E. **65**, 050801 (2002)
8. M.J. Ko, N. Waheed, M.S. Lavine, G.C. Rutledge, J. Chem. Phys. **121**, 2823 (2004)
9. T. Yamamoto, Polymer **54**, 3086 (2013)
10. T. Yamamoto, Macromolecules **47**, 3192 (2014)
11. T. Yamamoto, J. Chem. Phys. **139**, 054903 (2013)
12. T. Yamamoto, K. Sawada, J. Chem. Phys. **123**, 234906 (2005)
13. N.A. Romanos, D.N. Theodorou, Macromolecules **43**, 5455 (2010)
14. T. Yamamoto, in *Understanding Soft Condensed Matter via Modeling and Computation*, ed. by W. Hu, A.-C. Shi (World Scientific, Singapore, 2011)

Chapter 11
Polymer Blends: Bulk Property

Naoki Kobayashi

11.1 Introduction

The blending of a substance having rubbery properties into a polymer material has long been employed to improve the impact strength and other properties of polymer materials. Many models and expressions have been proposed regarding the "blending law," by which total physical properties are predicted from the additional amount of rubber in such material. Examples include a parallel model, series model, and mixed model [1] as the simplest models, and the Takayanagi model [2], which gives greater consideration to morphology. However, the morphology of the rubber-dispersed material of interest should be specified by microscopic observation or the like to determine which of the models or expressions should be applied. Results are thought to vary greatly depending on assumptions regarding the dispersed state, such as whether a dispersed phase is spherical or layered when the rubber-dispersed material has a sea–island structure and whether the rubber-dispersed material can actually have a sea–island structure [3]. In this chapter, the process from determining the morphology to predicting the elastic modulus is modeled using OCTA, thereby predicting the elastic modulus of the rubber-dispersed material without using the blending law. Instead, the prediction is based only on information on the primary structure and physical properties of each component molecule and the volume fractions of components and the elastic modulus of each component.

In the case of a ternary blend, for example, it is difficult to predict the elastic modulus using the aforementioned blending law. Thus, a ternary blend material of isotactic polypropylene (PP), ethylene-propylene rubber (EPR), and high-density

N. Kobayashi (✉)
Mitsui Chemicals Inc., Chiba, Japan
e-mail: Naoki.Kobayashi@mitsuichemicals.com

Japan Association for Chemical Innovation, *Computer Simulation of Polymeric Materials*, DOI 10.1007/978-981-10-0815-3_11

polyethylene (PE), which is a common rubber-dispersed polymer material and for which a large number of measured values have been reported, was selected as a material of interest. Simulation of the material's bulk morphology was then conducted using SUSHI. Using the obtained morphology, the elastic modulus of the bulk was calculated with MUFFIN5E_ELASTICA.

11.2 Method of Generating the Bulk Structure

A PP/EPR/PE system [4] that has been studied by Gahleitner et al. was simulated. SUSHI was used to create the system's morphology. The following three assumptions were made to compare experimental results of the injection-molded sample in a nonequilibrium state with the calculation results of the sample in an equilibrium state obtained with OCTA:

1. The elastic modulus is unaffected by the size of dispersed particles.
2. In a phase-separated process, there are changes in size but not shape in the morphology of the rubber-dispersed structure.
3. The elastic modulus of each component is not anisotropic, but rather exhibits only a linear response.

11.3 Calculation of the Morphology with SUSHI

The equilibrium structure of an incompatible multiphase polymer blend, which differs from the equilibrium structure of a block copolymer, is determined by the volume fractions, χ parameter difference, and system size; therefore, it can be said that, when the equilibrium structure is obtained through SUSHI's STATIC calculation, the molecular weight has little effect. The mesh size here is set to $32 \times 32 \times 2$ to conduct the MUFFIN5E_ELASTICA simulation in a pseudo two-dimensional manner. To reproduce the morphology of a core–shell structure obtained in experiments, three kinds of virtual polymers having $N = 20$ are used. The volume fractions are set as the literature values, A(PP):B(EPR):C(PE) = 0.85:0.10:0.05, and calculation is performed with a combination of $\chi_{AB} = \chi_{BC} = 0.2$ and $\chi_{AC} = 1.0$ to give a core–shell structure in the order of PP, EPR, and PE. Figure 11.1 shows the morphology (namely, the concentration of the EPR component) obtained by SUSHI under the aforementioned conditions. In the figure, the periphery is PP, the core in the center is PE, and the shell surrounding the core is EPR ("abc2d_out.udf").

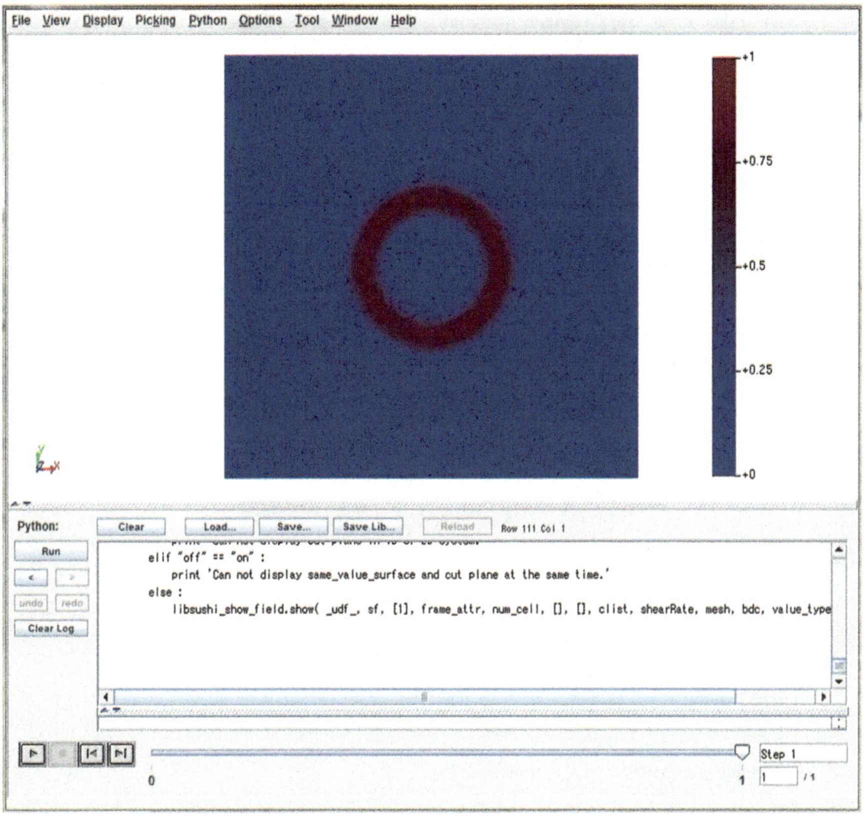

Fig. 11.1 Core–shell morphology (where the EPR concentration is illustrated) obtained by SUSHI

11.4 Procedure for Calculating Bulk Physical Properties

The volume fraction distribution data obtained by SUSHI are converted to the volume fraction distribution for the MUFFIN5E_ELASTICA module using OCTA's import function. First, the input user definable format (UDF) file ("EX05_in.udf") of application example 05 is read. The input UDF file reads a three-dimensional input of SUSHI from the samples of MUFFIN5E. From the action command displayed by right-clicking the file name, *import_field* is selected and the output UDF file of SUSHI described in the preceding section is assigned. Note that the number of grids designated by SUSHI is decreased by one in MUFFIN (Fig. 11.2). The data of the three components are input to *physical_data* as shown in Fig. 11.3. The elastic modulus of each component is input according to the measured values as shown in Table 11.1. The bulk modulus (K) is input to *BULK_MODULUS* and the shear modulus (G) to *SHEAR_MODULUS*. The relation between Young's modulus (E) and K and G is expressed by Eq. 11.1:

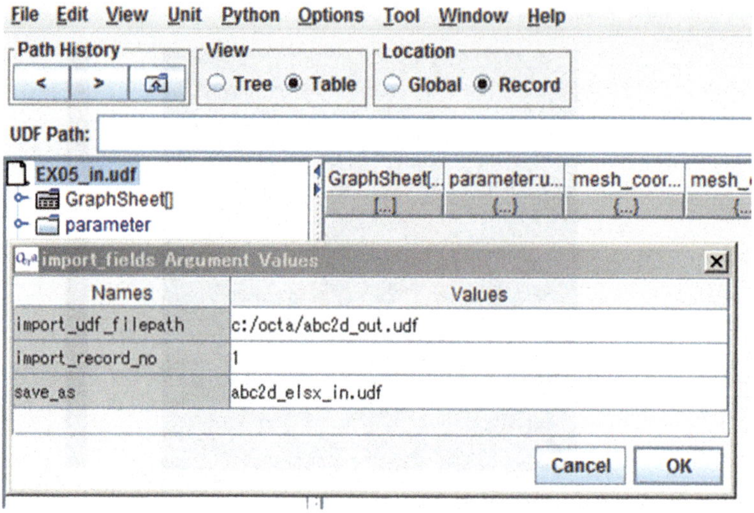

Fig. 11.2 Import of SUSHI data

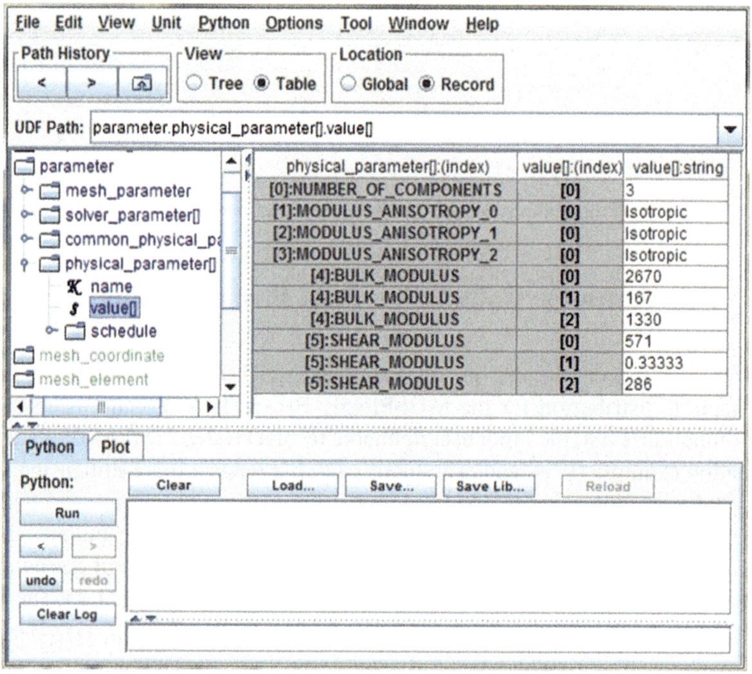

Fig. 11.3 Input of physical data

Table 11.1 Elastic modulus parameters of each component input to MUFFIN5E_ELASTICA

	PP	EPR	PE
Bulk modulus (MPa)	2670	167	1330
Shear modulus (MPa)	571	0.333	286

Table 11.2 Input of deformation conditions

		Partial region	Condition name	Value
Condition (1): shear (*x* axis)	1	BOUNDARY_VERTEX_YMIN	D_VEC	(−0.32, 0.0, 0.0)
	2	BOUNDARY_VERTEX_YMAX	D_VEC	(0.32, 0.0, 0.0)
Condition (2): shear (*y* axis)	1	BOUNDARY_VERTEX_XMIN	D_VEC	(0.0, −0.32, 0.0)
	2	BOUNDARY_VERTEX_XMAX	D_VEC	(0.0, 0.32, 0.0)
Condition (3): compression (*x* axis)	1	BOUNDARY_VERTEX_XMIN	D_VEC	(0.0, 0.0, 0.0)
	2	BOUNDARY_FACE_XMAX	N_LOAD	(−0.32, 0.0, 0.0)
Condition (4): compression (*y* axis)	1	BOUNDARY_VERTEX_YMIN	D_VEC	(0.0, 0.0, 0.0)
	2	BOUNDARY_FACE_YMAX	N_LOAD	(0.0, −0.32, 0.0)

$$E_{ave} = \frac{3G_{ave}}{1 + \frac{1}{3}\frac{G_{ave}}{K_{ave}}} \tag{11.1}$$

Next, the elastic modulus attributable to the morphology is determined by infinitesimally deforming the whole structure. In the case of an isotropic elastic material, the balance between the total strain free energy F and strain e can be expressed by Eq. 11.2:

$$F = G_{ave}\sum\left\{\left(e_{ij} - \frac{1}{d}\delta_{ij}e_{ll}\right)^2\right\} + K_{ave}\sum\left(\frac{1}{2}e_{kk}^2\right) \tag{11.2}$$

In this formula, G_{ave} is the whole shear modulus, K_{ave} is the whole bulk modulus, d is the number of dimensions, and δ is Kronecker's delta. Σ indicates the summation of coefficients at the nodal points. F in the formula, the coefficient of G_{ave}, and the coefficient of K_{ave} are calculated by MUFFIN5E_ELASTICA. Because there are two variables, K_{ave} and G_{ave} should be determined by two or more kinds of deformation. Here, infinitesimal shear deformation and compressive deformation are applied. Because an ideal linear elastic material is assumed, the tensile modulus and compressive modulus are equivalent.

Calculation conditions (1–4), comprising two kinds of shear deformation and two kinds of compressive deformation, are input as shown in Table 11.2, Figs. 11.4 and 11.5. The values given in Table 11.2 represent the force strength to be applied.

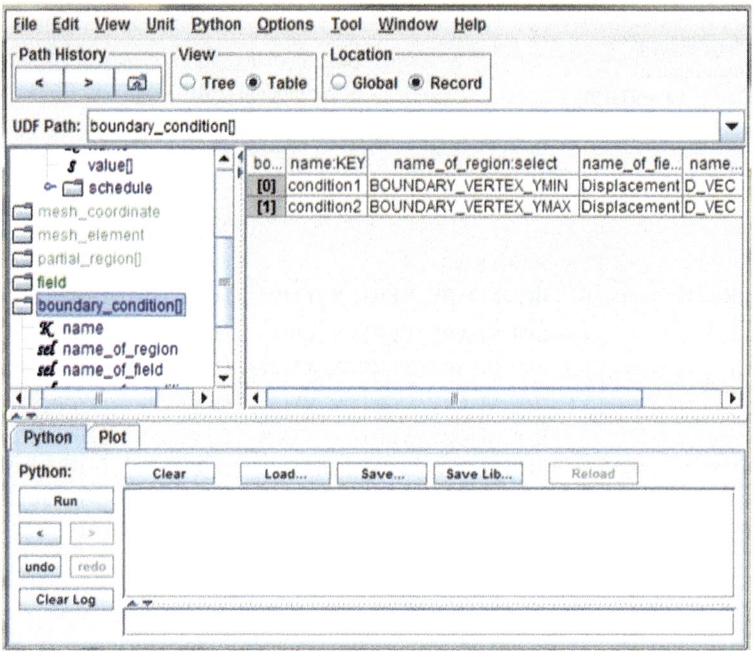

Fig. 11.4 Input of boundary conditions

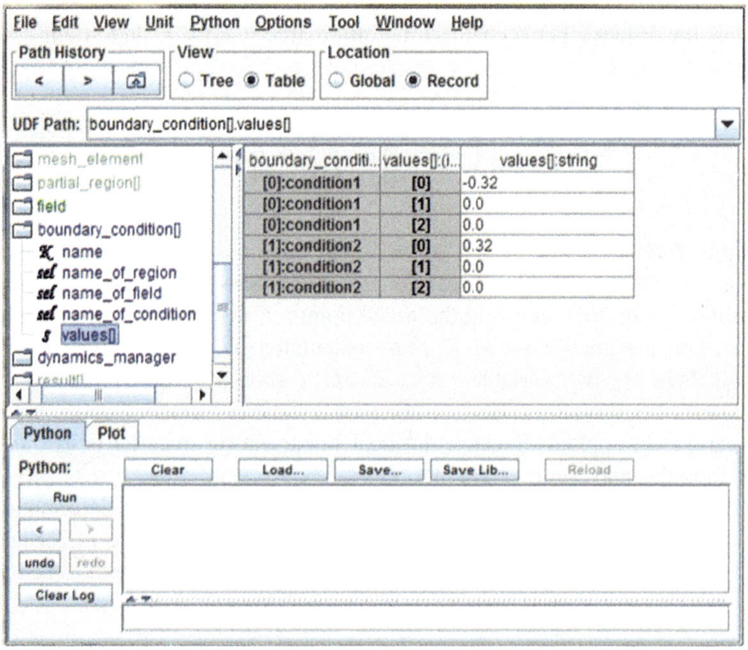

Fig. 11.5 Input of deformation amounts

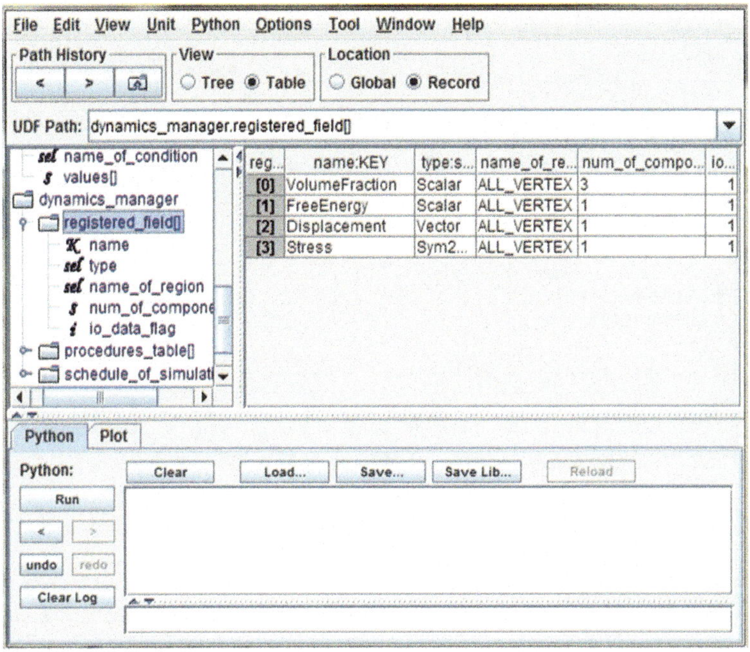

Fig. 11.6 Input of command conditions

Given that pseudo two dimensionality is assumed here, deformation is applied in the x and y directions only, and the anisotropy of morphology is canceled by calculating the average in the x and y directions. Input of the boundary condition can be made directly from a keyboard if the necessary condition does not appear upon pushing the select button.

Figure 11.6 shows the input of calculation command conditions in MUFFIN5E_ELASTICA. The number of components for *Volume_Fraction* is changed to three. For the remaining conditions, the default values may be used. Figure 11.7 shows the order of evolution in the calculation. The initializing command is changed to *MULTI_COMPONENT* because here the three-component system is handled. The input files are named "abc2d_elsx_in.udf", "abc2d_elsy_in.udf", "abc2d_elpx_in.udf", and "abc2d_elpy_in.udf".

11.5 Display of Calculation Results

Executing the aforementioned calculation using the MUFFIN5E_ELASTICA module generates output UDF files within a few seconds. Figure 11.8 shows an image of micro-shear deformation obtained by this calculation. Reading the generated UDF files with GOURMET reveals that a tag of *result*[] has been created, and

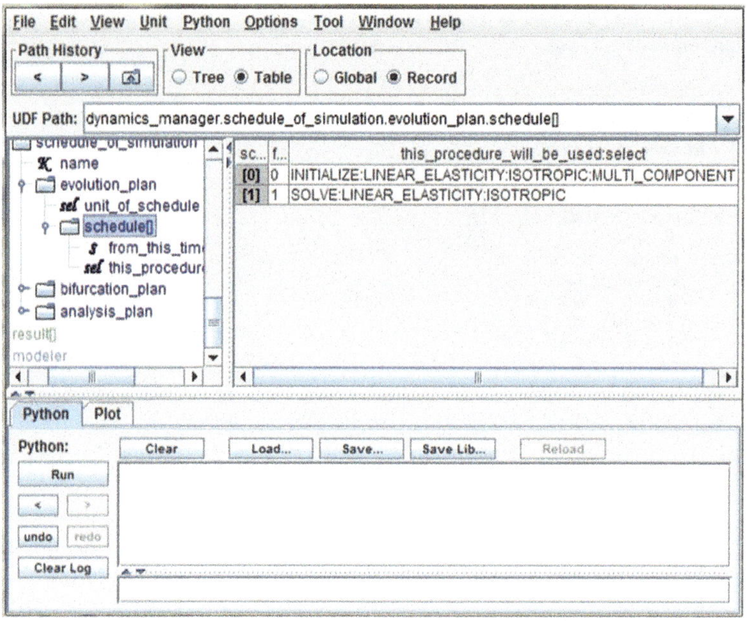

Fig. 11.7 Input of the evolution schedule

when the tag is opened, *F* of Eq. 11.2, the coefficient of *G* and the coefficient of *K* are displayed (Fig. 11.9). Figure 11.10 is a graph plotting the relation between *G* and *K* thus obtained. The intersections of the shear deformation and compressive deformation represent the bulk modulus and shear modulus of the bulk. Obtained results are given in Table 11.3. Nearly all of these values agree well with the literature values [4]. It is thought that, although the actual material has a three-dimensional structure, the effect of the core–shell structure obtained by encapsulating a material with another material having a different elastic modulus can be reproduced using a two-dimensional model. The output files are "abc2d_elsx_out.udf", "abc2d_elsy_out.udf", "abc2d_elpx_out.udf", and "abc2d_elpy_out.udf".

11.6 Concluding Remarks

The morphology at each volume fraction of a PP/EPR/PE polymer blend was calculated by a simulator that is based on the mean field method, SUSHI, using a self-consistent field method of the OCTA system. The elastic modulus at each volume fraction was calculated according to the results of morphology using a finite

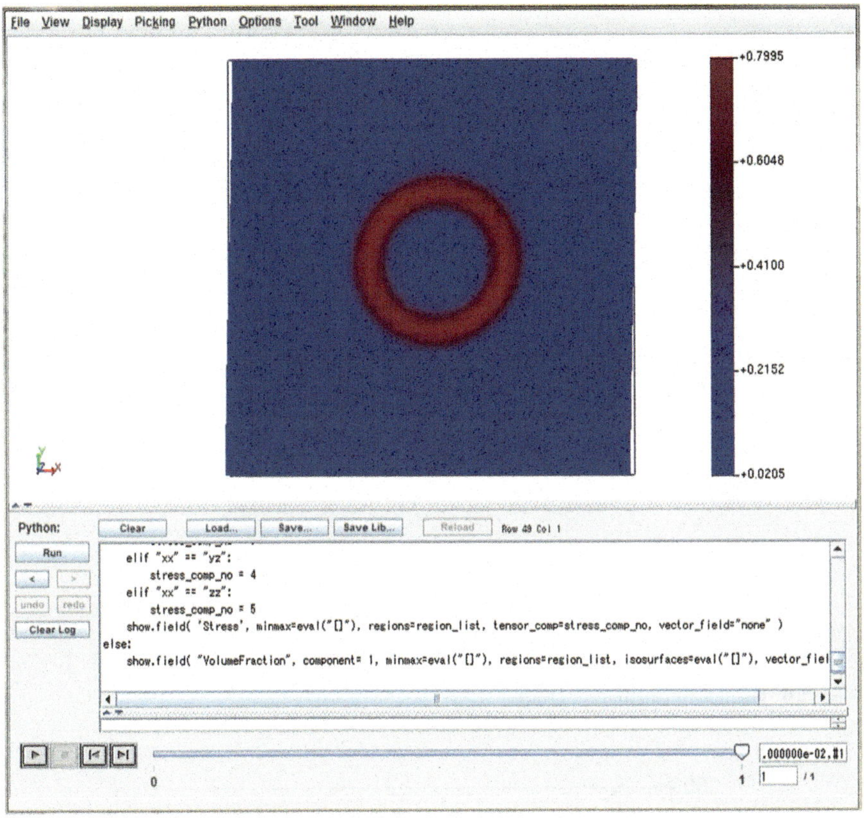

Fig. 11.8 Morphology after shear deformation

element method simulator for linear elastic materials, MUFFIN5E_ELASTICA, which is included in OCTA. The calculated values of the elastic modulus corresponding to the morphology at each volume fraction agreed well with the measured values, indicating that OCTA is sufficiently applicable to the design of real rubber-dispersed polymer materials.

11.7 Download Instruction

Sample files are available from the OCTA website. See the instruction on http://octa.jp/download_bookdata.html. Password for the sample (zip file) of this chapter is “Vt6238GZ”.

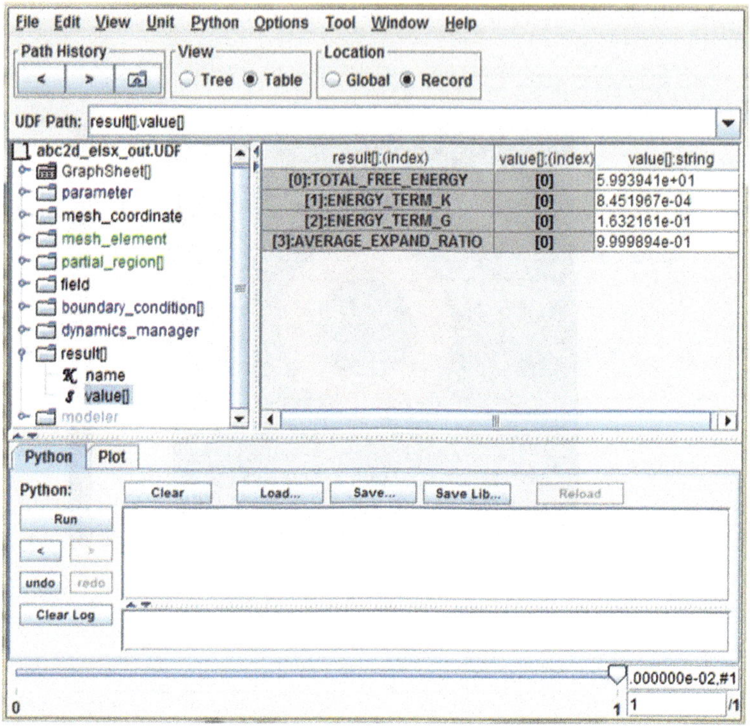

Fig. 11.9 Display of calculation results

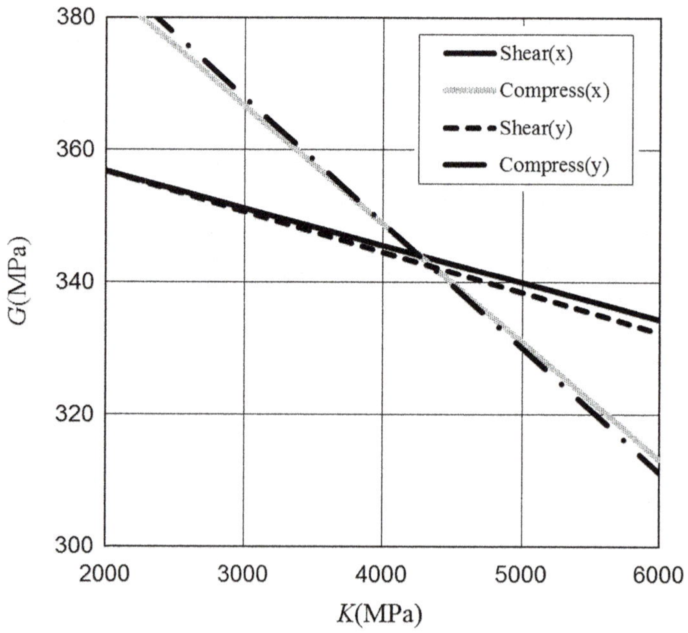

Fig. 11.10 *K* versus *G* for each deformation mode

Table 11.3 Elastic modulus of the ternary blend system obtained by MUFFIN5E_ELASTICA

K_{ave} (MPa)	G_{ave} (MPa)	E_{ave} (MPa)	$E_{literature\ value}$ (MPa)
4270	344	1000	1142

References

1. L.E. Nielsen, *Mechanical Properties of Polymers and Composites* (Marcel Dekker, New York, 1975). chapter 7
2. M. Takayanagi, H. Harima, Y. Iwata, J. Soc. Mater. Sci. Jpn. **12**, 389 (1963)
3. A. Veenstra, P.C.J. Verkooijen, B.J.J. Van Lent, J. Van Dam, A.P. De Boer, A.P.H.J. Nijhof, Polymer **41**, 1817 (2000)
4. M. Gahleitner, A. Hauer, K. Bernreitner, E. Ingolic, Int Polym Process **XVII**, 318 (2002)

Chapter 12
Polymer Blends: Interfacial Strength

Taku Ozawa

12.1 Introduction

An interface formed by different polymers strongly affects a material's mechanical properties, such as the elastic modulus and impact resistance. Generally speaking, the thickness of the interface varies according to the solubility of the polymers, greatly changing the interface properties. Moreover, interface properties have been improved by adding a compatibilizer. In this chapter, the effect of the interfacial structure on the interfacial fracture behavior is evaluated using self-consistent field theory (SCFT) and a coarse-grained molecular dynamics (CGMD) method. The specific procedures are described using SUSHI and COGNAC as engines of OCTA.

The thickness of the interface between polymers is generally known to be in the range of several tens of nanometers. It is thus unrealistic to use a full atomistic molecular dynamics method in which all motions of atoms are calculated, because the number of particles becomes enormous. Therefore, the CGMD method, which is capable of handling a larger system, is applied. However, it remains a challenge to create a fully relaxed initial structure. To solve this problem, the initial structure for CGMD simulation is created using a result of SCFT. This method is referred to as the "zooming function" in the OCTA system. Specifically, the volume fraction distribution of each segment constituting the polymers is used. Furthermore, to grasp universal characteristics, parameters characterizing a specific substance are not set for each polymer.

All input/output user-definable format (UDF) files are listed at the end of this chapter and can be referred to while reading this chapter.

T. Ozawa (✉)
JSOL Corporation, Tokyo, Japan
e-mail: ozawa.taku@jsol.co.jp

Japan Association for Chemical Innovation, *Computer Simulation of Polymeric Materials*, DOI 10.1007/978-981-10-0815-3_12

12.2 Calculation Model

Three different interfacial structures are evaluated. In each system, there is only one planar interface formed by two homopolymer components with $N = 100$ (A_{100} and B_{100}). Specifically, systems are constructed having a thick interface with $\chi N = 4.1$, a thin interface with $\chi N = 20$, and an interface with $\chi N = 20$, where compatibilizer molecules ($A_{50}B_{50}$) are localized. Here, χ represents the interaction between segments and N represents the number of particles that constitute the polymer. The strength of phase separation is estimated from χN, which is a multiplication of the interaction and number of particles; the interface becomes thinner as the value of χN increases.

First, an SCFT calculation in one dimension (where the system size is 48.0) is performed using SUSHI. A reflective boundary condition is set at each of the two boundaries, and the system is constructed so that only one interface is generated in the center of the system. The canonical ensemble is applied, and 50 vol.% each of A_{100} and B_{100} are involved in a two-component system. The conditions $A_{100} = B_{100} = 47.6$ vol.% and $A_{50}B_{50} = 4.8$ vol.% are set for a system containing a compatibilizer; these conditions match the number of polymer chains in COGNAC described later. Moreover, settings are made to output the segment volume fraction distribution (Fig. 12.1). In COGNAC, the initial structure in which the result of SCFT calculation is reflected can be created using the segment volume fraction distribution obtained by SCFT calculation employing the node density-biased Monte Carlo method. In addition, because it is difficult to manually set the parameters shown in Fig. 12.1, the simple Python script shown in Fig. 12.1 is created to set them automatically. Refer to the SUSHI manual for the meaning of each parameter.

Next, a CGMD calculation based on Langevin dynamics (i.e., the Kremer–Grest model) is carried out using COGNAC. FENE_LJ and Lennard–Jones potentials are set for bonding and nonbonding potentials, respectively. All distances at which the Lennard–Jones potential is zero are set to 1.0σ, and all cutoff values are set as $r_{\mathrm{cut\_off}} = 2.5\sigma$. With respect to ε and in accordance with [1], a potential well of 1.0ε is applied between particles of the same kind ($\varepsilon_{\mathrm{AA}}$ and $\varepsilon_{\mathrm{BB}}$), and a value consistent with the χ parameter in the SCFT calculation is applied between particles of different kinds. Specifically, ε is set as $\varepsilon_{\mathrm{AB}} = 0.986$ in the system with $\chi N = 4.1$ and $\varepsilon_{\mathrm{AB}} = 0.9$ in the system with $\chi N = 20$. In addition, regarding the relation between $\chi N = 20$ and $\varepsilon_{\mathrm{AB}} = 0.9$, [1] does not show perfect correspondence, and both are used as a parameter by which a sufficiently thin interface is formed.

For each homopolymer component, the chain numbers of A_{100} and B_{100} are both set to 40, and the chain number of the compatibilizer $A_{50}B_{50}$ is set to 4. The initial structure is created by applying the node density-biased Monte Carlo method using the calculation results from SUSHI. The parameters set when making the initial structure are shown in Fig. 12.2. Although the size of segments in SCFT and particles in the Kremer–Grest model does not coincide with each other, the initial structure is created here by setting parameters under the assumption of one-to-one correspondence for the sake of simplification.

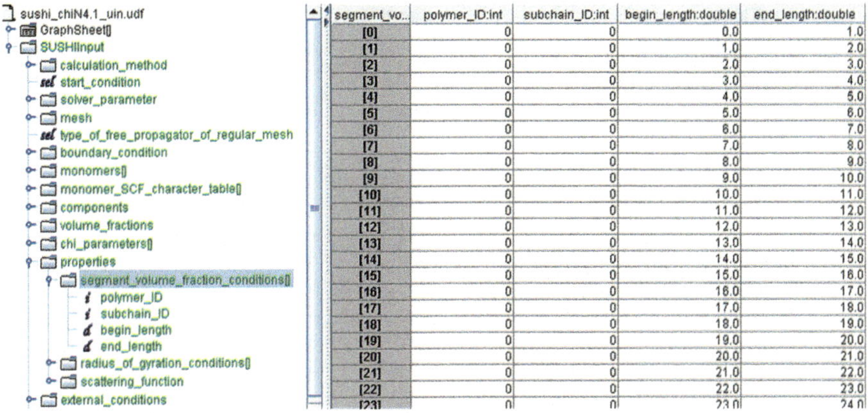


```
k = 0
for i in range(2):
        for j in range(100):
                $SUSHIInput.properties.segment_volume_fraction_conditions[k].polymer_ID   = i
                $SUSHIInput.properties.segment_volume_fraction_conditions[k].subchain_ID  = 0
                $SUSHIInput.properties.segment_volume_fraction_conditions[k].begin_length = float(j)
                $SUSHIInput.properties.segment_volume_fraction_conditions[k].end_length   = float(j+1)
                k = k + 1
```

Fig. 12.1 Settings for outputting the segment volume fraction distribution in SUSHI and a Python script for setting parameters (for a two-component system)

"REFLECTIVE2" is set for the boundary condition. By doing so, the reflective boundary condition is set at the two boundaries in the *Z* direction in the same manner as in the calculation in SUSHI. Note that the volume fraction distribution is reflected in the "Z direction" in COGNAC, when zooming using the one-dimensional calculation results of SUSHI.

The number density of the particles in the initial structure is set to 0.85, which is the default value in the Kremer–Grest model; the system size is $14.0\sigma \times 14.0\sigma \times 48.0\sigma$ in the two-component system and $14.35\sigma \times 14.35\sigma \times 48.0\sigma$ in the system containing a compatibilizer; here the length in the *Z* direction is the same in both systems, because the system size of SUSHI is set to 48.0 for each system. Moreover, considering that the mean square of the radius of gyration $<Rg^2>$ of the polymer chain with $N = 100$ in the bulk is about $25\sigma^2$, the system size is sufficiently large. When a number density of 0.85 and a length in the *Z* direction of 48.0σ are designated as parameters for creating the initial structure of COGNAC, each cell length in the *X* and *Y* directions is automatically set at the start of the calculation according to these values and the total number of particles.

More precisely, the value of stress in the *XY* direction (a direction parallel to the interface) is affected by placing the compatibilizer at the interface; therefore, it

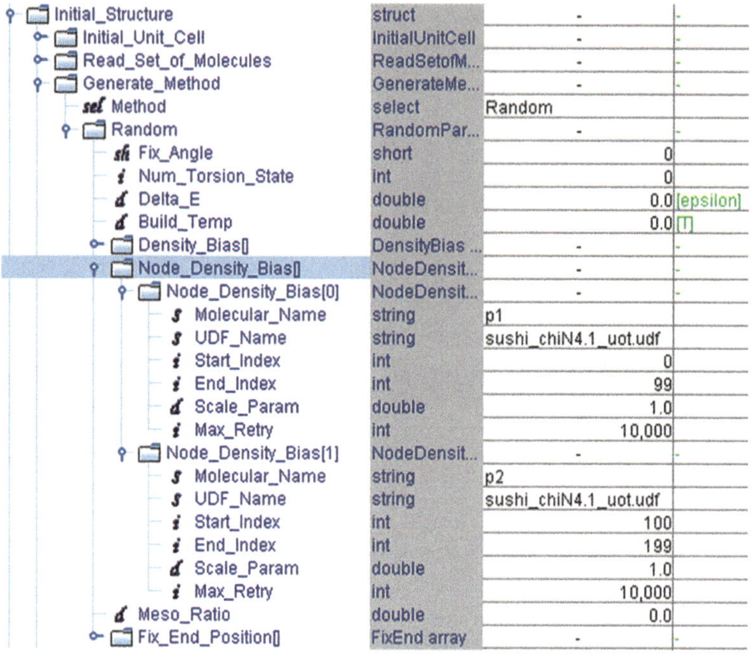

Fig. 12.2 Setting of zooming parameters

is necessary to study the adjustment of the system size depending on the density of the compatibilizer, though no special operation has been conducted in this case because the purpose of this chapter is simply to confirm the effect of introducing the compatibilizer.

After creating the initial structure using the zooming function, a calculation of relaxation of 600τ is performed. In COGNAC, it is possible to set an external field using the volume fraction distribution obtained by SUSHI, thereby allowing for adjustment of the interfacial structure. However, for simplicity, no special external field is set in this calculation, and the forces applied to each particle are only those of the Lennard–Jones potential and the bonding potential. Hereafter, the time step of the calculation is set to $dt = 0.006$ and the temperature to $T = 1.0$ in all calculations (where the unit of temperature is ε/k_B).

After calculating the relaxation, uniaxial elongation in the Z direction is calculated. In COGNAC, the deformation rate of cells is set; in this case, a constant value of $0.01\sigma/\tau$ is set (Fig. 12.3). The Poisson's ratio is set to 0.0, in which case the cell length in the XY direction does not change; in this way, a setting suitable for evaluating interfacial fracture behaviors is made.

The strain rate changes with time; the initial strain rate becomes approximately $2.08 \times 10^{-4}\tau^{-1}$. According to [1], when the potential above is used at $T = 1.0$, the system consisting of a polymer with $N = 20$ (without entanglement) is deformed under a strain rate of about $2.08 \times 10^{-4}\tau^{-1}$, and the whole polymer chain relaxes

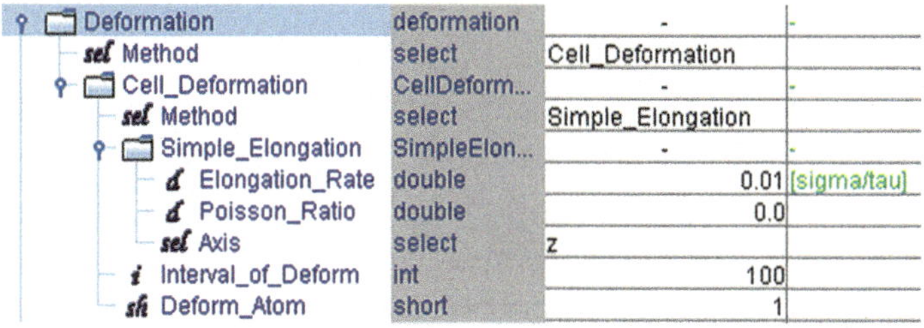

Fig. 12.3 Setting of uniaxial elongation conditions in COGNAC

against deformation (i.e., a melt state is achieved). However, when the system consisting of a polymer with $N = 100$ is deformed under the same strain rate, the deformation is faster than the relaxation of the whole polymer chain. In this way, it is important to confirm the relation between the relaxation time of a polymer chain and strain rate when calculating deformation. For example, even though a sufficiently high temperature is set as a condition of the molecular dynamics calculation for the purpose of evaluating the behaviors of polymer chains in a rubbery state, if high-speed deformation is applied, the relaxation cannot follow the deformation, meaning that the polymer is handled in a glass state. Meanwhile, the same strain rate is used in [2], and the interfacial fracture in a glass state is evaluated by setting the system temperature as $T = 0.3$.

12.3 Calculation Results

In both COGNAC and SUSHI, drawing and other various analysis tools are contained in actions performed on the results files. Calculation results are analyzed using these functions.

Figure 12.4 shows the calculation results of SUSHI. In (a) and (b), interfaces each having a different thickness due to the difference in χ parameters are obtained. In (c), the compatibilizer is concentrated at the interface, and a structure in which the compatibilizer molecule extends into each phase can be confirmed. Figure 12.5 captures part of the result files of SUSHI; it is seen that the volume fraction distribution of each segment is output according to the parameters set in Fig. 12.1. The node density-biased Monte Carlo method in COGNAC references these values. Figure 12.4 shows the volume fraction distribution of each component, not the volume fraction distribution of each segment.

Figure 12.6 shows the volume fraction distribution of each component in the Z direction after calculating relaxation using COGNAC for the system containing a compatibilizer (where the average is taken in the XY direction). A smooth distribution is not obtained because the evaluation is conducted on the basis of the snapshot in the last step, not the time-averaged snapshot; however, it is shown that the distribution has good agreement with the distribution of Fig. 12.4c.

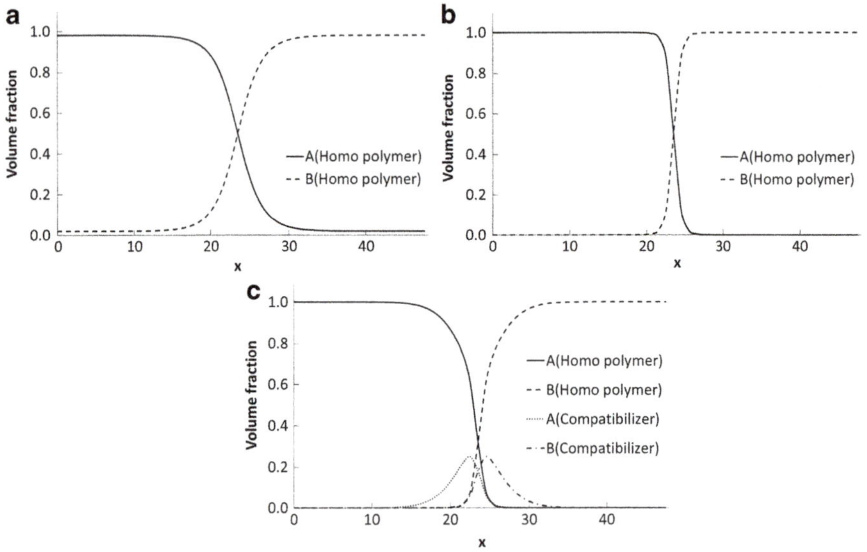

Fig. 12.4 Interfacial structures obtained by SUSHI (SCFT calculation). **a** $\chi N = 4.1$, **b** $\chi N = 20$, **c** $\chi N = 20$, with a compatibilizer

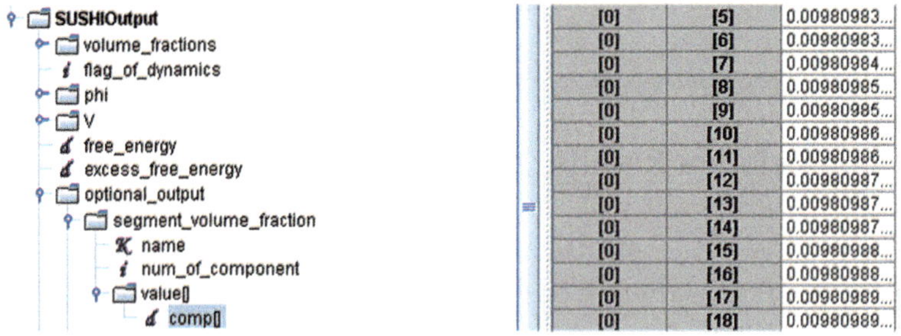

Fig. 12.5 Segment volume fraction distribution obtained by SUSHI (SCFT calculation)

Figures 12.7, 12.8, and 12.9 show snapshots at the initial state and at a strain of 0.5 obtained by uniaxial elongation using COGNAC. Given that the system shown in Fig. 12.7 has a thick interface with $\chi N = 4.1$, a fracture occurs in the bulk region. The result of SCFT calculation in Fig. 12.4a shows that the interface is thicker than the radius of gyration of the polymer chain, and because $\varepsilon_{AB} = 0.986$, the difference between the two components seems small. For these reasons, the effect of the interface is weak and the behavior of the polymer chain at the interface becomes similar to that in the bulk. Figure 12.8 shows the result for the system with $\chi N = 20$, where fracture behavior is observed at the interface because the interface is thin and molecular chains are not sufficiently entangled. Figure 12.9 shows the system containing a compatibilizer; the compatibilizer suppresses complete fracture of the interface.

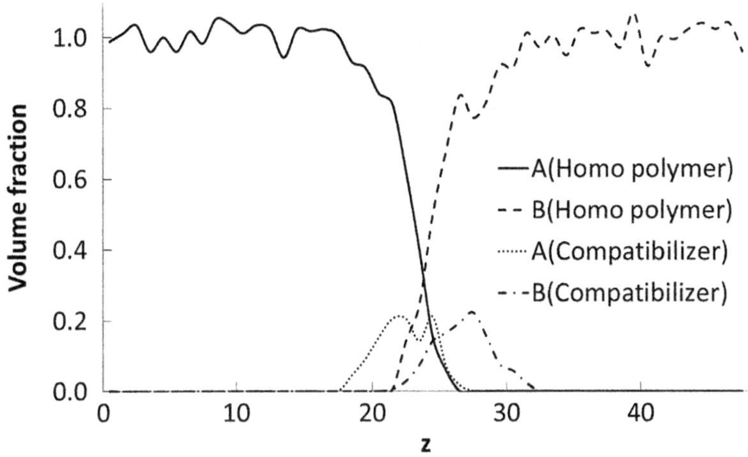

Fig. 12.6 Volume fraction distribution (system containing compatibilizer) in the *Z* direction after calculating relaxation with COGNAC

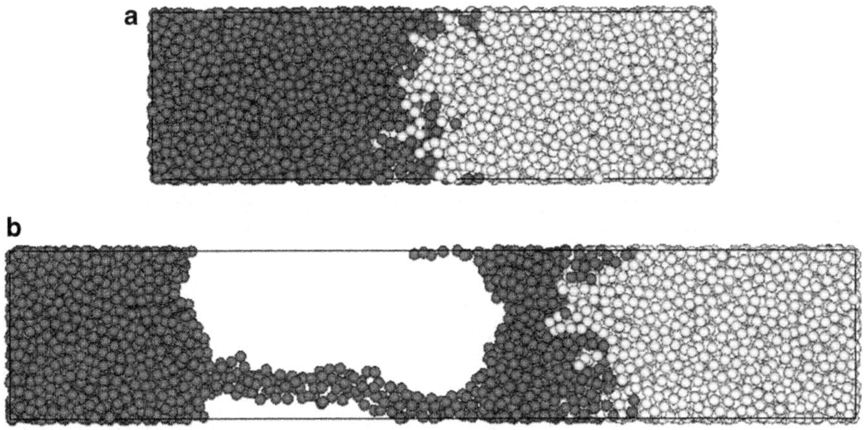

Fig. 12.7 Calculation of uniaxial elongation by COGNAC ($\chi N = 4.1$). **a** Initial state, **b** strain of 0.5

Figure 12.10 shows stress–strain curves of the respective systems. In all results, yield behavior appears at an early stage, and the stress falls after the yield. It can be thought that particularly the behaviors before and after yield depend on the initial structure; however, the calculation was only performed for one initial structure for each system. Nevertheless, with respect to behaviors after yield, characteristic results are grasped for the respective systems. In the system with $\chi N = 4.1$ and the system containing a compatibilizer, bundles of polymers form a fibril, and the stress is more maintained compared with the system with $\chi N = 20$. In [2], the fracture

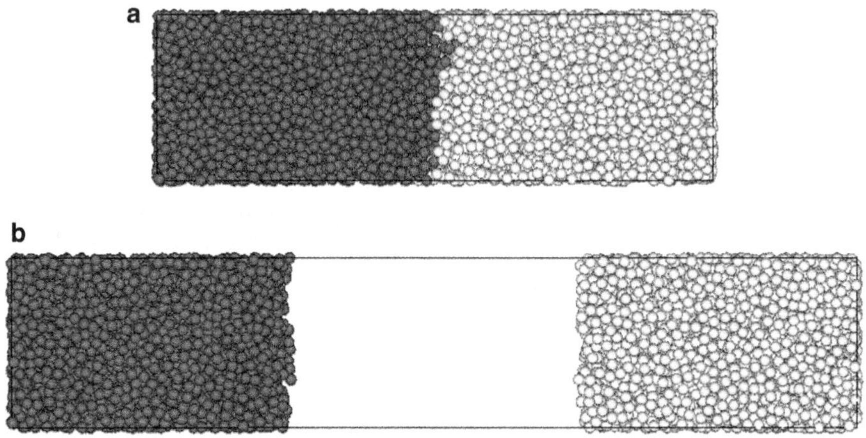

Fig. 12.8 Calculation of uniaxial elongation by COGNAC ($\chi N = 20$). **a** Initial state, **b** strain of 0.5

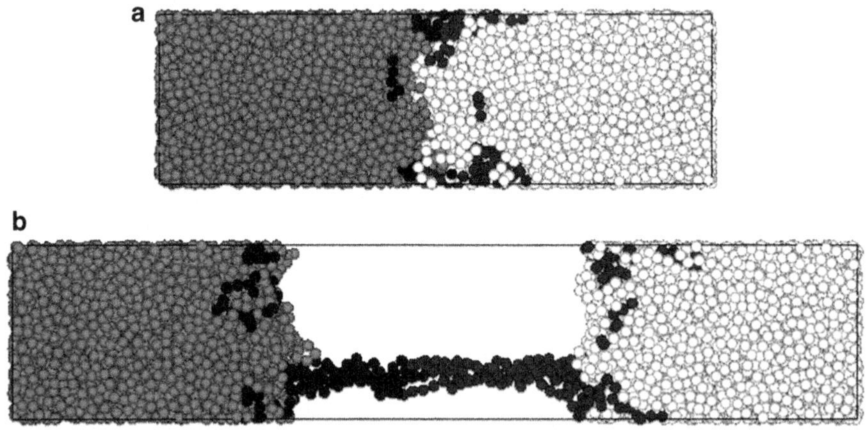

Fig. 12.9 Calculation of uniaxial elongation by COGNAC ($\chi N = 20$ with compatibilizer). **a** Initial state, **b** strain of 0.5

toughness (energy) is evaluated by integrating the stress value during the elongation. It can also be predicted from Fig. 12.10 that the fracture toughness becomes large when introducing the compatibilizer.

12.4 Conclusion

An SCFT calculation was performed using SUSHI to evaluate the interfacial structures of polymers in one dimension. The initial structures of CGMD simulation, in which the results of SCFT calculation were reflected using a zooming function,

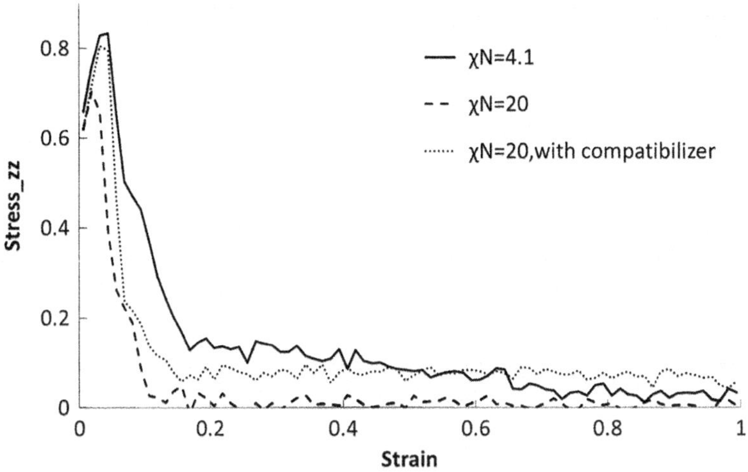

Fig. 12.10 Stress–strain curves obtained by COGNAC

were then created. From the calculation of uniaxial elongation, the effects of the thickness of an interface and the existence of the compatibilizer on interfacial fracture behaviors were evaluated.

12.5 Download Instruction

Sample files are available from the OCTA website. See the instruction on http://octa.jp/download_bookdata.html. Password for the sample (zip file) of this chapter is "w3VftXwd".

1. Input/output files of SUSHI
   - $\chi N = 4.1$
     Input: sushi_chiN4.1_uin.udf
     Output: sushi_chiN4.1_uot.udf
   - $\chi N = 20$
     Input: sushi_chiN20_uin.udf
     Output: sushi_chiN20_uot.udf
   - $\chi N = 20$ with compatibilizer
     Input: sushi_chiN20_block_uin.udf
     Output: sushi_chiN20_block_uot.udf
2. Input/output files of COGNAC
   - $\chi N = 4.1$
     Zooming input: cognac_chiN4.1_relax_in.udf

Zooming output: cognac_chiN4.1_relax_out.bdf
Uniaxial elongation input: cognac_chiN4.1_elongation_in.bdf
Uniaxial elongation output: cognac_chiN4.1_elongation_out.bdf

- $\chi N = 20$
  Zooming input: cognac_chiN20_relax_in.udf
  Zooming output: cognac_chiN20_relax_out.bdf
  Uniaxial elongation input: cognac_chiN20_elongation_in.bdf
  Uniaxial elongation output: cognac_chiN20_elongation_out.bdf
- $\chi N = 20$ with compatibilizer
  Zooming input: cognac_chiN20_block_relax_in.udf
  Zooming output: cognac_chiN20_block_relax_out.bdf
  Uniaxial elongation input: cognac_chiN20_block_elongation_in.bdf
  Uniaxial elongation output: cognac_chiN20_block_elongation_out.bdf

Among the above files, uniaxial elongation input files were created by saving zooming output files under different names to set calculation conditions.

References

1. T. Aoyagi, J. Takimoto, M. Doi, in *Proceedings of the International Conference on Advanced Polymers and Processing* (ICAPP, Yonezawa, 2001)
2. T. Aoyagi, Nihon Reoroji Gakkaishi **37**, 75 (2009)

Chapter 13
Composites: Morphology

Kenta Chaki and Taku Ozawa

13.1 Introduction

Nanocomposite materials comprising nanofillers dispersed in polymers have received increasing attention for their high functionality. Various nanofiller candidates, including carbon material, can be applied in such nanocomposite materials. The dispersed structure can be manipulated to improve the electric and thermal conductivities and mechanical properties of materials.

To evaluate and predict the disperse structure of nanofillers, simulations that take account of compatibility between fillers and polymers have been performed [1]. In this chapter, we conduct a dissipative particle dynamics (DPD) simulation [2] using COGNAC to observe the behavior of filler dispersion.

13.2 Modeling

A diblock copolymer is composed of two species of monomers. Changes in the ratio of each block length transform the phase structure in various ways. It has been reported [3] that a DPD simulation can reproduce the phase-separated structure of diblock copolymers. In addition, the interaction (conservative force) introduced in DPD simulation is related to the Flory–Huggins χ parameter, and this interaction affects the phase-separated structure. Therefore, DPD simulation has recently been applied to a variety of phase separation phenomena.

DPD simulation reproduces polymers as coarse-grained particles bonded by an entropic spring. In this chapter, we introduce two species of particles A and B with

K. Chaki • T. Ozawa (✉)
JSOL Corporation, Tokyo, Japan
e-mail: ozawa.taku@jsol.co.jp

Japan Association for Chemical Innovation, *Computer Simulation of Polymeric Materials*, DOI 10.1007/978-981-10-0815-3_13

different properties and compose a diblock copolymer by bonding 20 particles. The difference in the properties is introduced as interaction parameters of the conservative force. In detail, the parameters of the interactions between the same species are set to $a_{AA} = a_{BB} = 25$, which is equivalent to $\chi = 0$. Meanwhile, the parameter of the interaction between different species is set to $a_{AB} = 45$ according to [4]. In the COGNAC input file, we set *Interactions.Pair_Interaction*[].*Potential_Type* to "DPD". The cutoff is set to 1, which corresponds to the particle diameter.

The conservative force, as mentioned previously, is related to the χ parameter, which is also related to the solubility parameter. For example, the solubility parameter of a carbon nanotube (CNT) has been evaluated using full atomistic molecular dynamics and applied to the interaction parameter for a DPD simulation [5].

Bonds between two particles in the polymer are reproduced as the bonding potential of an entropic spring. Their equilibrium length and spring constant are 0 and 4, respectively. In inputting the COGNAC file, "DPD" is selected for *Molecular_Attributes.Bond_Potential*[].*Potential_Type*, and *R0* and the spring constant *C* are 0.86 and 4, respectively. Note that *R0* is not used in the dynamic calculation but used in setting up the initial structure of a random coil.

We first manipulate a CNT as a carbon material. A CNT is reproduced as a linear chain including 20 (C) particles. The parameter of interaction between two C particles is set to $a_{CC} = 25$ and that between A and C particles and that between B and C particles are set to $a_{AC} = 25$ and $a_{BC} = 35$, respectively. The interaction parameters are listed in Table 13.1. This configuration indicates that a C particle is more attracted to an A particle than a B particle. The bonding potential differs from that for the diblock copolymer, which is introduced as a harmonic potential and has an equilibrium length and spring constant of 0.75 and 20, respectively. In addition, an angle potential is applied to three continuous bonding particles to reproduce the rigidity of the CNT. In inputting the COGNAC file, "Theta" is selected for *Molecular_Attributes.Angle_Potential*[]. Its equilibrium angular *R0* and spring constant *K* are 0 and 40, respectively.

In addition to the CNT, we manipulate graphene as a carbon material. Graphene is reproduced as a disk. The disk is modeled using J-OCTA (http://www.j-octa.com), the commercial version of OCTA. The disk is hexagonal and the length of its side is 5σ. To stabilize its form, an angle potential is set. In this chapter, the interaction parameter set of graphene is the same as that of the CNT.

To visualize the results of the phase-separated structure and to use them in other simulation engines, we set up *Simulation_Conditions.Density_Output* in the COGNAC input file. This gives the distribution of the volume fraction in

Table 13.1 Interaction parameters

	A	B	C
A	25	45	25
B		25	35
C			25

Grid_Density of the COGNAC output file. To obtain the averaged structure, *Simulation_Conditions.Density_Output.Collect_Density_Steps* is set to 10.

13.3 Results and Discussion

Before manipulating carbon materials, we confirm the phase separation of diblock copolymers without fillers. We create 5000 polymers of A_2B_{18} or $A_{18}B_2$ (100,000 particles) in the bulk system with a periodic boundary condition and calculate for $10{,}000\tau$ ($\Delta t = 0.05$) to equilibrate the system. In the A_2B_{18} system, phase A is dispersed in phase B and there is phase separation. In the same way, in the $A_{18}B_2$ system, phase B is dispersed in phase A and there is phase separation. Figure 13.1 shows only particle A in the A_2B_{18} system.

Next, we confirm the dispersion of CNTs in the system of each homopolymer A_{20} and B_{20}. The system consists of approximately 100,000 particles including 1 vol.% CNTs. Figures 13.2 and 13.3 are snapshots of A_{20} and B_{20}, respectively. The CNTs are dispersed in the system of A_{20} and aggregated in the system of B_{20} because of the conservative force. A CNT is shown as a rod and we can visualize it in the same way using the Action command of COGNAC. The angle potential makes the configuration of the CNT rigid. The angle potential that we set up does not completely elongate the CNT but more or less winds the CNT. This roughly reproduces the real structure of the CNT.

We then simulate the system of a diblock copolymer including CNTs. Figures 13.4 and 13.5 show the distribution of CNTs in the phase-separated structure

Fig. 13.1 Phase-separated structure of A_2B_{18} (where only particle A is shown)

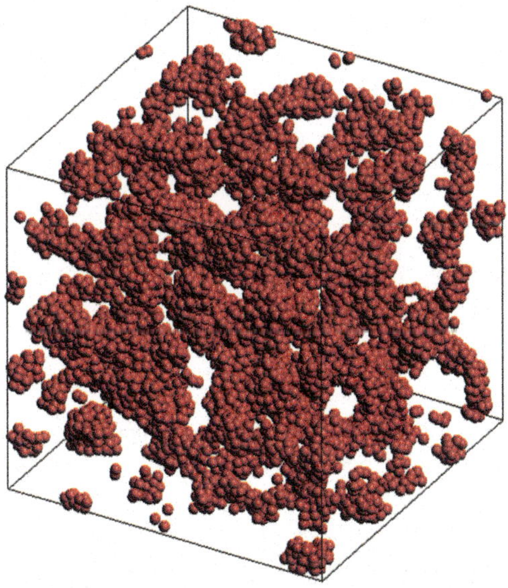

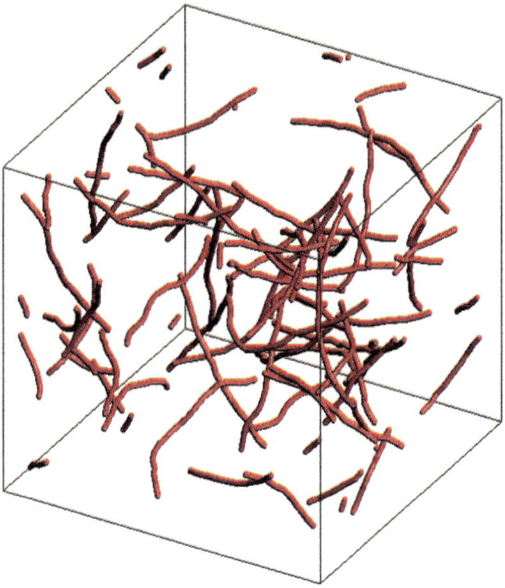

Fig. 13.2 CNT in the system of A_{20}

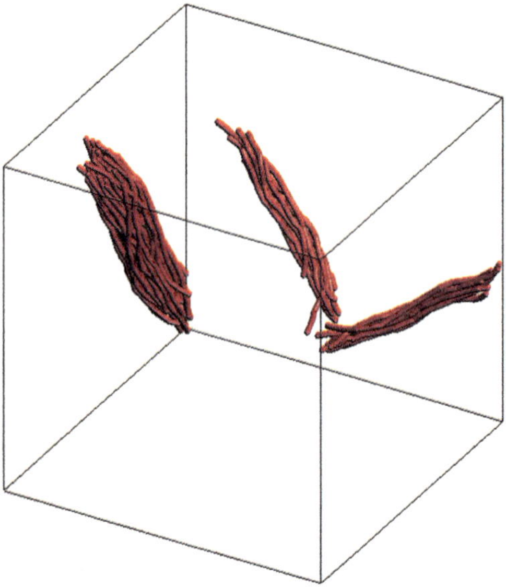

Fig. 13.3 CNT in the system of B_{20}

formed by A_2B_{18} and $A_{18}B_2$, respectively. The amounts of polymer and CNT are the same as before. Figures 13.4a and 13.5a show a smaller block ratio element particle of copolymers as a point. Figure 13.4a shows the isosurface when the local volume fraction of *Grid_Density* is 0.5. As mentioned previously, the CNT prefers particle A. In the $A_{18}B_2$ system, the CNTs avoid phase B and move into phase A. A

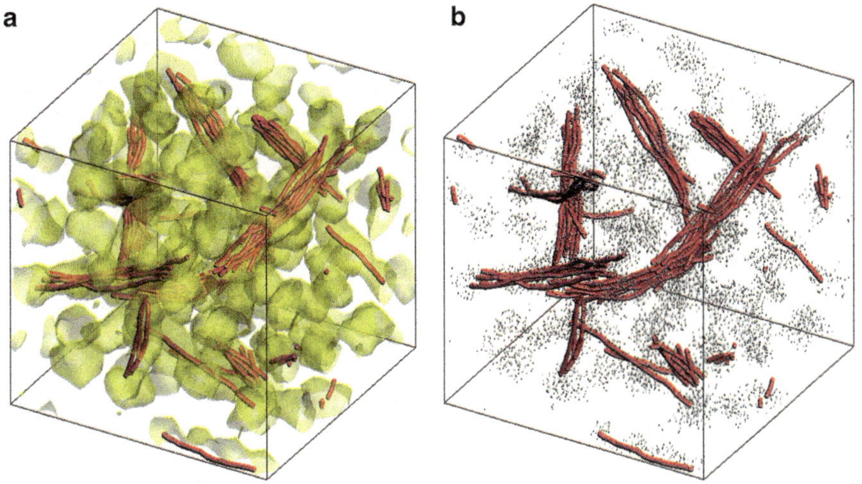

Fig. 13.4 CNT in the A_2B_{18} system: **a** phase A is shown as an isosurface of the volume fraction of 0.5, **b** particle A is shown as a point

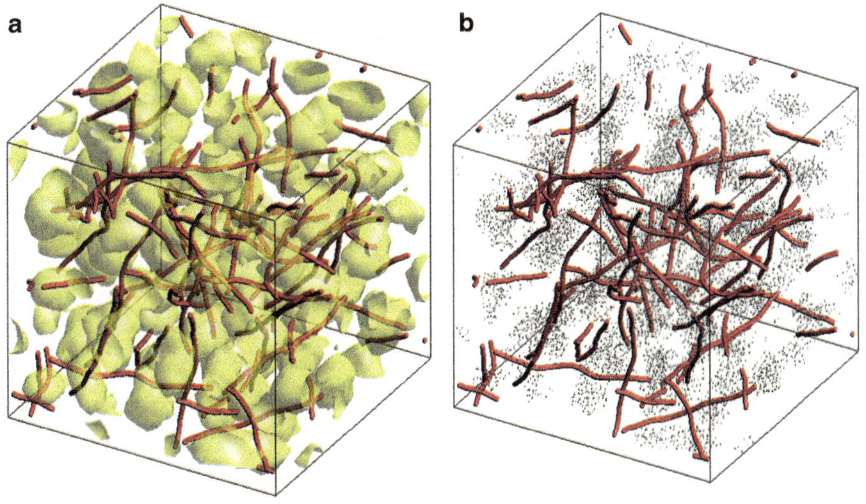

Fig. 13.5 CNT in the $A_{18}B_2$ system: **a** phase B is shown as an isosurface of the volume fraction of 0.5, **b** particle B is shown as point

distributed structure of CNTs consequently forms. Meanwhile, in the A_2B_{18} system, CNTs are confined in the smaller element of phase A. Partly owing to the rigidity of the CNTs, a bundle of CNTs forms. Compared with Fig. 13.1, the distribution of CNTs affects the phase separation of the polymer.

Finally, we simulate the system of a diblock copolymer including graphene. For comparison with the system including the CNT, the system consists of about

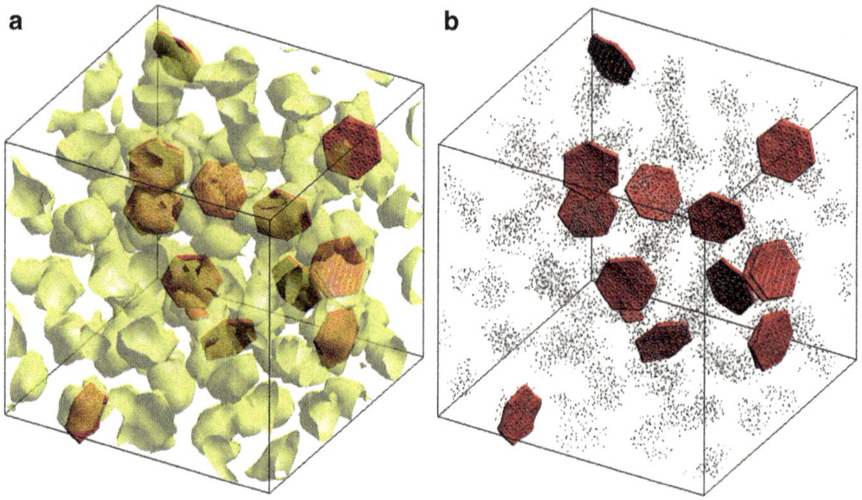

Fig. 13.6 Graphene in the A_2B_{18} system: **a** phase A is shown as an isosurface of the volume fraction of 0.5, **b** particle A is shown as a point

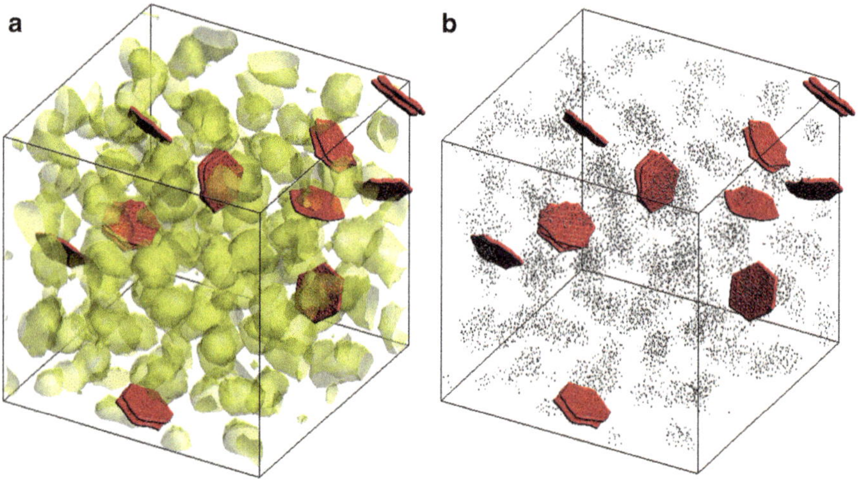

Fig. 13.7 Graphene in the $A_{18}B_2$ system: **a** phase B is shown as an isosurface of the volume fraction of 0.5, **b** particle B is shown as a point

100,000 particles including 1 vol.% graphene. We calculate for $50{,}000\tau$ ($\Delta t = 0.01$ so as not to collapse the form of graphene) to equilibrate the system. In contrast to the case for the CNT, graphene disperses and aggregates locally in both systems. Figures 13.6 and 13.7 are snapshots of the structure. It is seen that the shape of the filler affects the dispersion of the filler. Through the effect of the interaction, graphene disperses in phase A of both systems, which is a tendency similar to that of the system including CNTs.

We simulate the micro-phase separation of copolymers; such simulation will make it possible to manipulate the distribution of nanofillers. In such a case, DPD simulation allows us to predict the structure to a certain degree considering the solubility of two elements. In addition, DPD simulation can reveal the depletion effect [6] considering the sizes and shapes of obstacles. Meanwhile, we need to take care when considering the manufacturing process. For example, it is easy to apply shear flow using COGNAC but it is difficult to treat the real stirring condition. Moreover, DPD simulation cannot include an entanglement effect.

13.4 Application

Using the result for the distribution of fillers obtained in the DPD calculation, we can analyze and evaluate physical quantities. According to [4], finite element analysis (FEA) of the electric potential is performed, in which the structure including CNTs is converted to a distribution of the volume fraction, and the electric conductivity is assigned to each component. This analysis allows us to evaluate the electric conductivity of the whole system and percolation. Meanwhile, the percolation of fillers could be evaluated by tracing a path of contacts with particles. We should consider contacts with fillers in any event, especially in the case of electric conductivity.

MUFFIN in OCTA can conduct an analysis based on the continuum model by importing the distribution of the volume fraction from COGNAC. This is called the "zooming" function. Here, we briefly introduce the analysis of thermal conductivity with MUFFIN.

We use the distribution of the volume fraction obtained above. Heat conductivities are given as 1 for components A and B and 100 for component C. Figure 13.8 shows the temperature distribution on the CNTs in the $A_{18}B_2$ system. Figure 13.9 shows average heat conductivities of each system. Because the $A_{18}B_2$ system with CNTs is dispersive and percolative, the heat conductivity of this system would be larger than that of other systems.

13.5 Conclusion

We simulated filler dispersion in the system of a diblock copolymer. Through DPD simulation, we obtained the phase-separated structure of the polymer and the dispersion of fillers. In addition, we simulated heat conduction using the continuum model by "zooming" out the structure. The validity of multiscale analysis, a feature of OCTA, is clear.

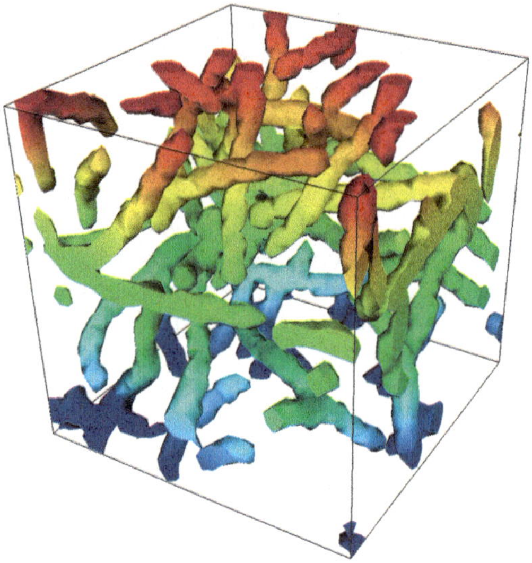

Fig. 13.8 Temperature distribution on the isosurface of the CNT volume fraction of 0.1 in the $A_{18}B_2$ system

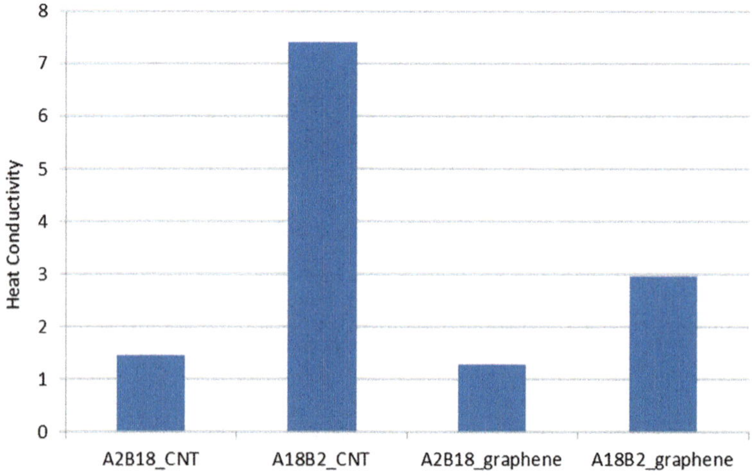

Fig. 13.9 Average heat conductivity

13.6 Download Instruction

Sample files are available from the OCTA website. See the instruction on http://octa.jp/download_bookdata.html. Password for the sample (zip file) of this chapter is "PautWz5s":

1. A_2B_{18} including CNT: "A2B18_in.bdf" (input) and "A2B18_out.bdf" (output)
2. $A_{18}B_2$ including CNT: "A18B2_in.bdf" (input) and "A18B2_out.bdf" (output)
3. A Python script that draws a CNT as a rod and the distribution of the volume fraction of element A (or B): "show.py"

References

1. Q.H. Zeng, A.B. Yu, G.Q. Lu, Prog. Polym. Sci. **33**, 191 (2008)
2. R.D. Groot, P.B. Warren, J. Chem. Phys. **107**, 4423 (1997)
3. R.D. Groot, T.J. Madden, J. Chem. Phys. **108**, 8713 (1998)
4. J.T. Wescott, P. Kung, A. Maiti, Appl. Phys. Lett. **90**, 033116 (2007)
5. A. Maiti, J.T. Wescott, P. Kung, Mol. Simul. **31**, 143 (2005)
6. S.-W. Hu, Y.-J. Sheng, H.-K. Tsao, Soft Matter **9**, 7261 (2013)

Chapter 14
Composites: Interfacial Strength

Taku Ozawa and Hiroya Nitta

14.1 Introduction

There has been much research on composites of polymers and nanofillers, which are considered a type of nanocomposite. The materials are expected to have high functionality (e.g., high electric or thermal conductivity, and/or appropriate mechanical strength or flexibility). There are two considerations in evaluating the functionality of nanocomposites. One is the dispersion structure and the resulting properties. The other is the interfacial phenomenon of polymers and nanofillers. Because the latter depends on the structure at an atomic scale, it is difficult to examine in experiments. Nanoscale simulations are thus conducted for the study of the interfacial phenomenon.

While the dispersion structure has been analyzed using dissipative particle dynamics described in the previous chapter, the interfacial strength is evaluated using molecular dynamics (MD). Here we present a normal separation (traction) simulation at the interface between polyethylene and a graphene.

14.2 Simulation Conditions

The model used here is taken from the literature [1]; however, we simplify the model and simulation steps to facilitate the process of following the simulation. We use the united atom (UA) model to model the polyethylene, while the polyethylene modeled in the literature is a full atomistic model. In the UA model, each group of CH_2 or CH_3 is treated as one particle. The choice of the UA model makes modeling easier

T. Ozawa (✉) • H. Nitta
JSOL Corporation, Tokyo, Japan
e-mail: ozawa.taku@jsol.co.jp

Japan Association for Chemical Innovation, *Computer Simulation of Polymeric Materials*, DOI 10.1007/978-981-10-0815-3_14

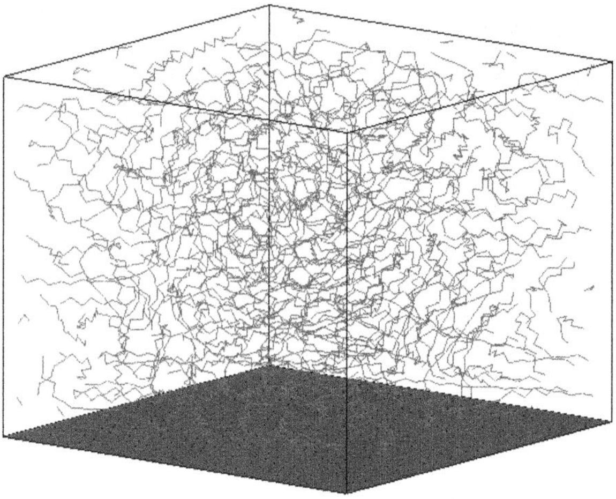

Fig. 14.1 The graphene–polymer interfacial system considered in this chapter (*black*: graphene; *gray*: polyethylene)

and reduces the processing load of a computer in the simulation. The carbon atoms that constitute graphene are fixed at given points, and their role is only that of a wall that interacts with polymers. For easy reference, the user definable format (UDF) files used in the simulation are listed at the end of the chapter.

We consider that the model described here is adequate for qualitatively analyzing the problem because the polymer has small polarizations and the graphene is electronically neutral. It is noted that, in the case of polymers with polar functional groups or particles having finite charges, models need to be more sophisticated than the model described here.

We consider linear-chain polyethylene. Each chain comprises 60 monomers and both ends are terminated by methyl groups. Although the model is shorter than real polymers, it is sufficiently long for evaluating interfacial properties, given that the size of the simulation cell described later is about 5 nm.

The system modeled for the simulation is shown in Fig. 14.1. The simulation cell is filled with amorphous polyethylene, and the graphene is located at the bottom of the cell ($Z = 0.0$). The system dimensions X and Y are determined from the unit length of graphene, and the size in the Z direction is determined from the density of the polyethylene.

Figure 14.2 shows screenshots of the unit and force-field parameter settings for the simulation. The parameters used for the polyethylene are taken from [2], while those used for the graphene are taken from [3]. Some molecular modeling software programs have parameter sets similar to those described here and provide a graphical user interface through which molecular structures can be constructed. The polyethylene can be constructed with "potential_map.udf" and SILK, which are appended to COGNAC. Information of the units, such as units of mass, energy, and length, are listed in *Unit_Parameter*, and the dimensionless quantities can be dimensionalized using this information. C_33, C_32, and CB listed in Fig. 14.2 represent CH_3, CH_2, and a carbon atom in graphene, respectively.

a

Unit_Parameter	struct	-	-
Name	string		
Comment	string		
Mass	double	1.0	[amu]
Energy	double	4.1855	[kJ/mol]
Length	double	0.1	[nm]

b

Pair_Interaction[]:(index)	sigma:double	epsilon:double
[0]:C_33-C_33	3.69937	0.25
[1]:C_33-C_32	3.66164	0.222711
[2]:C_32-C_32	3.62391	0.1984
[3]:CB-C_32	3.51546	0.11649
[4]:CB-C_33	3.55319	0.13077

c

Molecular_Attributes	struct	-	-
Atom_Type[]	Atom_Type ...	-	-
Atom_Type[0]	Atom_Type	-	-
Name	KEY	C_33	
Mass	double	15.01	[mass]
Atom_Type[1]	Atom_Type	-	-
Name	KEY	C_32	
Mass	double	14.01	[mass]
Atom_Type[2]	Atom_Type	-	-
Name	KEY	CB	
Mass	double	12.01	[mass]
Bond_Potential[]	Bond_Poten...	-	-
Bond_Potential[0]	Bond_Poten...	-	-
Name	KEY	C_32-C_33-1-1	
Potential_Type	select	Harmonic	
R0	double	1.53	[sigma]
Harmonic	harmonic	-	-
K	double	700.0	[epsilon/(sigma^2)]
Bond_Potential[1]	Bond_Poten...	-	-
Name	KEY	C_32-C_32-1-1	
Potential_Type	select	Harmonic	
R0	double	1.53	[sigma]
Harmonic	harmonic	-	-
K	double	700.0	[epsilon/(sigma^2)]
Angle_Potential[]	Angle_Pote...	-	-
Angle_Potential[0]	Angle_Pote...	-	-
Name	KEY	X-C_32-X-X-X	
Potential_Type	select	Theta	
theta0	double	70.529	[degree]
Theta	Kconst	-	-
K	double	100.0	[epsilon/rad^2]
Torsion_Potential[]	Torsion_Pot...	-	-
Torsion_Potential[0]	Torsion_Pot...	-	-
Name	KEY	X-C_32-C_32-X-X-1-X	
Potential_Type	select	Cosine_Polynomial	
Cosine_Polynomial	CosinePoly...		-
K	double	1.0	[epsilon]
N	int	4	
p[]	double array	-	-
p[0]	double	1.0	
p[1]	double	3.0	
p[2]	double	0.0	
p[3]	double	-4.0	

Fig. 14.2 (**a**) Units defined in the simulation. (**b**) Lennard-Jones potential parameters defined for the nonbonding interactions. (**c**) Bond potential parameters used in the simulation

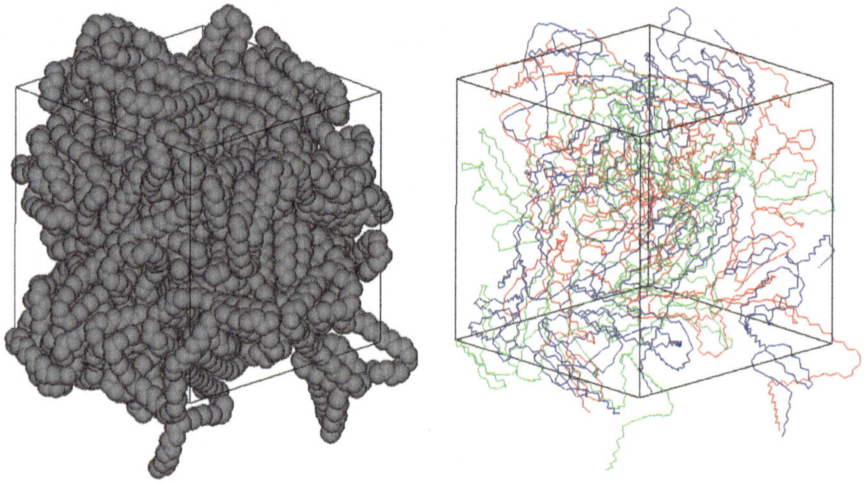

Fig. 14.3 The bulk structure of polyethylene; the right-hand image shows molecules in line view, with colors showing individual chains

We next describe how to build the system. The polymer system consists of 32 polyethylene chains. In [1], the polymers are initially dilute, and an NPT ensemble dynamics simulation is then performed, and finally, the system is cooled to 100 K. We adopt a simpler approach in this chapter. Polyethylene chains are located with density of 1.0 g/cm^3 in the periodic cell by setting *Initial_Structure.Generate_Method.Method* to "Random". An MD simulation with an NVT Nose–Hoover thermostat is then performed until the time evolution reaches 0.5 ns at a temperature $T = 100$ K. The obtained structure is shown in Fig. 14.3. In the line view, we see the spatial distribution of the chains and the alignment of some chains like that in a crystal.

We next add graphene to the system. We shift the Z-coordinates of the polymers and enlarge the Z length of the cell so that polymers exist in the central part of the cell. The boundary condition for the Z direction is changed to "reflective" and "none" at $Z = Z_{\mathrm{max}}$ and $Z = 0$, respectively. This is done by setting *Simulation_Conditions.Boundary_Conditions.c_axis* to "Reflective1". At $Z = 0$, a crystal structure corresponding to graphene is prepared (Fig. 14.4). At this time, we modify the cell lengths in X and Y directions to construct a periodic structure in X and Y directions. Although modifying the cell lengths in X and Y directions produces excess forces between polymers, the structure is allowed to relax before MD simulation. The structure before the relaxation simulation is depicted in Fig. 14.5. As described above, the carbon atoms that constitute the graphene are fixed at initial positions. This is done by registering all constraint atoms with the array *Simulation_Conditions.Constraint_Conditions.Constraint_Atom[ ]*. Because the above procedure involves an enormous amount of manual input, we run a Python script for this setting. The script is included in sample files.

Fig. 14.4 The graphene structure that we constructed

We then perform an NPT ensemble dynamics simulation at pressure $P = 1$ atm and temperature $T = 100$ K to create an interfacial structure between polyethylene chains and the graphene. For this simulation, we set *Simulation_Conditions.Solver.Dynamics* to "NPT_Parrinello_Rahman_Nose_Hoover", and *Fix_Angle* and *Fix_Cell_Length* to "1" and "xy", respectively. Finally, we obtain the structure shown in Fig. 14.1, where the cell sizes in *X*, *Y*, and *Z* directions are 4.68, 4.69, and 3.88 nm, respectively. We now have an initial structure for the simulation, and we present the settings for the normal separation simulation in the following. The carbon atoms in the graphene are again restrained in this simulation. For the cell deformation condition, the upper 40 % of atoms in the *Z* direction are set to have the velocity $v_z = 25$ m/s (where v_x and v_y are zeros). This corresponds to the deformation of the material with a zero Poisson's ratio. The settings in the UDF file are shown in Fig. 14.6. The velocity for the deformation is set to ten times higher than that in [1]. We apply the condition for the reduction of the load on a computer and that the polymers are expected to behave like glass under the simulated temperature.

14.3 Results and Discussion

Figure 14.7 shows snapshots taken from the normal separation simulation. The upper particles move away from the interface at a constant velocity and the polyethylene chains close to the interface separate gradually. Voids form in the

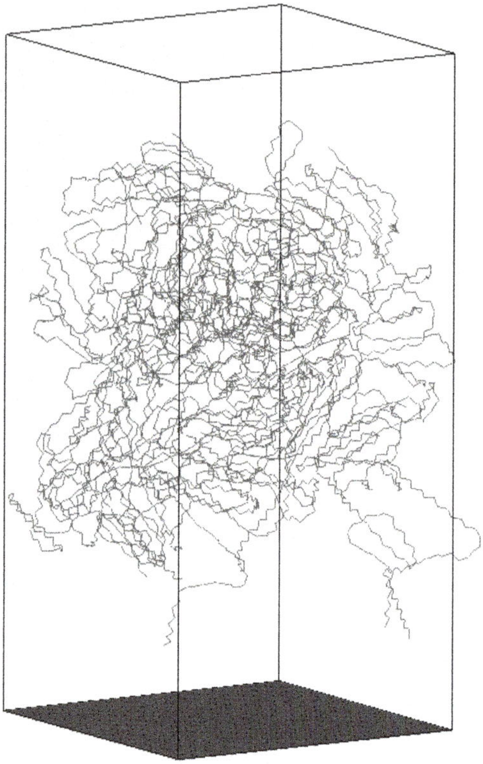

Fig. 14.5 Polyethylene and graphene have been combined to generate an interface

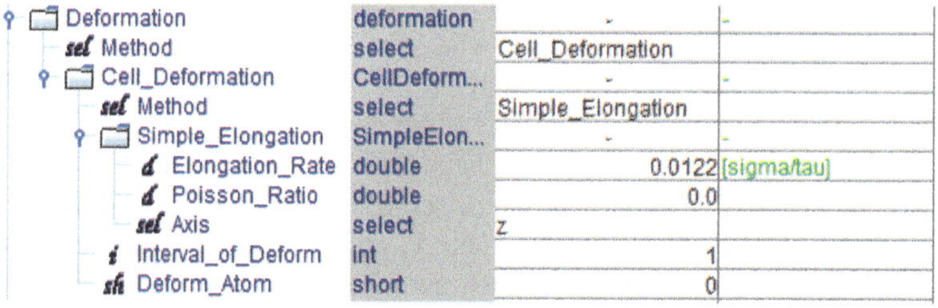

Fig. 14.6 Settings for the cell deformation used in the normal separation simulation

interfacial region because the Poisson's ratio is zero. Some parts of the chains remain adsorbed on the interface in the simulation. The chains that connect the absorbed and upper polyethylenes form a fibril, and the chains in the fibril align like a crystal. In Fig. 14.8, average forces in the *Z* direction, which act on each graphene carbon, are plotted on a graph with the cell deformation shown on the horizontal axis. Relatively large forces are recognized immediately after the simulation starts, after which the forces weaken gradually. Even after deformation exceeding 8 nm, the forces have finite value because of the fibril and the absorbed particles described above.

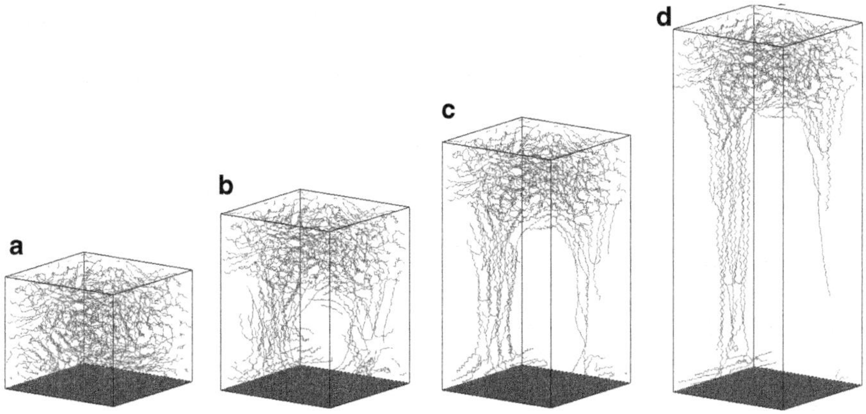

Fig. 14.7 Snapshots obtained during the normal separation simulation: (**a**) the initial state and the state where the cell deformation reaches (**b**) 2.5 nm, (**c**) 5.0 nm and (**d**) 8.5 nm

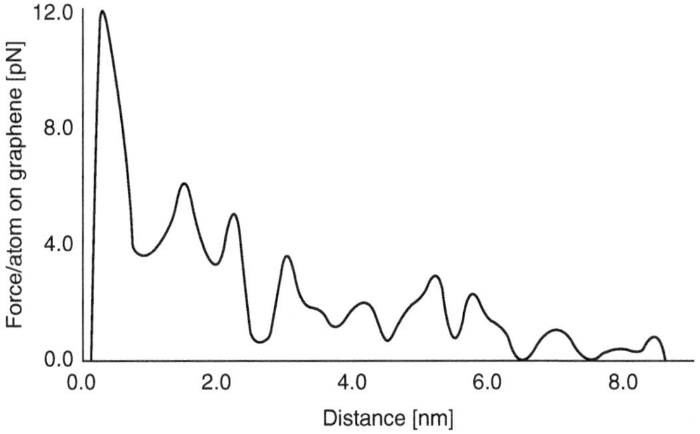

Fig. 14.8 The force applied to the graphene during the normal separation simulation

14.4 Conclusion

We modeled the interface between polymers and graphene and simulated the normal separation of polymers from the graphene surface. Applying the simulations described in this chapter should facilitate an understanding of phenomena and allow a qualitative evaluation of the physical properties at the interface. In the literature [1], the shear strain (sliding separation) simulation is performed, and the simulation gives values comparable to experimental values obtained for polyethylene absorbed on carbon nanotubes.

It is difficult to compare the simulation performed here with experiments directly and quantitatively for several reasons. The simulation conditions such as the polymer chain length or the deformation speed are far from the conditions of experiments. It is noted that detailed and realistic models are needed for the treatment of some functional groups and polarizable polymers. Poisson's ratio will have to be determined more precisely when comparing simulation results with experiments. Furthermore, the inhomogeneity of the surface has not been taken into account in the example presented here, which is a problem that needs to be addressed in future work.

14.5 Download Instruction

Sample files are available from the OCTA website. See the instruction on http://octa.jp/download_bookdata.html. Password for the sample (zip file) of this chapter is "8wh2KGfs".

1. Construction for the bulk polymer: bulk_relax.bdf and bulk_relax_out.bdf
2. Construction for the interface: interface.bdf and interface_out.bdf
3. Simulation for normal separation: separation.bdf and separation_out.bdf
4. Script that adds a graphene structure to the given UDF file: python_set_graphene.py
5. Script for the constraint setting: python_move_wall.py

References

1. A.P. Awasthi, D.C. Lagoudas, D.C. Hammerand, Model. Simul. Mater. Sci. Eng. **17**, 015002 (2009). doi:10.1088/0965-0393/17/1/015002
2. S.L. Mayo, D. Olafson, W.A. Goddard III, J. Phys. Chem. **94**, 8897 (1990). doi:10.1021/j100389a010
3. J.P. Lu, X.-P. Li, R.M. Martin, Phys. Rev. Lett. **68**, 1551 (1992). doi:10.1103/PhysRevLett.68.1551

Chapter 15
Cross-Linked Rubber

Hiroshi Shima

15.1 Introduction

Typical rubber materials are usually used in a cross-linked state. It is therefore important to understand the effect of cross-linking when developing rubber materials. In particular, considering that rubber materials are often used in a deformed state, it is important to grasp the effect of cross-linking under deformation.

For usual non-cross-linked polymers, there are various theoretical models and calculation methods based on such models, and viscoelasticity can be predicted from the molecular weight, structure (e.g., branches), molecular weight between entanglement points, and relaxation time of entanglement using simulators such as PASTA or NAPLES.

For cross-linked materials, among theoretical models, the results of Edwards–Vilgis slip-link model [1] have been shown to agree with those of experiments conducted by Urayama [2]. Among simulators, PASTA cannot handle cross-linked materials whereas NAPLES can; however, because NAPLES simulation involves a large degree of coarse graining, it has a limited range of applications. A calculation based on coarse-grained molecular dynamics, the Kremer–Grest model [3], has the following merits:

- Physical properties can be predicted with a model of arbitrarily defined molecular weight between cross-linking points and its distribution.
- The observation of behaviors of molecular chains under large deformation is possible.

Furthermore, the calculation is considered to have a wide range of application despite its large calculation load.

H. Shima (✉)
BRIDGESTONE Corporation, Tokyo, Japan
e-mail: hiroshi.shima@bridgestone.com

Japan Association for Chemical Innovation, *Computer Simulation of Polymeric Materials*, DOI 10.1007/978-981-10-0815-3_15

COGNAC can easily handle deformation and cross-linking reactions in addition to the basic functions of the calculation made in the coarse-grained molecular dynamics method; therefore, COGNAC is suitable for the analysis of cross-linked materials. The following describes the formation of cross-linked structures and elongational physical properties in the analysis of cross-linked materials using COGNAC.

15.2 Formation of Cross-Linked Structures

15.2.1 Formation Method

The main methods of forming cross-linked structures include a method for random cross-linking and a method for end cross-linking. The following outlines these two methods (Fig. 15.1):

1. Method for random cross-linking

   Random cross-linking is conducted according to the following procedure, as described in the literature [4]:

   - A system containing a plurality of polymers is prepared.
   - The desired number of cross-linking points is placed randomly in the system.
   - A polymer particle at the position closest to the cross-linking points is considered to be an activated particle.
   - A particle that is within a distance of 1.3 σ from the activated particle is randomly selected, and a bond is created between the selected particle and the activated particle. In this procedure, particles located near the already-reacted particle by a distance of one or two particles in the initial stage of a polymer are excluded from the reaction when selecting the activated particle and the particle to react with the activated particle.

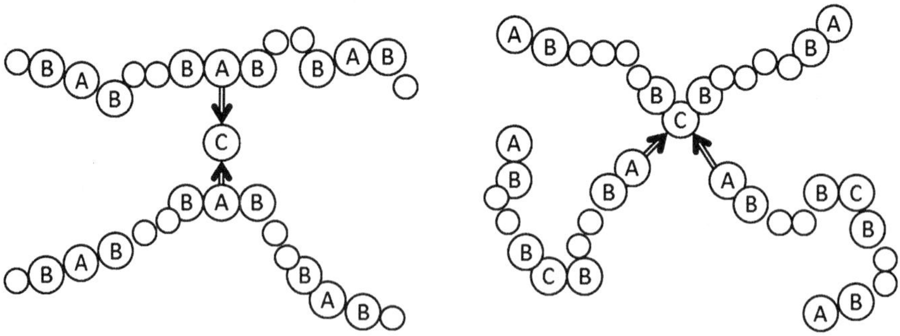

Fig. 15.1 Method of creating a cross-linked structure. *Left*: Random cross-linking using cross-linking particles. *C* cross-linking particle; *A* cross-linkable polymer particle. *Right*: End cross-linking. *C* four-functional particle; *A* end reactivated particle

Regarding the abovementioned distance (within 1.3 σ), Kremer and Grest note that "We chose this value of r_x since it equaled the persistence length, though the results are not sensitive to the precise choice of r_x" [4]. Additionally, this reaction condition of within 1.3 σ has been adopted in many papers.

According to this method, random cross-linking can be conducted in a manner similar to that of actual cross-linking.

Modifications of the aforementioned method resembling sulfur cross-linking in rubber include a method [5] employing cross-linking particles in place of the cross-linking points. A system containing cross-linking particles and polymers is created to react with polymer particles close to the cross-linking particles. To prevent cross-linking of adjacent particles, an approach in which a polymer is considered to be made of alternately connected cross-linkable and non-cross-linkable particles or the like is used.

2. Method for end cross-linking

   End cross-linking is conducted according to a previously reported method [6] by which four-functional cross-linking particles are used to react with the ends of polymers:

   - A system containing polymers, the molecular weight of which is equivalent to the molecular weight between cross-linking points, is prepared.
   - Four-functional particles are attached to 1/4 of all polymer end particles.
   - A bond is created when a four-functional particle and an end particle are closer to each other than a predetermined distance.

Employing this method, the molecular weight between the cross-linking points can be set to a predetermined value (e.g., a value of a uniform or particular distribution). However, the calculation for the reaction takes a long time. Moreover, it is not desirable that all ends react with the functional group, resulting in a system in which, for example, there are approximately 10 % dangling chains.

15.2.2 *Procedure for Creating a Cross-Linked Structure*

The following describes the procedure for creating (1) random cross-linking and (2) end cross-linking. The following procedure creates a cross-linked structure and calculates elongational deformation:

- Creation of an input user definable format (UDF) file
- Calculation of equilibration (1)
- Calculation of the cross-linking reaction
- Calculation of equilibration (2)
- Calculation of uniaxial elongation

The method of using cross-linking particles is applied to random cross-linking, and the method of placing four-functional particles in the center to react both ends with the reactivated particle is applied to end cross-linking.

Hereinafter, specific operations are described. The same can be performed by copying the random_crosslink (end_crosslink) folder to execute run_crosslink+elong.bat.

15.2.3 Creation of an Input UDF File

First, an input UDF file containing polymers (and cross-linking particles) is created. The UDF file for the present calculation requires the following items in COGNAC. For details, refer to the manual (Cognac90_eng.pdf, section 5.2):

1. Simulation conditions

   Setting of general calculation conditions (e.g., temperature, pressure, time step, and ensemble)

2. Initial structure

   Selection of an initial structure (e.g., random/restart)

3. Molecular attributes

   Setting of atomic species, bonding potentials, and other attributes to be used in simulation

4. Interactions

   Setting of nonbonding potentials, electrostatic potentials, and external fields

5. Setting of molecules

   Setting of molecular structures (giving information on bonding and assignment of coordinates and potentials)

Python scripts have been prepared to apply the aforementioned configuration.

When 3_2_1\sample\make_polymer_RC.py (in the case of random cross-linking) or make_polymer_EC.py (in the case of end cross-linking) is copied to input >python make_polymer_RC.py (or make_polymer_EC.py) from the command line in GourmetTerm (a terminal configured in the case of Linux), the input UDF file is created.

To change the molecular structures, the range (Fig. 15.2) at the top of the Python script is changed.

The molecular structures are represented such that the atomic species are lined up in the Python list. In Python, ["A"]*1 + ["B"]*4 is the same as ["A", "B", "B", "B", "B"].

```
################## Polymer and cross-linker
polymers = [
[5,"N1000", (["A"]*1+["B"]*4)*200],
[50,"Crosslinker", ["C"]]
]
'''
[5,"N1000", (["A"]*1+["B"]*4)*200] Making 5 polymer chains named "N1000" with 200 repetition of
ABBBB
[50,"Crosslinker", ["C"]]  Making 50 molecules named "Crosslinker" with 1 "C" atom
'''
```

Fig. 15.2 Molecular structure settings in make_polymer_RC.py

```
############### COGNAC: Fatal Error ###############
bond length of FENE potential exceeded maximum bond length !!
##################################################
```

Fig. 15.3 Example error message from COGNAC

Besides molecular structures, the main content set here is as follows:

- Calculation time (time step, total step, and output step)
  The time step *delta_T* is set to 0.006. The literature value of the time step was initially 0.006 but changed to 0.012 in subsequent papers. It is preferable for the time step to be shorter when a long molecule is randomly created as an initial structure. The error seen in Fig. 15.3 sometimes occurs depending on the initial structure that is randomly created; however, this type of error can be avoided by making *delta_T* small. Another way to avoid this error is to generate initial structures several times.
- Temperature and pressure
  Temperature $T = 1.0$ and pressure $P = 0.0$. The pressure P is invalid in NVT calculations where the volume is constant.
- Ensemble
  NVT_Kremer_Grest is selected. Friction is set to 0.5 in accordance with the literature.
  The following ensemble is selected in the case of a calculation made under constant pressure (e.g., $P = 0.0$).
  NPT_Andersen_Kremer_Grest
  NPT_Parrinello_Rahman_Kremer_Grest
  The former isotropically deforms, while the latter anisotropically deforms. The setting of *cell_mass* is necessary in both cases. When *cell_mass* is large, the pressure control becomes moderate.
- Setting of periodic boundary conditions
  Periodic boundary conditions are set in all three directions; however, it is essential to set *Periodic_Bond* to 1 in the analysis of cross-linked materials. This creates a bond of the image distance between particles under periodic boundary conditions.

For configuration details, check the Python file while referencing "5.2 Description of input UDF" in the manual. Moreover, when an appropriate value for a parameter (e.g., *Simulation_Conditions.Dynamics_Conditions.Max_Force*) to be input is unclear, the values in the sample file (e.g., \OCTA8\ENGINES\COGNAC90\sample\block\A20B40A10_in.udf) should be referenced.

When the input UDF file is created, the initial structure is generated by executing the following command from the command line.

```
>cognac90 -I XXX_in.udf -O XXX_out.udf > XXX_in.log
```

If no changes are made, XXX is "5-A1B4x200+50-C1" for random cross-linking and "50-A1B49C1B49A1" for end cross-linking.

15.2.4 Calculation of Equilibration (1)

The equilibration is calculated using the created input files. It is desirable to perform the calculation for as long as possible within the allowable range. In [3], the calculation time is defined as the time required for the molecule's center of mass to move a distance about twice the radius of gyration *Rg*. However, it is basically impossible to perform the calculation according to this definition in a random cross-linking system because of the large molecular weight; researchers should therefore determine the calculation time while confirming the feasibility.

Specific operations are listed as follows.

The aforementioned XXX_in.udf file is copied to create XXX_equil1_in.udf, which is opened in GOURMET and configured to the following settings (Fig. 15.4):

1. Settings for time (time step, total step, and output step):

   In *Simulation_Conditions.Dynamics_Conditions*, *delta_T* is set to 0.01.
   Total_Steps is set to, for example, 500,000.
   Output_Interval_Steps is set to, for example, 50,000.
   There is no problem with *delta_T* being 0.012.
   It is desirable that *Total_Steps* be as large as possible within the allowable range as mentioned above.

2. Settings for the creation of the initial structure

   Settings are arranged so that the calculation restarts from the aforementioned XXX_out.udf:

   Initial_Structure.Generate_Method.Method is set to Restart.
   Initial_Structure.Generate_Method.Restart.UDF_Name is set to XXX_out.udf.
   Initial_Structure.Generate_Method.Restart.Record is set to -1. Here, the value of -1 represents the final record. In the case of restarting from the middle state, *Ini-tial_Structure.Generate_Method.Restart.Record* is set to the number of the middle state.

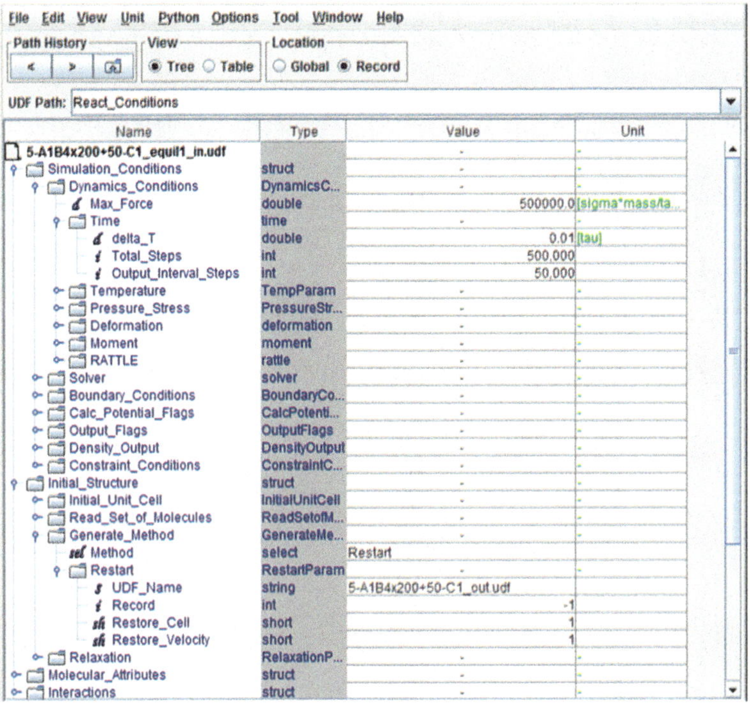

Fig. 15.4 Settings for the calculation of equilibration (1)

Initial_Structure.Generate_Method.Restart.Restore_Cell is set to 1.

When *Initial_Structure.Generate_Method.Restart.Restore_Velocity* is set to 1, the velocity in the previous calculation is set to be the initial velocity, and when *Initial_Structure.Generate_Method.Restart.Restore_Velocity* is set to zero, the velocity is newly set.

When the input UDF file is created, we perform the calculation by executing the following command from the command line.

```
>cognac90 -I XXX_equil1_in.udf -O XXX_equil1_out.udf
> XXX_equil1_in.log
```

15.2.5 Calculation of the Cross-Linking Reaction

1. Performing the calculation

   Specific operations are given in the following.

   The aforementioned XXX_in.udf file is copied to create XXX_react1_in.udf, which is opened in GOURMET to configure the following settings (Fig. 15.5):

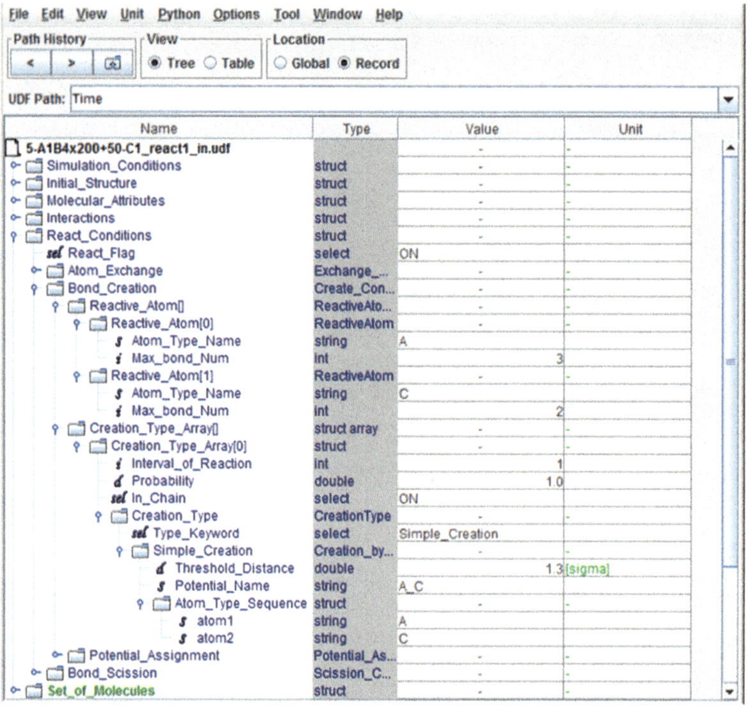

Fig. 15.5 Settings for the calculation of the cross-linking reaction

- Settings for time (time step, total step, and output step):

  In *Simulation_Conditions.Dynamics_Conditions*, delta_T is set to 0.01.
  Total_Steps is set to, for example, 10,000.
  Output_Interval_Steps is set to, for example, 100.

  It is necessary to make *Total_Steps* large in end cross-linking. Figure 15.6 shows the elapsed time and the number of cross-linking reactions in a sample model; however, some reacting particles remain unreacted.
- Settings for the creation of the initial structure

  Settings are arranged so that the calculation restarts from XXX_equil1_out.udf, which is the previous calculation result.
- Settings for the reaction

  In the settings for the reaction, whether the reaction occurs, the number of bonds that can be made by each particle, the reaction distances, the kinds of bonds created in the reaction, and the types of reacting particles are input:

  React_Conditions.React_Flag is set to 1.
  Because the array in *React_Conditions.Bond_Creation.Reactive_Atom*[] is initially empty, two arrays are created by clicking *Reactive_Atom*[]

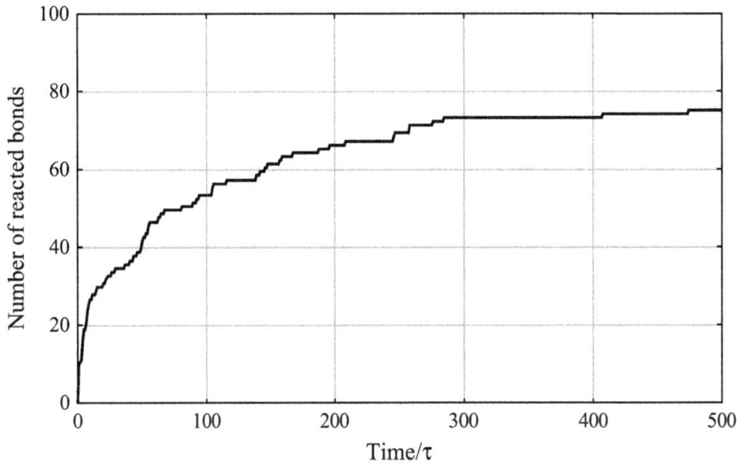

Fig. 15.6 Elapsed time and number of reactions (example of end cross-linking)

positioned in the second position of the tree view. Then, after inputting Ctrl + I or Ctrl + E twice, the values shown in Fig.15.5 are input. *React_Conditions.Bond_Creation.Creation_Type_Array*[*0*] is created in the same manner.

Figure 15.5 shows the case of random cross-linking; the case of end cross-linking is as follows:

Bond_Creation.Reactive_Atom[*0*].*Atom_Type_Name* is set to A.
Bond_Creation.Reactive_Atom[*0*].*Max_bond_Num* is set to 2.
Bond_Creation.Reactive_Atom[1].*Atom_Type_Name* is set to C.
Bond_Creation.Reactive_Atom[1].*Max_bond_Num* is set to 4.

When the input UDF file is created, we perform the calculation by executing the following command from the command line.

```
>cognac90 -I XXX_react1_in.udf -O
   XXX_react1_out.udf
> XXX_react_in.log
```

2. Confirmation of calculation results

After the calculation is completed, the reaction is confirmed.

First, "XXX_react1_out.udf" is opened from GOURMET. *Set_of_Molecules* is displayed in green at record 0 (the record number is displayed at lower right), and when the slider below is shifted, *Set_of_Molecules* turns black, which indicates *Set_of_Molecules* has changed. It is understood that the number of molecules decreases owing to the cross-linking reaction.

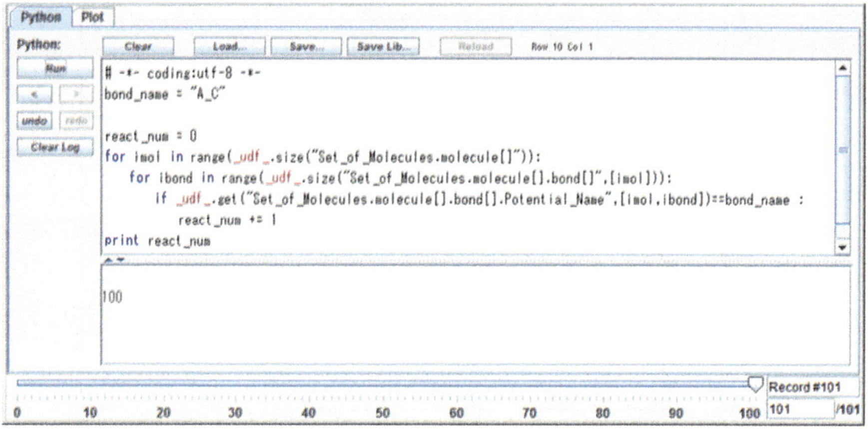

Fig. 15.7 Calculation of the number of cross-linking reactions

Next, the number of bonds of potential name *A_C* produced by the reaction is counted. When search_bonds_on_gourmet.py is pasted into the Python Scripting Window in GOURMET and the Run button is pushed, the number of bonds is displayed in the Python Log Window below (Fig. 15.7). In the case of random cross-linking, when the number of A_C bonds is twice the number of C particles in the system, it is concluded that all particles have reacted.

Moreover, a Python script that displays *Atom_Id* of reacted "A" and "C" particles ("crosslinked_atoms.py") is prepared to check where the reaction occurs. The script is used by pasting it into the Python Script Window in GOURMET in the same manner as for "search_bonds_on_gourmet.py".

15.2.6 Calculation of Equilibration (2)

The aforementioned "XXX_equil1_in.udf" file is copied to create "XXX_equil2_in.udf", which is opened in GOURMET to configure the following settings (Fig. 15.8):

- Settings for time (time step, total step, and output step)
- Settings for the creation of the initial structure:

1. *Initial_Structure.Read_Set_of_Molecules*

   Because the molecule is changed by the cross-linking reaction, it is necessary to read *Set_of_Molecules* after the reaction. *UDF_Name* is set as the output "XXX_react1_out.udf" of the calculation of the cross-linking reaction, and the value of *Record* from which the record should be read is specified. However, if the final record should be read, the value can be set as a value larger than the actual number of records. When −1 is input, the initial structure

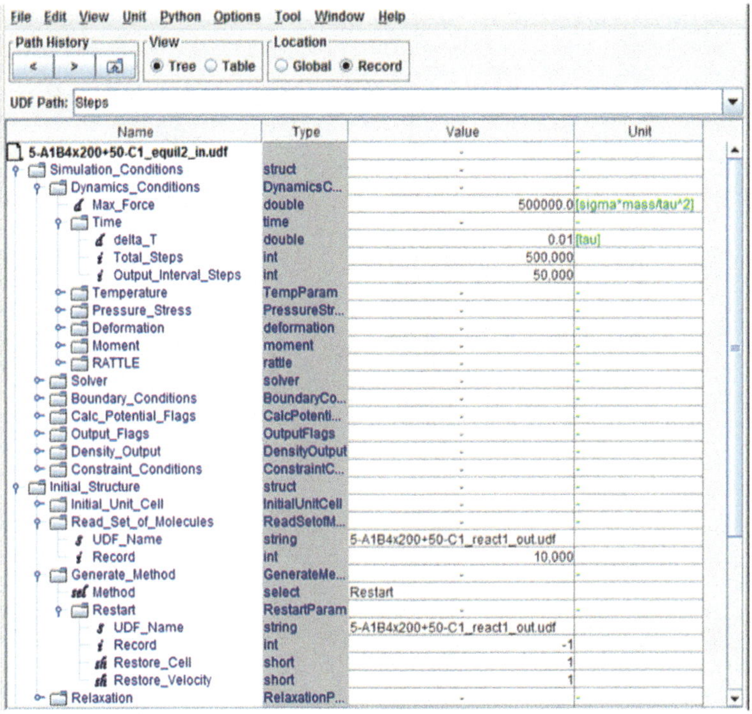

Fig. 15.8 Settings for the calculation of equilibration (2)

is read. Note that in the case of reading positions of particles, the final record is read in COGNAC, when −1 is input for *Generate_Method.Restart_Record*.

2. *Initial_Structure.Generate_Method*

   Settings are arranged so that the calculation restarts from XXX_react1_out.udf, which is the previous calculation result.

   When the input UDF file is created, we perform the calculation by executing the following command from the command line.

```
>cognac90 -I XXX_equil2_in.udf -O XXX_equil2_out.udf
> XXX_equil2_in.log
```

15.3 Elongational Physical Properties

15.3.1 Calculation of Elongational Deformation

In calculating elongational deformation, conditions such as uniaxial or biaxial elongation, the elongation rate, and Poisson's ratio are set in addition to the aforementioned settings of time and the creation of the initial structure. Here, the input file name is XXX_elong1_in.udf (Fig. 15.9).

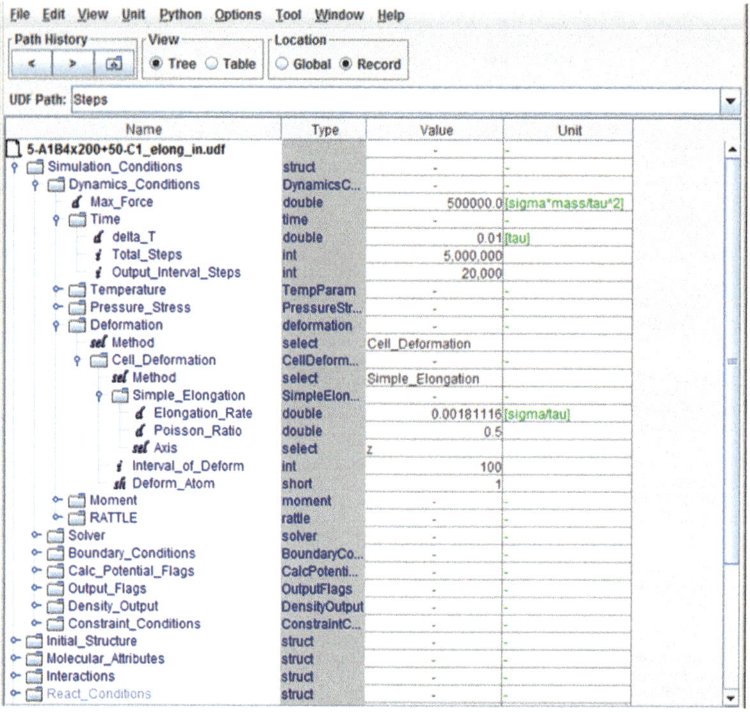

Fig. 15.9 Settings for the calculation of elongational deformation

In COGNAC's *Simple_Elongation*, deformation is input in terms of the deformation rate of a cell. In the sample model, the size of the cell is 18.1116; therefore, *Elongation_Rate* in the case of a deformation rate at which, for example, elongation of 100 % occurs in 10,000 τ becomes 0.0018116. In the case of the NVT calculation, Poisson's ratio is set to 0.5 for incompressible rubber. Moreover, *Axis* is set as "z" in the case of uniaxial elongation and "xy" in the case of biaxial elongation. When *Interval_of_Deform*, which is the interval of the time step for performing cell deformation, is set to 1, the calculation time is long because lists of neighboring particles are generated every step in which the cell size is deformed. *Deform_Atom* is a flag used to determine whether particle positions should be changed according to affine deformation during deformation of the cell. When cross-linking particles are used, polymer molecules do not greatly move during the first equilibration process; however, cross-linking particles greatly move and the *z* coordinate thereof sometimes becomes, for example, 5,000. At this time, if a cell having a length of 20 deforms by 0.01 % (0.002 σ), the cross-linking particle having a *z* coordinate of 5,000 moves by 0.5 σ from the periodic boundary, while a bonded polymer particle does not move. This distance is not negligible relative to the maximum length of the FENE-LJ bond potential of 1.5, and it is preferable to set *Deform_Atom* to 1.

In the case of an NPT calculation, *NPT_Parrinello_Rahman_Kremer_Grest* is used as a solver. When using this, *Fix_Angle* is set to 1 in addition to setting *Cell_mass* and *Friction*.

As mentioned above, deformation in COGNAC only occurs in the *z* direction or *xy* direction. Because much time is spent calculating the equilibration and the cross-linking reaction, in the case of a cubic cell, it may be more practical to allow deformation to occur in three directions and take the average value. Python scripts "udf_z2x.py" and "udf_z2y.py" were prepared to change the axis of coordinates. Using these, the coordinates of particles [x, y, z] can be changed to [−z, y, x] and [x, −z, y], respectively, by inputting ">python udf_z2x(y).py udf file name" from the command line.

When the input UDF file is created, we perform the calculation by executing the following command from the command line.

```
>cognac90 -I XXX_elong_in.udf -O
XXX_elong_out.udf>
XXX_elong_in.log
```

15.3.2 Observation of the Deformed State

To observe the deformed state, a file is opened and the file name in GOURMET is right-clicked to select VIEW_show... from the menu displayed. Once the menu is displayed, *type*, *bc*, and *color* are selected. Given that *Periodic_Bond* is set, *bc* is set as "atom".

Figure 15.10 shows an example of a snapshot taken during 300 % elongation. In this case, *type* is set as "ball-stick", *bc* is set as "atom", and *color* is set as "atom"; by increasing the radius with *Draw_Attributes.Atom_Type*[], only the C particles are displayed.

15.3.3 Calculation of the Stress–Strain Property

To draw a stress–strain curve, the file name is right-clicked as is done to select **PLOT_SS_curve**.... *P_0* is for pressure correction. In the present calculation,

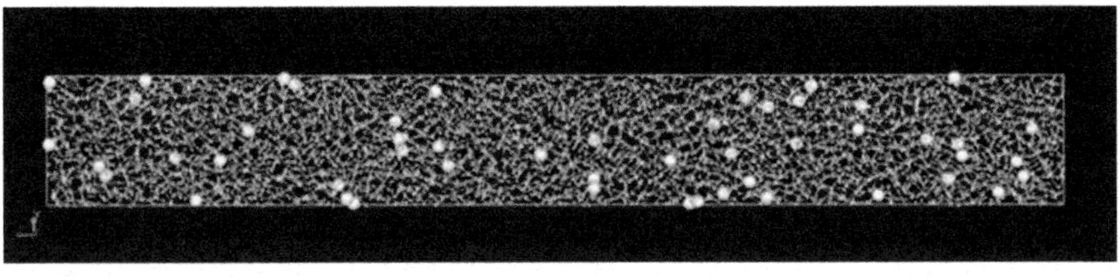

Fig. 15.10 Example of a snapshot taken during elongation

(xx + yy)/2 is selected, although such selection is not necessary when P is set as $P = 0$ in the NPT calculation. By this setting, $\sigma_{zz} - (\sigma_{xx} + \sigma_{yy})/2$ is plotted. When *Output_Fraction* is set to "no", the stress is displayed; when *Output_Fraction* is set to "yes", the nonbonding potential component of the stress and bonding potential component of the stress are displayed in addition to the stress. Figure 15.11 shows an example of the plot.

What is output in this case is the stress, not the industrial stress that is measured in usual experiments and used for stress–strain curves. To obtain the industrial stress, it is necessary to make a correction in terms of the change in cross-sectional area associated with the deformation and to process the data separately. Reading the data from XXX_elong_out.dat is one method to that end. Another method is to add the following two lines to the end of the Python script used in **PLOT_SS_curve** when it appears in the Python Scripting Window, to the execute Run, and finally to display the output in the Python Log Window so that it can be copied (Fig. 15.12):

```
For i in range(len(dataList[0]))
Print dataList[0][i],dataList[1][i]
```

The attached "read_records_of_files.py" can also be used.

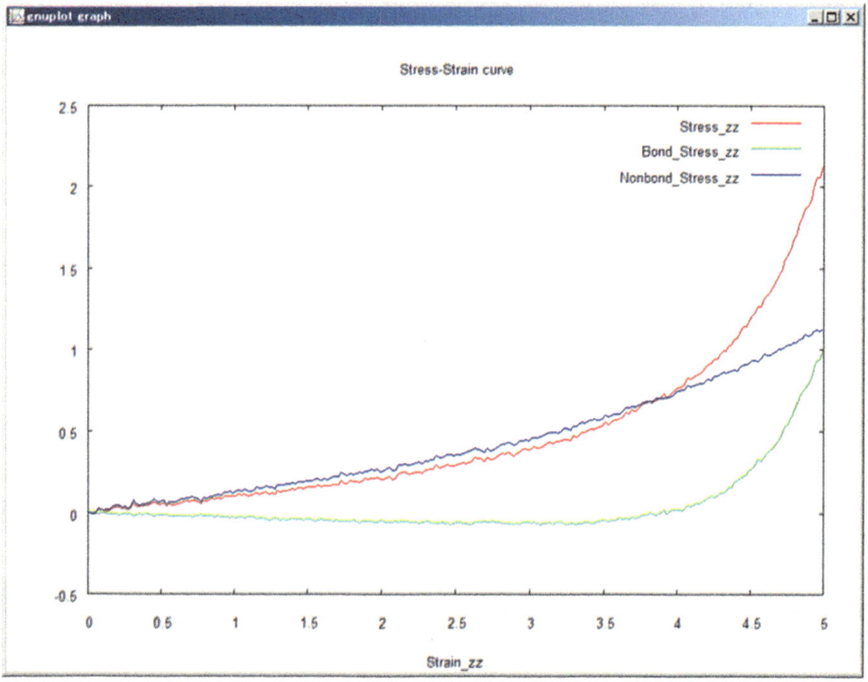

Fig. 15.11 Example of a plot for a stress–strain curve

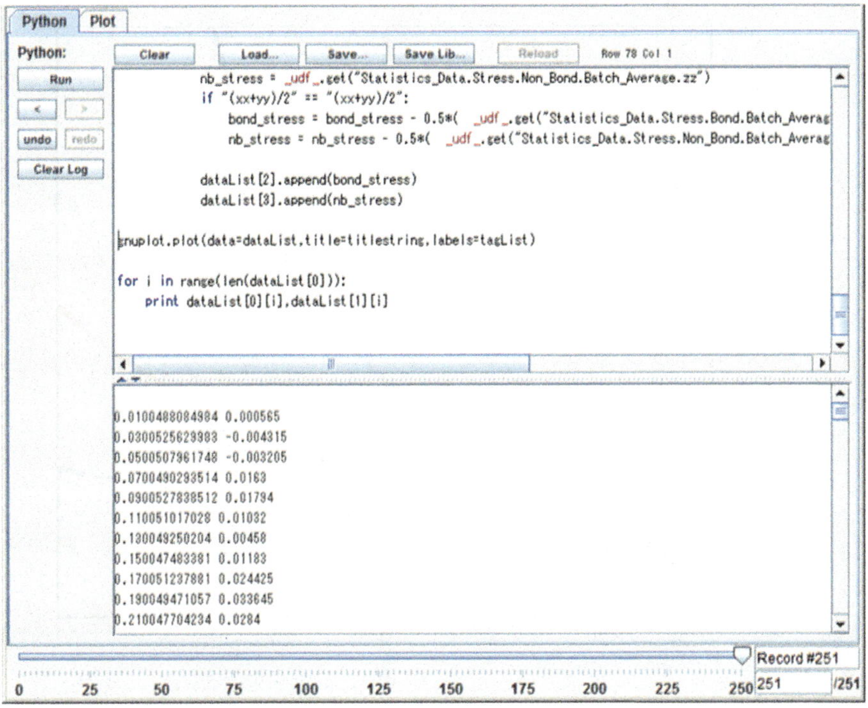

Fig. 15.12 Numerical data output of stress and strain

15.3.4 Example of Calculating the Elongational Physical Property (1): Method Using Cross-Linking Particles

The following shows stress–strain calculation results for a system that contains cross-linking particles with 20 molecules having a molecular weight of 2,000. In the present example, NVT deformation is conducted with a Poisson ratio of 0.5. Moreover, considering the calculation time, COGNAC is used to create the initial structure and to perform the calculation of the cross-linking reaction; however, with respect to other calculations, VSOP (a parallelizing solver created by JSOL Corporation) is used. A single molecule is constituted by repeating ABBBBBBBBB 200 times. Moreover, the time required for equilibration was set to 200,000 τ both before and after cross-linking.

Figure 15.13 shows stress–strain curves for 100, 200, and 400 cross-linking particles. The deformation rate in this case is $10 - 5\lambda/\tau$, such that 600 % elongation is achieved in 600,000 τ.

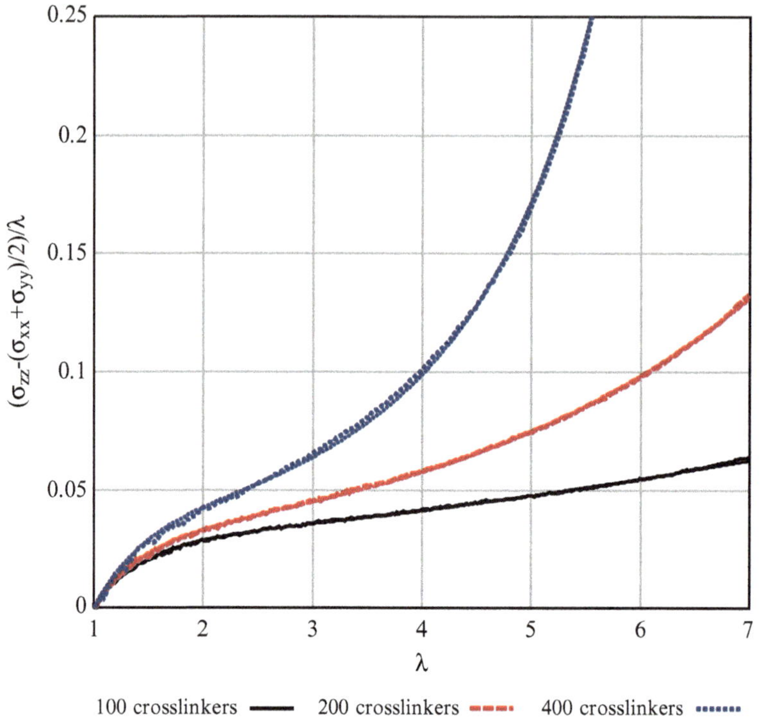

Fig. 15.13 Cross-link density and stress–strain curve. *Thick line*: coarse-grained MD; *thin line*: Edwards–Vilgis model

The results obtained by fitting with the Edwards–Vilgis slip-link model [1] as a theoretical model are also shown; good agreement is seen.

It has been confirmed that the model agrees well with experiments [2]. This model is one in which the entanglement effect of polymer chains is handled as a virtual link that moves on the polymer chains; the cross-link is handled as a link that does not move on the polymer chains. The model is also one that takes the finite extended effect into consideration. The model, as shown by Eq. 15.1, is represented by four parameters, namely, the cross-linked chain density *Nc*, the entanglement point density *Ns*, a coefficient η representing friction, and a coefficient α representing extension; the cross-link-derived component and entanglement-derived component are separately handled. The model's free energy is computed using the Eq. 15.1. The first term on the right side represents the cross-link-derived component, while the second term on the right side represents the entanglement-derived component. In addition, the stress (e.g., repulsion between beads) is included in the entanglement component-derived stress component obtained in this analysis:

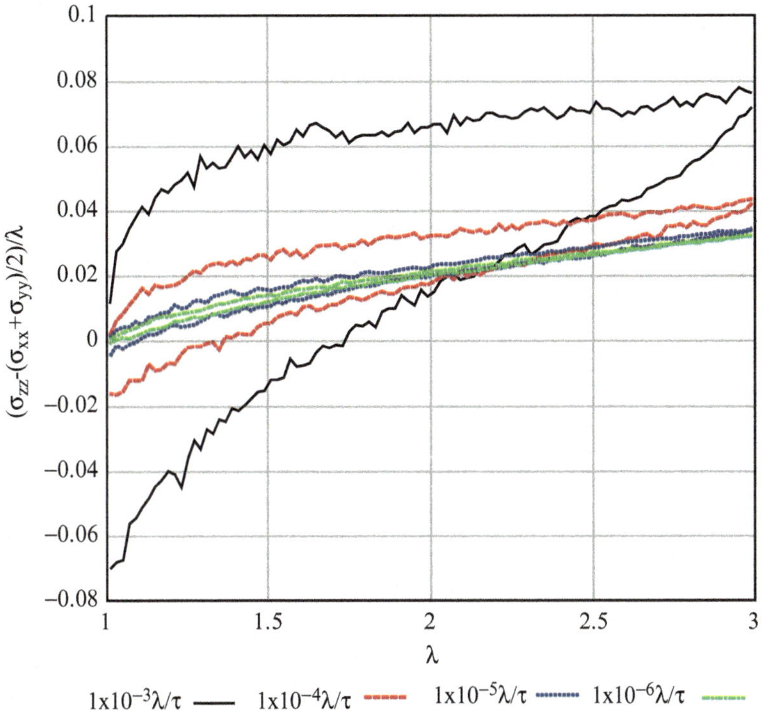

Fig. 15.14 Deformation rate and stress–strain curve

$$\frac{F}{k_B T} = \frac{1}{2} N_c \left\{ \frac{\sum \left(1-\alpha^2\right) \lambda_i^2}{1-\alpha^2 \sum \lambda_i^2} - \log\left(1-\alpha^2 \sum \lambda_i^2\right) \right\}$$
$$+ \frac{1}{2} N_s \left[\sum \left\{ \frac{\lambda_i^2 \left(1+\eta\right)\left(1-\alpha^2\right)}{\left(1+\eta\lambda_i^2\right)\left(1-\alpha^2\sum\lambda_i^2\right)} + \log\left(1+\eta\lambda_i^2\right) \right\} - \log\left(1-\alpha^2\sum\lambda_i^2\right) \right] \tag{15.1}$$

Figure 15.14 shows the effect of the deformation rate. The system with 200 cross-linking particles is elongated by 200 % and contracted by 200 % under four conditions at deformation rates of 10^{-3}, 10^{-4}, 10^{-5}, and 10^{-6} λ/τ. The hysteresis greatly differs depending on the deformation rate; when the deformation rate is fast, the hysteresis loss becomes extremely large. When evaluating the stress–strain relation of a cross-linked material, it is important to set an appropriate deformation rate to solve the problem. Deformation should be set so as to be slower than 10^{-5} λ/τ for the problem of usual rubber deformation.

Figure 15.15 shows the results of separating the stress into the cross-link-derived component and entanglement-derived component in the Edwards–Vilgis model to

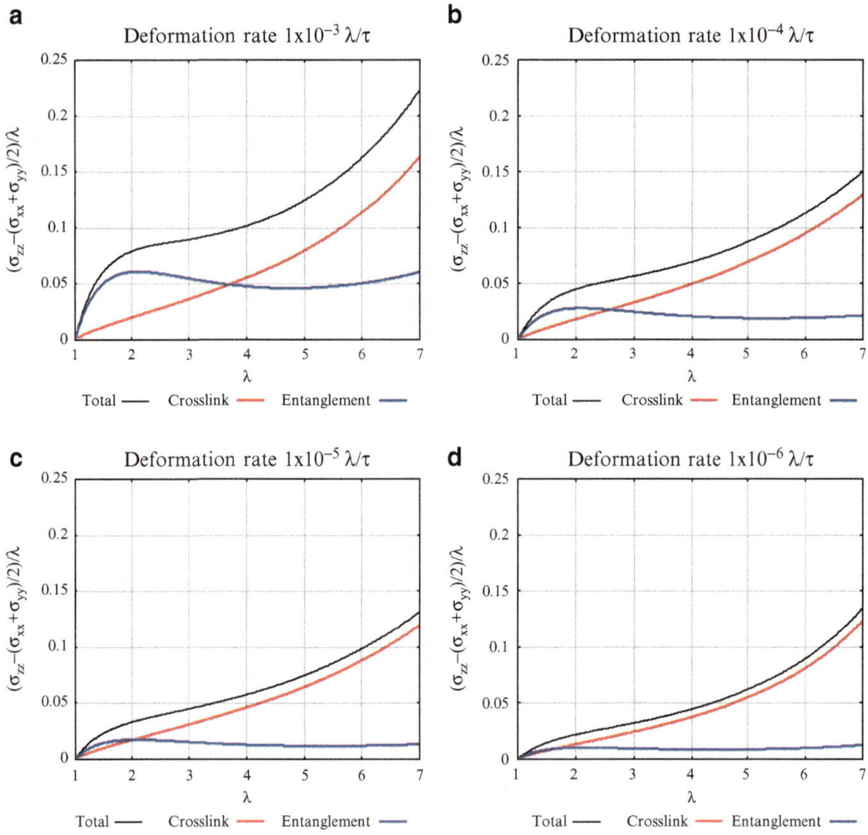

Fig. 15.15 Changes in the cross-link-derived (*dotted line*) and entanglement-derived (*broken line*) components of stress associated with changes in the deformation rate

understand the effect of the deformation rate. It is understood that the entanglement-derived component becomes extremely large during fast deformation, and the results indicate that deformation should preferably be slower from the standpoint of evaluating the effect of cross-linking.

15.3.5 Example of Calculating the Elongational Physical Property (2): Method for end Cross-Linking

The following shows an example of calculating end cross-linking in which the molecular weight between cross-linking points is set to be uniform. In the present example, NPT deformation is conducted at zero pressure for evaluating the stress in the non-elongational direction by conducting biaxial deformation and deformation

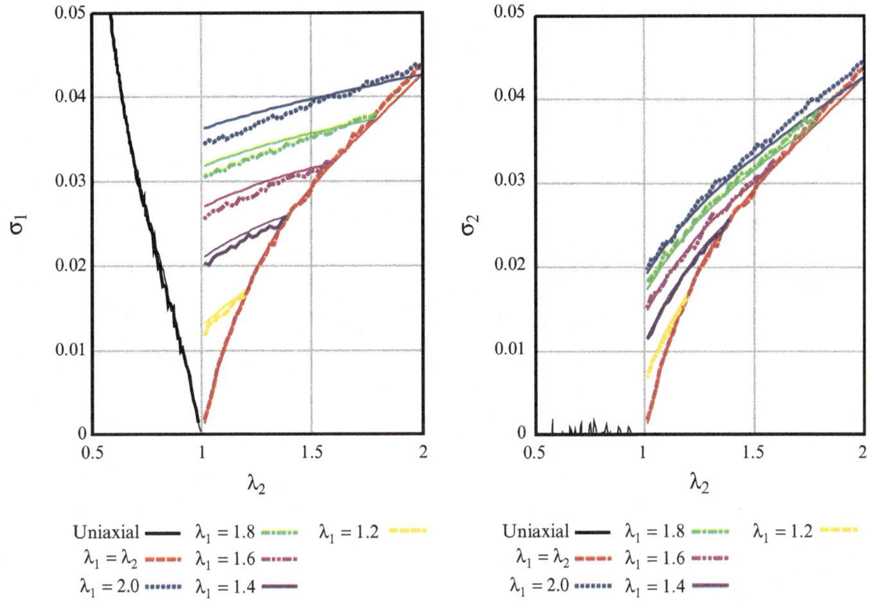

Fig. 15.16 Stress–strain in end cross-linking. *Thick lines*, calculation results; *thin lines*, Edwards–Vilgis model

with one axis fixed. The cutoff distance of the nonbonding interaction is set to 2.0 to achieve density of about 0.85 at zero pressure. Moreover, considering the calculation time, COGNAC is used to create the initial structure and to calculate the cross-linking reaction; however, LAMMPS [7] (an MD solver for large-scale parallel calculation) is used for the other calculations.

Deformation was conducted under seven conditions: (1) $\lambda_1 = 1$ to 4 (uniaxial elongation); (2) $\lambda_1 = \lambda_2 = 1$ to 2 (biaxial elongation); (3) $\lambda_1 = 1.2$, $\lambda_2 = 1.2$–1.0 (one axis-fixed, uniaxial deformation, the same applies hereinafter); (4) $\lambda_1 = 1.4$, $\lambda_2 = 1.4$–1.0; (5) $\lambda_1 = 1.6$, $\lambda_2 = 1.6$–1.0; (6) $\lambda_1 = 1.8$, $\lambda_2 = 1.8$–1.0; and (7) $\lambda_1 = 2.0$, $\lambda_2 = 2.0$–1.0. The calculation was performed according to the result of (2) under conditions (3) to (7). The deformation rate was set to 10^{-6} λ/τ.

Figure 15.16 shows the stress–strain relation for a system in which the molecular weight between cross-linking points is set to 50 (total number of particles: 40,400) together with the results obtained by fitting the stress strain relation with the Edwards–Vilgis model. It is confirmed that the stress–strain relation agrees with the Edwards–Vilgis model. Table 15.1 gives the parameter values obtained by fitting a system having a molecular weight between cross-linking points of 50 and a system having a molecular weight between cross-linking points of 100 (total number of particles: 40,200) with the Edwards–Vilgis model. Results were obtained in which the *Nc* equivalent to the cross-link density corresponded well to the cross-link density in the calculation model.

Table 15.1 Fitting parameters of the Edwards–Vilgis model

	η	α	Ns	Nc(E-V)	Nc(CGMD)
Mc50	0.2	0.120	0.0117	0.0111	0.0160
Mc100	0.2	0.113	0.0114	0.0051	0.0077

Only the calculation results for the stress–strain relation are shown above. Although not shown here, it is useful to obtain the changes associated with the elongation of molecular chains, such as changes in bond orientation or entanglement using methods [8, 9] of the primitive path. This will facilitate deeper understanding of the phenomenon occurring during the deformation of cross-linked materials.

15.4 Download Instruction

Sample files are available from the OCTA website. See the instruction on http://octa.jp/download_bookdata.html. Password for the sample(zip file) of this chapter is "GawXCHSV".

References

1. S. Edwards, T. Vilgis, Polymer **27**, 483 (1986)
2. K. Urayama, T. Kawamura, S. Kohjiya, Macromolecules **34**, 8261 (2001)
3. K. Kremer, G. Grest, J. Chem. Phys. **92**, 5057 (1990)
4. G. Grest, K. Kremer, Macromolecules **23**, 4994 (1990)
5. T. Hayakawa, K. Hashimoto, H. Morita, M. Doi, Investigation of stress–strain relationship for polymer networks with crosslink types by molecular dynamics simulation (OCTA), Paper presented at the international rubber conference 2005, Pacifico Yokohama, Yokohama, 24–28 October 2005
6. E. Duering, K. Kremer, G. Grest, J. Chem. Phys. **101**, 8169 (1994)
7. S. Plimpton, J. Comp. Phys. **117**, 1 (1995)
8. R. Everaers, S. Sukumaran, G. Grest, C. Svaneborg, A. Sivasubaramanian, K. Kremer, Science **303**, 823 (2004)
9. S. Sukumaran, G. Grest, K. Kremer, R. Everaers, J. Polym. Sci. Pol. Phys. **43**, 917 (2005)

Chapter 16
Thermoplastic Elastomers

Takeshi Aoyagi

16.1 Introduction

A thermoplastic elastomer is a block copolymer of glassy blocks and rubberlike blocks. Block copolymers show plasticity at temperatures above the glass transition Tg of the glassy component and elasticity when cooled below Tg. Typical examples of thermoplastic elastomers include styrene–butadiene–styrene and styrene–isoprene–styrene triblock copolymers. As shown in Fig. 16.1, triblock copolymers possess a microphase-separated structure. The styrene block domains that are in a glassy state at room temperature act as physical crosslinks, allowing the material to exhibit rubberlike elasticity.

When a thermoplastic elastomer is elongated, the morphology and fraction of bridge chains, which bridge different glassy domains, of the soft component both greatly influence the physical properties of the material because rubber elasticity is exhibited by the bridge chains, as shown in Fig. 16.1.

Molecular dynamics simulation is a useful tool to study such elastic behavior during elongation. However, considering the time and length scale required, it is necessary to use a coarse-grained molecular model, such as a bead–spring model that has a high degree of coarse graining. To understand the influence of microphase-separated structure on the physical properties of a material, it is of course necessary to deal with the length scale of the microphase-separated structure. Because styrene-based elastomers used as industrial products typically have a domain spacing of approximately 20 nm, it is necessary to handle a length scale of several tens of nanometers, so a million or more atoms are needed to generate a full atomistic model. Furthermore, it is necessary to elongate the system so that the soft segments

T. Aoyagi (✉)
Research Center for Computational Design of Advanced Functional Materials,
National Institute of Advanced Industrial Science and Technology, Ibaraki, Japan
e-mail: aoyagi.t@aist.go.jp

Japan Association for Chemical Innovation, *Computer Simulation of Polymeric Materials*, DOI 10.1007/978-981-10-0815-3_16

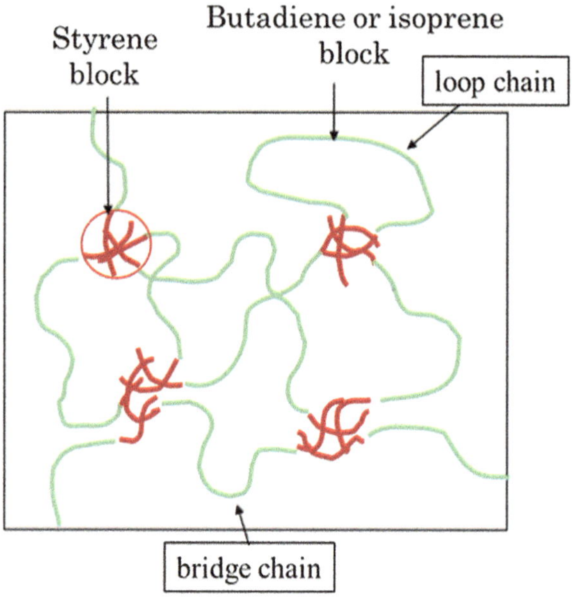

Fig. 16.1 Example of the microphase-separated structure of a styrene-based thermoplastic elastomer

can fully relax to reproduce the rubberlike elasticity. If the deformation rate is too high, even the soft segments behave like a glassy state. Therefore, to study the rubberlike elastic behavior of thermoplastic elastomers, it is a realistic choice to analyze qualitative behavior using the bead–spring model.

In addition, it is also necessary to consider a method to generate the initial structure of microphase separation. Even with the bead–spring model, it is extremely difficult to cover a timescale with a molecular dynamics simulation during which the microphase-separated structure is formed from a uniform state. Therefore, in this example, the coordinates of molecular chains to be used for coarse-grained molecular dynamics are generated from the distribution of segment volume fraction obtained from a self-consistent field (SCF) calculation by SUSHI using a density-biased Monte Carlo (DBMC) method proposed by Aoyagi et al. [1].

In this example, a simulation procedure based largely on the article by Aoyagi et al. [2] is introduced. Therefore, it is recommended to refer to [2] when reading this example. A few changes were made to efficiently compute this example using the present version of OCTA and more advanced computer hardware than in [2].

16.2 Initial Structure Preparation

The first step, involving generation of a microphase-separated structure in a body-centered cubic (BCC) lattice of an ABA triblock copolymer by SUSHI and generation of a bead–spring model by COGNAC using the BCC microphase-separated structure from SUSHI, is described in this section.

16.2.1 Selection of a Calculation Model

The volume fraction of each type of segment is important when generating a molecular model of a block copolymer in SUSHI. Because the miscibility is determined by the parameter χN, where N is the total number of segments in a chain and χ corresponds to the miscibility, arbitrary N can be taken if the parameter χ is scaled at the same time.

The question is how to determine N. It is realistic to take both the desired resolution of molecular weight to be calculated and the calculation time into consideration when determining N. For example, when the effect of M of 90,000; 100,000; and 110,000 must be predicted, N may be set so that $N = 9$, 10, and 11, respectively, by assuming that $M = 10{,}000$ is one segment. However, if the difference between $M = 98{,}000$ and 100,000 must be predicted, it is necessary that the calculation be performed by setting N as $N = 49$ and 50, respectively, so $M = 2000$ is one segment. In the case of a block copolymer, the resolution relates not only to the total chain length but also to the block fraction. For example, when a block copolymer of A3B44A3 with $N = 50$ is selected, as in [2], the volume fraction of A can be changed by a pitch of 4 % considering the symmetry of the ABA triblock. If smaller differences in volume fractions are to be investigated, it is necessary to make N larger.

In this case, the objective is to analyze the behavior of the polymer when a hard domain is spherical; therefore, it is necessary to select the volume fraction and χ of the spherical phase in the phase diagram. Because it is understood from the phase diagram reported by Matsen et al. [3] that the BCC phase is exhibited when $\chi > 1.6$ with $N = 50$ and volume fraction of segment A $\phi_A = 0.12$, χ was set as $\chi = 2.4$ in this case (note that Matsen et al. [3] set N of a symmetrical triblock to half the actual chain length to compare it with the diblock). The objective here is to analyze the behavior of the ideal BCC structure; therefore, it is not necessarily essential to use a styrene fraction in an actual thermoplastic elastomer, or the unmodified χ of actual styrene–butadiene or styrene–isoprene.

Consequently, here the calculation is performed using a molecular model of A3B44A3 ($N = 50$ and $\phi_A = 0.12$) and $\chi_{AB} = 2.4$ in which the BCC phase is expected to be stable.

16.2.2 Generation of BCC Structure

Next, the BCC structure of A3B44A3 is generated using SUSHI. First, the unit cell size must be determined as an input condition of SUSHI. In this case, because the objective is to generate a coarse-grained molecular dynamics model and analyze elastic behavior during elongation based on the BCC phase structure obtained by SUSHI, it is desirable that the unit cell size be as large as possible. Here, a cell size of 2 × 2 × 2 of the BCC unit lattice is used in accordance with [2]. Then, the size of this unit lattice needs to be determined. In the current version of SUSHI, the cell size can be optimized using a technique called system size optimization (SSO), which is described below. However, it is necessary to estimate an initial value of the cell size that is close to that of the stable structure.

In general, the domain spacing of a microphase-separated structure of a block copolymer has a value somewhat larger than the average distance between the ends of the molecular chain <*R*>. Because the molecular chain used in a SCF is an ideal chain model, $\sqrt{\langle R^2 \rangle} = \sqrt{(N-1)}$, and $\sqrt{\langle R^2 \rangle} = 7.0$ in the case of $N = 50$.

The BCC unit lattice becomes $l = 7.0 \times 2/\sqrt{3} \cong 8.1$. Accordingly, it is considered that a value of about 17 may be assigned as the initial value of the unit cell to be used for calculation.

In this manner, the initial value of the unit cell is determined. However, when a simulation is started using a uniform state as the initial structure, the structure will converge to a locally stable structure with low symmetry, and the BCC phase, which is the most stable structure and highly symmetrical, is not obtained in most cases. To avoid such results, a bias should be applied to the chemical potential field by domain specification at the start of the calculation.

Hereinafter, the aforementioned points are confirmed with reference to the actual UDF file. In addition, a navigation-type V2 format (refer to SUSHI manual appendix A) is used here as the SUSHI input UDF file. First, as shown in Fig. 16.2, system

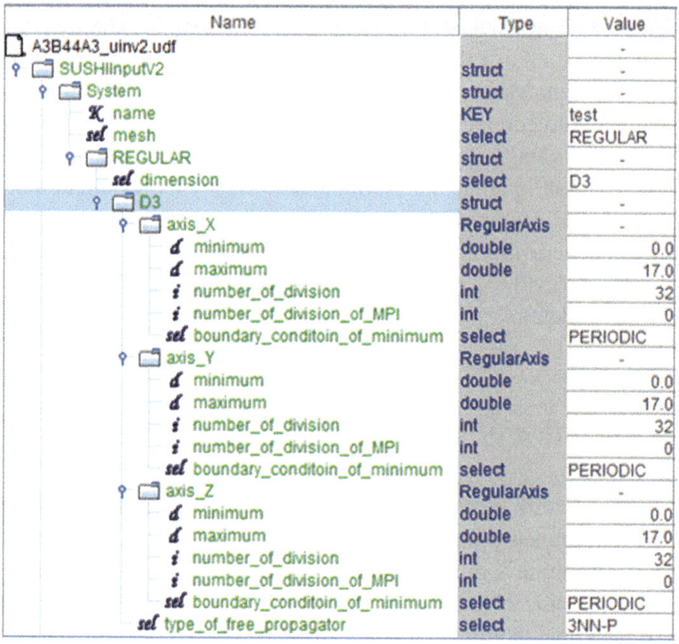

Fig. 16.2 Specification of system size

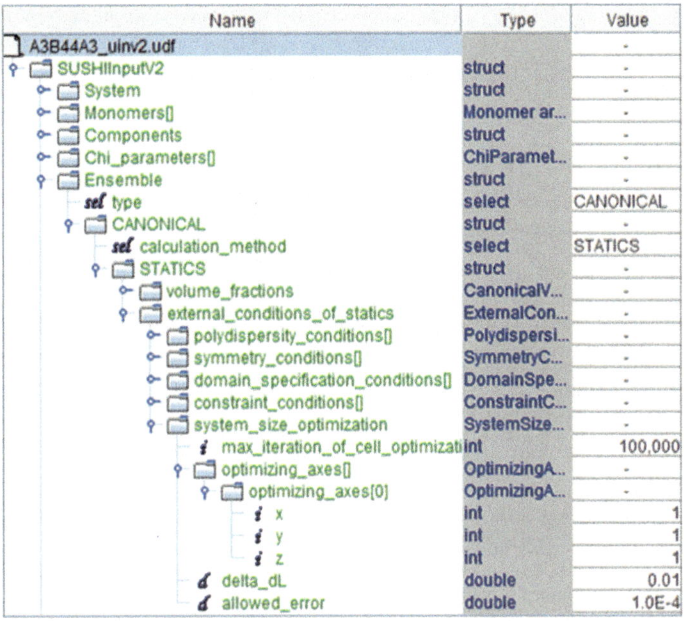

Fig. 16.3 Specifications of SSO

size is set as minimum = 0.0 and maximum = 17.0 for each of the *x*, *y*, and *z* axes. The number of divisions of the mesh is set to 32 so that the mesh width of each axis *dx*, *dy*, and *dz* is about 0.5.

Next, Fig. 16.3 shows an example of SSO settings. When a numerical value larger than zero is set for *max_iteration_of_cell_optimization*, SSO is executed. Moreover, when *x*, *y*, and *z* are set as $x = y = z = 1$ in *optimization_axes*[*0*], each of the *x*, *y*, and *z* axes is changed in the same proportion, thereby maintaining the cubic structure. Here, note the value of *delta_dL*. A very small value is set for *delta_dL* by default, and when SSO is executed without changing this value, sometimes there is almost no change in cell size when the cell size converges. The value *delta_dL* corresponds to the magnitude of change used to vary the cell size in SSO. Therefore, when *delta_dL* is too small, sometimes the cell size is regarded as converged because the difference in energy between before and after changing the cell size is sufficiently small. Thus, when a calculation is completed with little change in cell size, it is recommended to set *delta_dL* larger and retry the calculation. Here, the calculation was performed by setting *delta_dL* as a large value of 0.01.

Finally, bias is applied to the initial volume fraction distribution using domain specification. For a three-dimensional simulation, specifying the domain where the bias is applied is somewhat complicated; however, the domain is basically specified for each axis, and the bias is applied in the region specified in all axes. Figure 16.4 provides an example of an actual UDF file, while Fig. 16.5 shows domains where the bias is applied according to specifications.

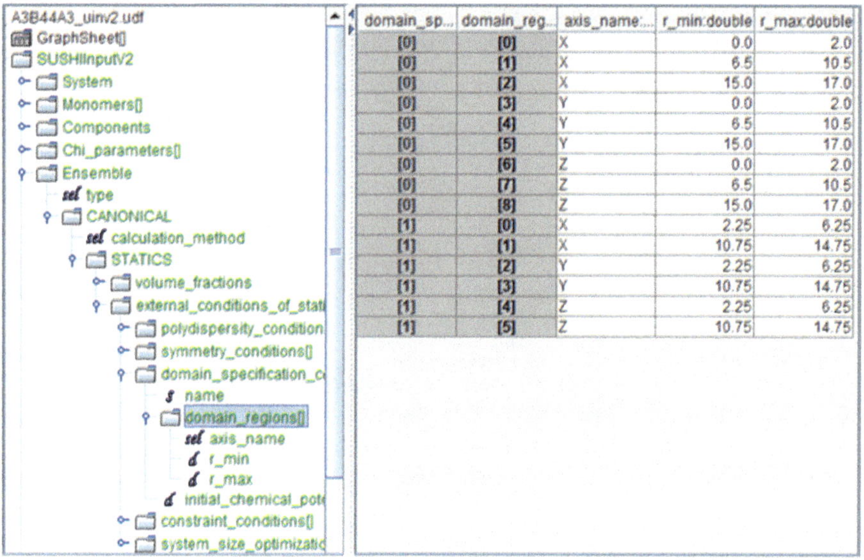

Fig. 16.4 Setup for domain specification

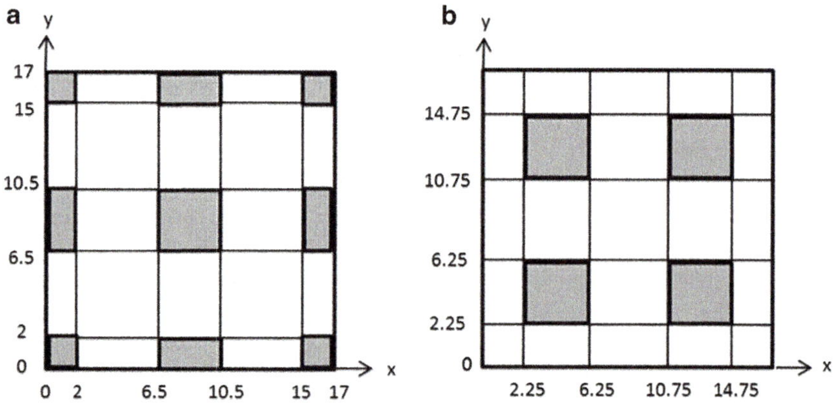

Fig. 16.5 Domain specification: (**a**) domains specified by *domain_specification_conditions*[*0*], ranges from 0 to 2, from 6.5 to 10.5, and from 15 to 17 are also specified for the z axis; (**b**) domains specified by *domain_specification_conditions*[1], ranges from 2.25 to 6.25 and from 10.75 to 14.75 are also specified for the z axis

When a calculation is performed using the settings described above, a BCC structure like that shown in Fig. 16.6 is obtained. The calculation time will be approximately 10–15 min on a single core of the CPU of a middle-range PC.

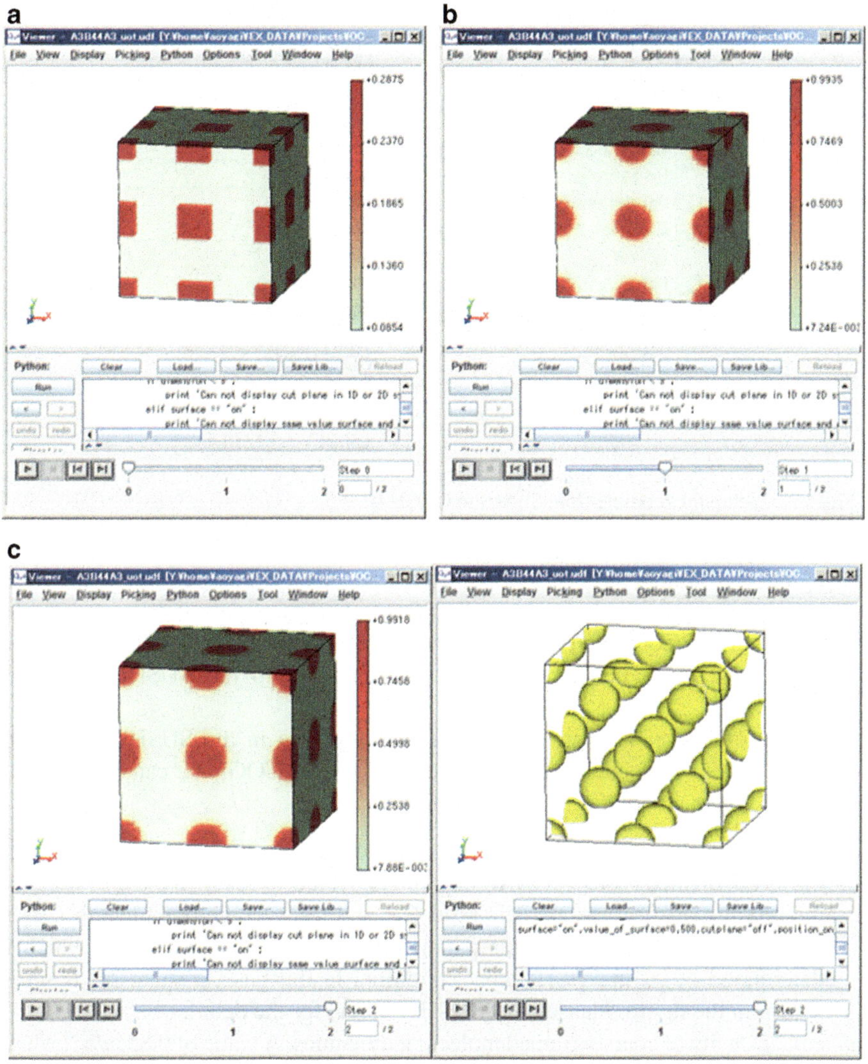

Fig. 16.6 Results for A3B44A3 BCC structure: (**a**) initial structure (note the difference in scales of volume fractions), (**b**) converged structure before SSO, and (**c**) converged structure after SSO

16.2.3 Setup for Zooming

Next, a setup for zooming from SUSHI to COGNAC using the DBMC method is described. Because the current version of SUSHI has a function to generate the output of *segment_volume_fraction* necessary for the DBMC method as well as create COGNAC input files corresponding to such an output, the setup for zooming

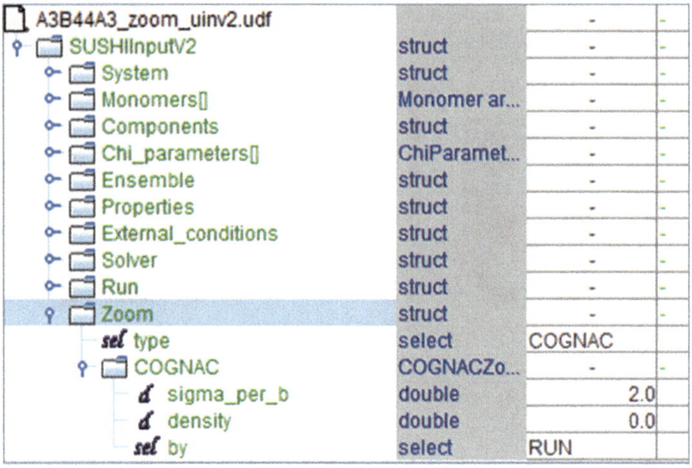

Fig. 16.7 Setup for zooming from SUSHI to COGNAC

using the new functions is introduced here. Specifically, SUSHI is launched using the results of the BCC phase obtained in Sect. 16.2.2 as restart data, and a SUSHI output UDF file containing *segment_volume_fraction* and a COGNAC input file with data on the molecular topology corresponding to the SUSHI output are generated.

Figure 16.7 displays part of the setup for zooming in the SUSHI input UDF file. COGNAC is selected as the type (at present, only COGNAC can be selected). COGNAC parameters are configured as follows.

sigma_per_b: This is a setting for the ratio of the segment length in SUSHI to the bead diameter in COGNAC. Minimum value is given by $C_\infty l$, which is the product of the characteristic ratio C_∞ and the average bond length l, as is described in [2]. However, this parameter must take discrete value, which is determined by *ds* in SUSHI input condition. In this case, a value 2.0 is set as shown in the figure. Please refer SUSHI manual for the detail.

density: Density of coarse-grained molecular dynamics. A value of 0.85 m/σ^3, which is usually used in the Kremer–Grest model, is set as the default value. Therefore, this value may be used without modification.

by: "RUN" is selected to calculate *segment_volume_fraction* after performing a SCF calculation once from the restart data.

Another point that should be noted regarding restart settings is that cell size is changed by setting SSO in the calculation in Sect. 16.2.2. Thus, it is necessary to read the cell size of the final structure at the time of restart, and therefore *SUSHIInputV2.Run.type* must be set to "RESTART_READMESH".

After configuring the settings in this manner, SUSHI is restarted by giving the following argument (although the file name is arbitrary, the file name of the example is used here).

```
> SUSHI -IA3B44A3_zoom_uinv2.udf -RA3B44A3_uot.udf
-OA3B44A3_zoom_uot.udf    -Zrelax_in.udf         >
A3B44A3_zoom_uot.log
```

Here, the file designated by -Z is a COGNAC template file that is found in "ENGINES\SUSHI10.5\Susi\def_udf" in the OCTA8 folder. This file must be copied into the execution directory.

Once the calculation is complete, files such as SUSHI archive UDF file "A3B44A3_zoom_uinv2_uar.udf", which contains *segment_volume_fraction*, and COGNAC input UDF "relax_A3B44A3_zoom_uinv2_in.udf" are created.

16.2.4 Setup for Simulation Conditions of COGNAC

In the COGNAC input UDF file created by the SUSHI zooming function, the initial settings for various conditions contained in the "relax_in.udf" are included. These conditions are generally set in accordance with the conditions in [2]. However, there are several points to note, as follows:

Pair Interaction The parameters of nonbonding interactions are important when analyzing the elastic behavior of thermoplastic elastomers. The model here consists of two kinds of particles, A and B, and A–A, B–B, and A–B all interact via Lennard–Jones potentials. Although the interaction between particles is determined by Lennard–Jones parameters in the original formulation, length parameter σ and energy parameter ε are all set to 1.0 in this model. Instead, differences of interaction are represented by changing cutoff distances. More specifically, if A particles are intended to have a high Tg and be in a glassy state under the elongation conditions, then a long cutoff distance of 2.5σ is set for the particles so that they interact via a strong attractive force, thereby representing the physical properties of particles with high Tg. Conversely, a cutoff distance of $2^{1/6}\sigma$ is set for B particles so that they interact via only a repulsive force, thereby representing their physical properties as elastic. Furthermore, the cutoff distance for A–B was adjusted to maintain the morphology created by SUSHI. Try-and-error-based optimization gave a value of 1.5σ.

Setting the cutoff distances as described above to reproduce the characteristics of each kind of particle helps to minimize computational time compared with controlling ε with the same long cutoff distance. However, the difference between the results obtained by controlling the cutoff distance and ε has not yet been fully verified and should be studied in the future.

Interaction Site Range As described in Sect. 5.2.4 of the COGNAC manual, the optimal value of range for each interaction site depends on the cutoff distance. When the cutoff distances are different for each particle pair, as in the calculation model used here, it is efficient to set a different value for the range. In this case, because the cutoff distance of A–A is set to 2.5σ, the range of A is set to 1.7σ, including a somewhat larger buffer. Meanwhile, because the cutoff distance of B–B is $2^{1/6}\sigma$, the range of B is set to 1.0σ. In this case, the sum of the ranges in A–B is 2.7σ, which is acceptable because the range is sufficiently long relative to the cutoff distance of 1.5σ.

The current version of COGNAC is capable of parallel computing, and the optimized value of range depends on the degree of parallelization. However, the basic idea is that, as described above, it is necessary to set the range so as to satisfy the necessary conditions between the cutoff distance and range for every pair of A–A, B–B, and A–B.

Temperature As described under the settings for pair interaction, a simulation must be conducted at a temperature T lower than Tg for A particles. Tg of this bead–spring model is known to be 0.45–0.5 ε/k_B when the cutoff distance is set to 2.5σ. In this case, T is set as $T = 0.4\ \varepsilon/k_B$.

Degree of Polymerization and Number of Segments In [2], the number of particles N_{MD} of the molecular chain in COGNAC is set as $N_{MD} = C_\infty N_{SCF}$ from the number of segments in SUSHI, N_{SCF}, and C_∞ of the bead–spring model. However, when *Set_of_Molecules* in COGNAC is created by SUSHI zooming, the number of particles in a chain becomes a value close to the designated scale among the numbers that are two, four, five, eight, and ten times the number of segments in SUSHI because of the restriction of the pitch width in the path integral of the chain. Accordingly, because N_{MD} is twice as large as N_{SCF} here, the structure of A6B88A6 is exported to *Set_of_Molecules*.

DBMC Settings The SUSHI output UDF file name and corresponding molecular species and indices of particles are specified by *Node_Density_Bias* in *Initial_Structure.Generate_Method.Random*. In accordance with the aforementioned scale, the positions of particles that have indices 0–99 in an A6B88A6 molecule are generated using the DBMC method. It is noted that the output data is read from the archive UDF file (in this case, "A3B44A3_zoom_uinv2_uar.udf") that writes out the final result, not the output UDF file specified at the time of launching SUSHI.

Time Step In "relax_in.udf", which is used in the SUSHI zooming function and is a COGNAC input template, *delta_T* and *Total_Steps* are set as *delta_T* = 0.012τ and *Total_Steps* = 10,000; therefore, the simulation is conducted with a total time t of 120τ. Here, *Total_Steps* is set to 1,000,000, while t is set as $t = 12{,}000\tau$ to allow structural relaxation after generating the initial structure. The total time required for relaxation is estimated from the autocorrelation function of the normal coordinate and mean-square displacement (MSD) of the center of gravity of chains in the case of homopolymer melts. Specifically, the time required for the autocorrelation function to decay to 1/e or the time required to become MSD = $2\langle Rg^2 \rangle$ represents relaxation time.

However, the thermoplastic elastomer being studied here has physical cross-linking, so relaxation of the entire molecular chain generally does not occur. Accordingly, the relaxation time for soft segment B88 may be used as a rough estimate of the whole elastomer. Strictly speaking, it is desirable that the necessary simulation time be determined after calculating the relaxation time by simulating a homopolymer of B88 at $T = 0.4\varepsilon/k_B$. However, here the relaxation time is set to 12,000τ based on [2].

Output_Interval_Steps is also changed in accordance with *Total_Step*. In the sample, because *Output_Interval_Steps* is set to 10,000, 100 trajectories will be output.

16.2.5 COGNAC Execution

COGNAC is executed with the input UDF file created in Sect. 16.2.3 and 16.2.4 to generate the relaxed structure with a BCC structure using the bead–spring model. As a SUSHI output UDF file for DBMC, "A3B44A3_zoom_uinv2_uar.udf" is specified in the COGNAC input UDF file, "relax_A3B44A3_zoom_uinv2_in.udf". Therefore, as shown below, both files are placed in the same folder to execute COGNAC. Although the output file name is arbitrary, the file name here is the same as that in the sample.

```
> cognac90 -I relax_A3B44A3_zoom_uinv2_in.udf -O
  A6B88A6_eq.bdf > A6B88A6_eq.log
```

In addition, in the output UDF file, the extension is specified as "bdf", not "udf". When using this extension, a binary UDF file is automatically generated to decrease the file size and improve the access speed in GOURMET. Therefore, it is recommended to use a "bdf" file when generating a large UDF file. Moreover, because it takes a long time to perform each calculation, it is recommended to perform parallel computation if possible using the -n option.

Figure 16.8 shows the initial structure and that after relaxation generated by the DBMC method. When the boundary condition is changed to atom, particles will be displayed in the cell, which makes the morphology easy to understand.

However, when all particles are displayed, it becomes difficult to see the distribution of hard blocks. To clarify the domains, it is possible to display only the hard particles (A particles). To do so, hierarchical structures of *Set_of_Molecules* are traced as *molecule*[*0*] → *atom*[*0*]; then, in the window that appears by right-clicking on *atom*[*0*], *show_same_type* should be selected. This means that all particles of the same type as designated by *molecule*[*0*].*atom*[*0*] (in this case, all A particles) are displayed as shown in Fig. 16.9. These results confirm that the BCC structure calculated by SUSHI is reproduced.

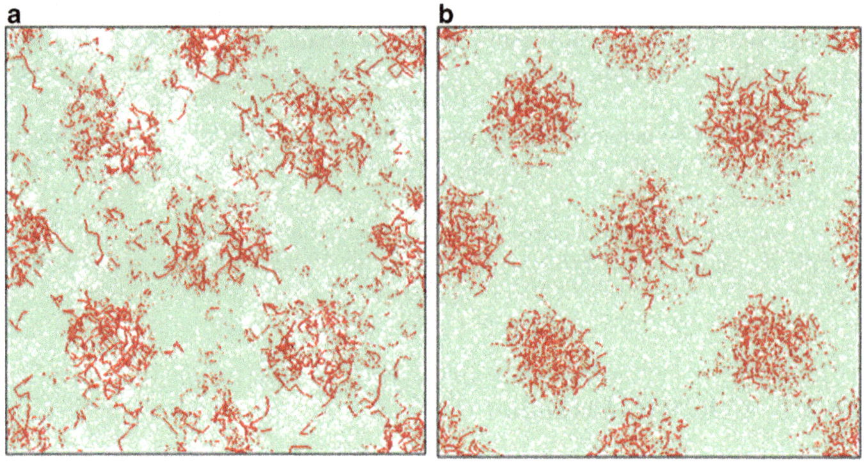

Fig. 16.8 BCC structure obtained by coarse-grained MD: (**a**) initial structure and (**b**) relaxed structure

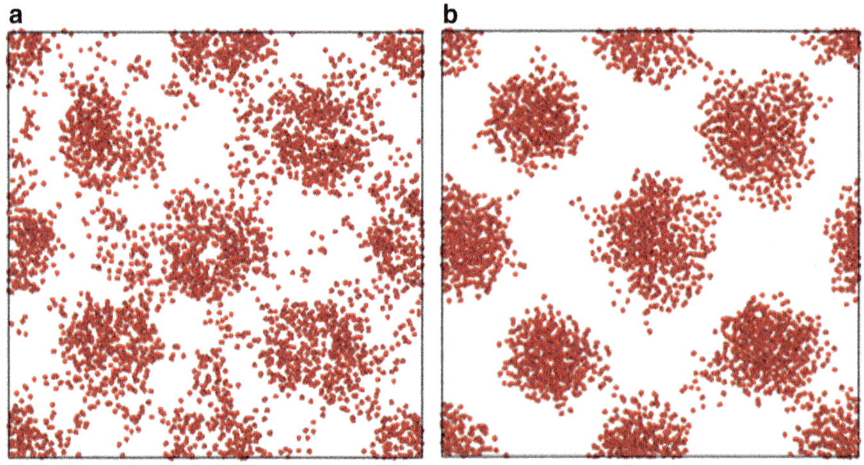

Fig. 16.9 BCC structure obtained by coarse-grained MD (only hard segments are shown): (**a**) initial structure and (**b**) relaxed structure

16.3 Elongational Properties

Next, the procedure to study the elastic behavior of the obtained microphase-separated structure of the ABA triblock copolymer is described.

16.3.1 Setup for Elongational Deformation

Elongational deformation is conducted here by cell deformation, which is a function of COGNAC. With cell deformation, elastic behavior is analyzed from the change in stress when the unit cell is forced to deform during the simulation. As the input UDF file for elongation, "relax_A3B44A3_zoom_uinv2_in.udf", which contains the relaxed structure, is used. In accordance with the sample, "relax_A3B44A3_zoom_uinv2_in.udf" is copied as "A6B88A6_deform_in.udf". The file will be edited to set the conditions for elongation. Here, the following three parts should be changed: (1) the initial structure (*Initial_Structure*), (2) conditions of deformation (*Simulation_Conditions.Dynamics_Conditions.Deformation*), and (3) the total time (*Simulation_Conditions.Dynamics_Conditions.Time*).

First, the relaxed structure, which was generated previously, should be used as restart data. Therefore, *Initial_Structure.Generate_Method.Method* is changed to "Restart". Furthermore, because relaxation has already been performed, the flag for *Intial_Structure.Relaxation.Relaxation* is set to 0 so that further relaxation is not performed.

Next, with respect to settings for deformation, *Cell_Deformation* is selected in *Simulation_Conditions.Dynamics_Conditions.Deformation.Method* and then "Simple_Elongation" is selected in *Method* of *Cell_Deformation*. The deformation rate (*Elongation_Rate*) and Poisson's ratio (*Poisson_Ratio*) are set as conditions of *Simple_Elongation*. When the deformation rate is too fast, the B block, which should exhibit rubberlike elasticity as a soft segment, cannot be relaxed, and it shows glass-like behavior. Therefore, it is desirable that deformation should be conducted over a timescale longer than the relaxation time of the B block. Specifically, the deformation rate may be set so that the reciprocal of the strain rate is equal to or longer than the relaxation time. In this case, the initial strain rate is set to $1/12{,}000\tau^{-1}$ based on the estimated relaxation time of $12{,}000\tau$. In fact, because the input of *Elongation_Rate* is the amount of deformation (in units of σ/τ) per unit time τ, the input value is calculated by the following equation using the strain rate and unit cell size (34.1323σ):

$$\text{Elongation Rate} = 34.1323 \times {}^{1}\!/_{12,000} = 0.002844358\ \left[\sigma\tau^{-1}\right] \tag{16.1}$$

Poisson's ratio is a physical property obtained as an output. There is a method to obtain Poisson's ratio in which pressure is controlled only in the axis perpendicular to elongation. However, this calculation was performed by fixing Poisson's ratio to a constant value of 0.5.

The default values can be used for the other elongation conditions. For details, refer to the COGNAC manual.

The total time step of the simulation is set from the total strain. Because the total strain is 100 % at $12{,}000\tau$ at the defined deformation speed, the total time step should be set to $12{,}000\tau$, in other words, 1×10^6 steps to study the deformation until

100 % strain. At the same time, the output interval is set to 20,000 in the sample. Therefore, 50 trajectories will be stored during the process of elongation to a strain of 100 %.

16.3.2 Preparation of the Initial Structure

The BCC structure generated in this example is highly symmetrical, so the x, y, and z axes are equivalent. Therefore, the average value of stress obtained from the elongation to each of the x, y, and z axes can be used to decrease statistical error. However, the *Simple_Elongation* function of COGNAC only supports elongation in the z-axis direction. Accordingly, the axes of coordinates in the initial structure need to be converted to perform elongation in these three directions.

First, the structure of the last record in the relaxation is exported to another UDF file. When the UDF file for a relaxed structure is read by GOURMET and then moved to the last record and **EXPORT_data** is selected from the Action tool, the window shown in Fig. 16.10 appears.

Here, the command **EXPORT_data** is used to export only the selected record to another UDF file. With the default *file_type* and *file_name*, pressing **OK** will create a file in which the record numbers have been added (here, "A6B88A6_eq_12.udf").

Next, to convert coordinates, an exported UDF file is read by GOURMET, and the axes of coordinates are converted using a Python script ("swap_axis.py"), as displayed in Fig. 16.11.

The positions (*pos*) and velocities (*vel*) of all particles are edited with the script by iterating a "for" loop for the molecules and atoms in each molecule.

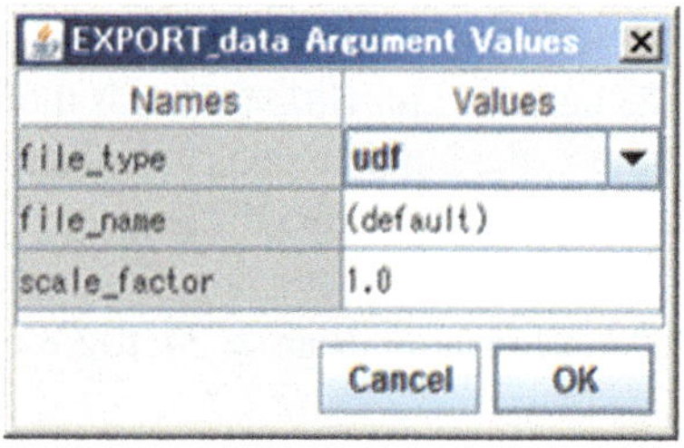

Fig. 16.10 Export_data parameters

```
for i in range(0,size('Structure.Position.mol[]')):
        for j in range(0,size('Structure.Position.mol[].atom[]', [i])):
                pos = $Structure.Position.mol[i].atom[j]
                vel = $Structure.Velocity.mol[i].atom[j]
                $Structure.Position.mol[i].atom[j]  = [pos[1], pos[2], pos[0]]
                $Structure.Velocity.mol[i].atom[j]  = [vel[1], vel[2], vel[0]]
```

Fig. 16.11 Python script for converting axes of coordinates

The coordinates of positions and velocities are converted so that coordinates (x, y, z) that are originally (x, y, z) = ([0], [1], [2]) are converted to ([1], [2], [0]) = (y, z, x), as shown in the last two lines in Fig. 16.11. Three different structures with converted coordinates are obtained by storing the three structures in different files: one with the original coordinates, one with coordinates obtained by executing the script once, and one with coordinates obtained by executing the script twice. In the example, the three kinds of files thus generated are denoted as "A6B88A6_eq_x.udf", "A6B88A6_eq_y.udf", and "A6B88A6_eq_z.udf".

16.3.3 Performing Uniaxial Elongation

Simulation of uniaxial elongation along the z axis is conducted using the input file generated in Sect. 16.3.1 and restarting from the three different relaxed structures generated in Sect. 16.3.2. The simulation is started using the command shown below.

```
>      cognac90-IA6B88A6_deform_in.udf
-RA6B88A8_eq_{x|y|z}.udf
-OA6B88A6_deform_{x|y|z}_out.bdf                    >
A6B88A6_deform_{x|y|z}_out.log
```

In the example, elongation is limited to 100 %. However, if attempting to perform a calculation involving larger elongation, it will require a longer time than calculating 100 % elongation, so it is recommended to perform parallel computation using the -n option. Moreover, because the trajectory data size also becomes large, it is recommended to output to a binary format.

16.3.4 Analyzing Results

Displaying the Structure When the calculation results of elongation are read by GOURMET and the molecular structure is displayed, deformation of unit cells like that in Fig. 16.12 is observed.

It is also effective to display some of the soft segments, as depicted in Fig. 16.13, to observe the structure of soft segments during elongation.

To do so, the Python script ("show_partialB.py") shown in Fig. 16.14 can be used.

With this script, all structures for only the ten molecules with index = 0–9 are shown as lines, and all A particles are shown as balls.

Stress–strain curve analysis PLOT_SS_Curve, an Action tool, is used to analyze the change in stress during elongation. When **PLOT_SS_Curve** is started using the default parameters, a graph like that shown in Fig. 16.15 is produced.

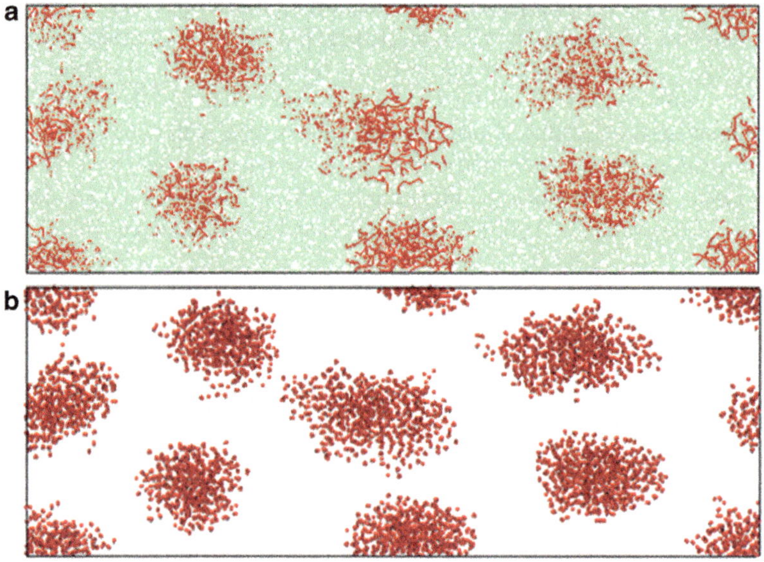

Fig. 16.12 Simulation results of elongation: (**a**) all particles and (**b**) only A particles

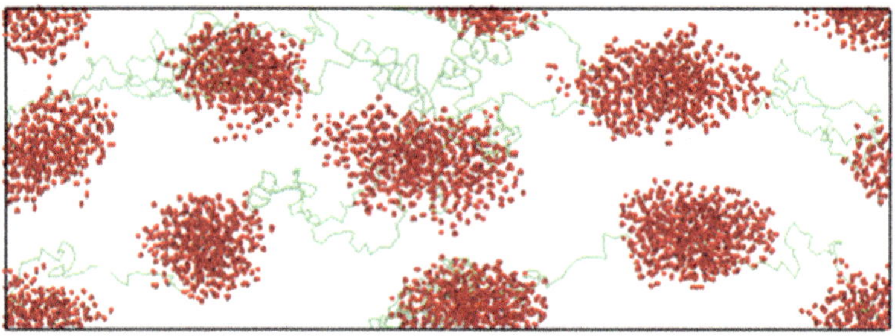

Fig. 16.13 Illustration of the A particles and some of the B particles

Fig. 16.14 Python script for displaying only the A particles and part of the B particles

```
import CognacShowLib
current = currentRecord()
if current < 0:
    jump(0)
    current = 0
show = CognacShowLib.CognacShow(_udf_)
for i in range(0,10):
    show.molecule(i,"line", "atom", "atom")
show.atom("atomA","ball","atom")
show.cell([0,0,0,1])
```

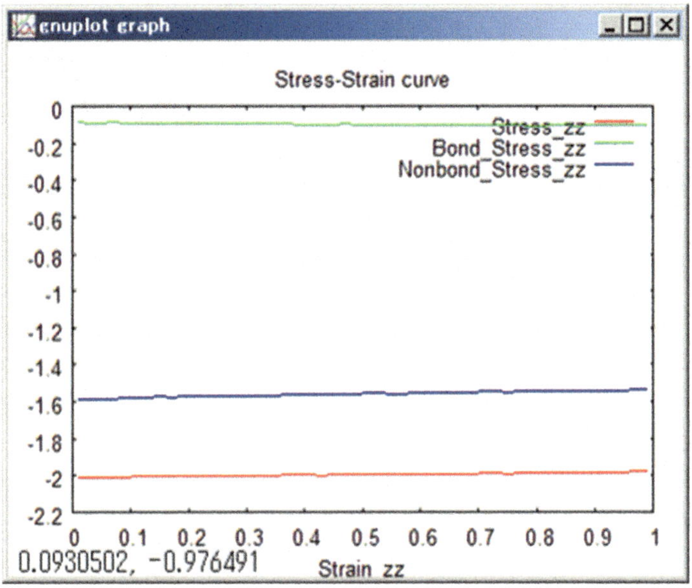

Fig. 16.15 Results of PLOT_SS_Curve

This command shows stress by dividing it into components of bond potential and nonbonding potential. Because the absolute values of each term are very different, it is difficult to see the changes when displayed on the same scale. In addition, to obtain an average for three different directions, numerical data should be extracted from the results to perform analysis with other software. The numerical data used by the plot is exported to a file named "plot.dat", which is created in either the same folder as the read UDF file or in "C:\OCTA8\GOURMET\tmp" according to the procedure for starting GOURMET. It is recommended to copy "plot.dat" into another file every time when a plot is displayed because the file is overwritten when new data is plotted.

Figure 16.16 shows an example of the averaged stress–strain relation for three directions of elongation. The plot was obtained by calculating and plotting the change from the initial stress.

Figure 16.17 displays the stress components of bond potential and nonbonding potential.

It is clear from the results that the stress from nonbonding potential is dominant and that stress derived from bond stretching becomes negative when the elongation becomes large. In the calculation conditions used here, the stress has a negative value in the equilibrium state. The absolute value of the stress in the orientation direction becomes large because of molecular orientation. Therefore, the negative value is considered to be large. This is, in a sense, an artifact of calculation. However, it can be confirmed that bond stretching is sufficiently relaxed in the elongation process, and the positive stress caused by unnatural expansion of bond

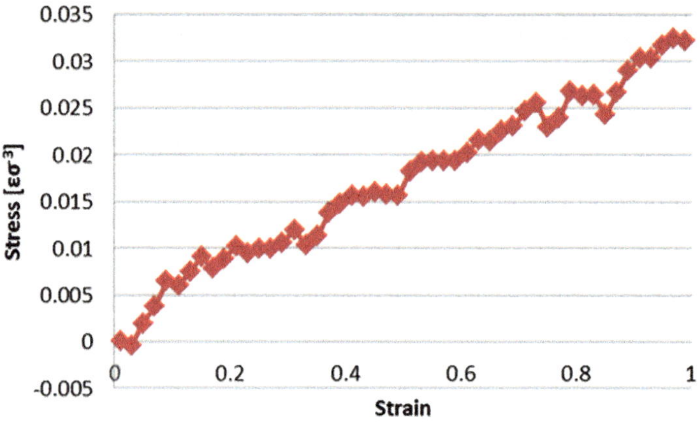

Fig. 16.16 Stress–strain curve based on total stress

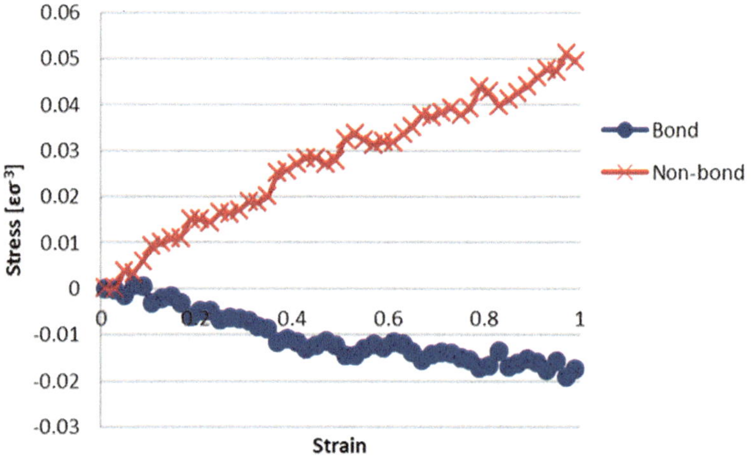

Fig. 16.17 Components of stress

length as rubberlike elasticity does not appear. Meanwhile, the fact that the stress component of nonbonding interaction becomes positive can be explained as follows. The stress derived from the interaction between nonbonding particles is also negative in this system, so the reason it turns into positive stress may be because the number of interactions in the orientation direction decreases or a strong attractive force occurs. With respect to the former, the bonds are oriented with elongation, and it is logical that the number of nonbonding particle pairs in the orientation direction decrease. Moreover, with respect to the latter, it is possible that the hard segments are subjected to elongation and are extended to a distance where a stronger attractive force occurs. Regardless, the results indicate a source of stress apparently different from the original rubberlike elasticity of a single chain. Further detailed analysis is necessary to determine the origin of this stress.

16.4 Download Instruction

Sample files are available from the OCTA website. See the instruction on http://octa.jp/download_bookdata.html. Password for the sample (zip file) of this chapter is “GEisDXbh”.

References

1. T. Aoyagi, F. Sawa, T. Shoji, H. Fukunaga, J. Takimoto, M. Doi, Comput. Phys. Commun. **145**, 267 (2002)
2. T. Aoyagi, T. Honda, M. Doi, J. Chem. Phys. **117**, 8153 (2002)
3. M.W. Matsen, R.B. Thompson, J. Chem. Phys. **111**, 7139 (1999)

Chapter 17
Filler-Filled Rubbers

Hiroshi Morita

17.1 Introduction

Although pure polymer materials are not mechanically strong, they can be reinforced when mixed with a solid object. A typical example of adding a solid filler is rubber materials used for tires. To reinforce a rubber material, carbon black particles, which are nanoscale particles made of carbon only, are introduced in a technique that has surprisingly been used since the early 1900s. The technique in which silica is mixed into nano-filler was also developed recently. Recent experimental analysis has revealed the relationship between the dispersed structure of fillers and mechanical properties. However, in some cases, it is difficult to analyze experimental observations of the dispersed structures in the mixing process or physical properties of those materials. Simulation studies are thus expected.

It is also important to consider the models of the materials used in the simulations for analysis. This is because the applicability of the analysis strongly depends on these models. In the filler-filled system, major contents are fillers and polymers, and polymers are cross-linked. In this system, there are several specific length scales, such as sizes of filler and polymer, sizes of cross-linked network and filler network, etc., and the differences of those sizes make our simulations complicated. Thus, the selection of the minimum unit and total size in the model is one of the critical issues of constructing the analysis system and the application of those analyses. For example, using a coarse-grained molecular dynamics (coarse-grained MD) method, the filler dispersion system in the small size can be analyzed in the polymer chain level [1]. By use of the finite element method, the stress concentration at small strain can be analyzed [2]. There are also models for analyzing dispersion of fillers in polymer blend systems [3].

H. Morita (✉)
National Institute of Advanced Industrial Science and Technology, Ibaraki, Japan
e-mail: h.morita@aist.go.jp

Japan Association for Chemical Innovation, *Computer Simulation of Polymeric Materials*, DOI 10.1007/978-981-10-0815-3_17

Against the background described above, this chapter describes a model and simulation of filler-filled materials. The applicability of analysis based on this simulation is also described. In coarse-grained MD simulation, the bead–spring chain is used as a minimum unit of a polymer chain, and a filler, which is larger than the polymer, is also modeled by connected beads. A filler-filled system is modeled as a mixture of bead–spring chains and model fillers. Note that the appropriate scale at which to simulate the filler-filled system has not yet been decided. There is also a size limitation of User Definable Format (UDF) files. As a result, this book gives examples of smaller systems.

Section 17.2 describes the creation of the model structure of the filler-filled system using coarse-grained MD, while Sect. 17.3 describes the structure and mechanical properties of the filler-filled system undergoing elongation. Polymers are not cross-linked in Sect. 17.2 but are cross-linked in Sect. 17.3.

17.2 Filler Dispersion Structure

This section describes how to make the filler-dispersed structures in coarse-grained MD simulation and employs the models used in [1] using COGNAC. The structures of filler-filled materials are generally classified into two groups: dispersed and aggregated structures. Various factors, such as the filler size, molecular weight, cross-linked network size, filling fraction of the filler, and cohesive force of the filler and polymer themselves, are involved in the formation of these structures. In this section, the structures of filler-filled materials are examined while changing the interaction between filler and polymer as described in [1]. In addition, calculations are made while changing the molecular weight.

17.2.1 Calculation Model

Several models for fillers in coarse-grained MD simulation have previously been proposed, and this chapter uses the models developed by Starr et al. [1]. One particle is set as a core particle, and 12 other particles are placed around the core at the vertices of a regular icosahedron. A bonding interaction is added between the core and each outer particle. There are also bonding interactions between neighboring outer particles. These bonding interactions ensure that the model filler remains in the structure. The 12 outer particles interact with polymer particles in a nonbonding interaction. In this model, the radius of the filler is almost 1.5 times the radius of the original particle, and the filler is therefore considered to be slightly larger than a polymer radius of gyration. In [1], the aggregated structures can be found in the system of a polymer chain of 40 particles, and the change in the dispersion structure of the filler with a change in the molecular weight (i.e., length of the polymer chain) would be interesting. Figure 17.1 shows the model structure of the filler.

Fig. 17.1 Starr's filler model

Section 17.2 conducts the following topics:

- The filler-filled system of a polymer melt is simulated, and the dispersed structures are examined.
- The filler-dispersed structures are examined with a change in the interaction between filler and polymer.
- The filler-dispersed structures are examined with a change in the number of particles in a polymer chain (i.e., a change in the chain length).

Note that a smaller system is used as an example owing to several limitations. In the verification of the aggregated structure, the system size and observation time strongly affect the results. When examining the aggregated structure, care must be taken with these parameters. For research purposes, it is recommended to use a system having at least ten times the number of particles used here.

17.2.2 Tips in Creating Input Data

The following describes the specific procedure of the calculation. In the calculation procedure, how to put fillers in the system is an important problem to solve. One method uses the automatic generation of COGNAC. In another method, fillers replace on the alternate particles that are placed in advance at the supposed location during the generation of polymers. The problem with the former method is that the filler may deform from the icosahedron structure owing to the collapse of polymer particles in the generation steps. To avoid these collapses, the initial structure has been generated under a low-density condition. However, this method is nonsensical in that it generates a less entangled structure. Primitively, a polymer chain naturally tends to curl into a clew due to the entropic effect. Even if there is compression afterward, the only result is the generation of compressed clews, and there is less entanglement. Obviously, the structure created in this manner will be mechanically weak.

Here, a modified method using the auto-generation function of COGNAC is adopted. First, the calculation for the generation of the dispersion structure is conducted by COGNAC under a normal density condition, and the calculation is stopped immediately after the first step. (In most cases, the calculations fail owing to a collapse.) Using the obtained structure, the filler is modified and reshaped to an icosahedron form. After modification, a relaxation simulation is performed. This method is adopted here because it is the easiest way of generating the structure.

For the first step of the calculation, an input file must be prepared. In the calculation using COGNAC, a UDF file that only includes *Set_Of_Molecules* for the polymer must be created. How to create *Set_Of_Molecules* is discussed in previous chapters and is therefore not discussed here. Next, *Set_Of_Molecules* for the filler must be added to the previous UDF file. Here, four UDF files for 8, 16, 32, and 64 fillers are provided; the file names are "Filler-ich8_in.udf," "Filler-ich16_in.udf," "Filler-ich32_in.udf," and "Filler-ich64_in.udf," respectively. The UDF file for the filler is merged into the UDF file for the polymer only using the action command of *EDIT_Merge_Set_of_Molecules*. The total number of fillers can be changed by adding or reducing the number of fillers in the UDF files. Here, a UDF file containing *Set_Of_Molecules* for the mixture of fillers and polymers can be prepared. As a sample file, the UDF file for the mixture of the 64-filler system and polymers is "Poly40-Filler64-ichRX_in.udf."

The initial structure is created using the prepared UDF file. To obtain the initial structure including fillers and polymers, only the first step of the calculation using COGNAC is performed. In most cases, the calculation will stop once the maximum force is surpassed, and most of fillers do not take icosahedron forms and thus crash. The fillers are reshaped using the Python script "FillerReconst.py." Using this script, fillers are fixed and the UDF file can be used as an initial configuration for a long-period relaxation simulation.

In the original Kremer–Grest model, the interaction between the beads of polymers is repulsive, and in the present simulation, a repulsive interaction is added between the polymers, and the system is constrained according to the setting of the density or pressure. In a polymer melt system, under the condition of density of 0.85, the pressure is 5.27. To set a pressure of 5.27, a simulation using an NPT ensemble is performed to optimize the volume of this system. Here, three UDF files for the relaxation simulation are prepared; "Poly40-Filler64-ichRX0_in.udf," "Poly100-Filler64-ichRX0_in.udf," and "Poly200-Filler64-ichRX0_in.udf" correspond to polymer lengths of 40, 100, and 200, respectively.

Next, the relaxation simulation using the NVT ensemble is performed. In this relaxation simulation, a repulsive interaction is added between polymers, between polymer and fillers, and between fillers. In particular, the cutoff distance is set as 1.12246. To follow [1], it is necessary to change the interaction and observe the corresponding dispersion structure by controlling the coefficient ε of the

Lennard-Jones potential between the polymer and filler. Furthermore, in the next section, an attractive interaction between the polymer and filler is added in the calculation of mechanical properties. Therefore, after the relaxation simulation using the NVT ensemble, a further relaxation simulation is performed using the changed ε. Here, six files are prepared for the simulation with a change in the length of the polymer and a change in the interaction between the polymer and filler. The files are "Poly40-Filler64-ichRX2_in.udf," "Poly100-Filler64-ichRX2_in.udf," and "Poly200-Filler64-ichRX2_in.udf" for $\varepsilon = 1.0$ and "Poly40-Filler64-ichRX2r_in.udf," "Poly100-Filler64-ichRX2r_in.udf," and "Poly200-Filler64-ichRX2r_in.udf" for $\varepsilon = 2.0$.

17.2.3 Analysis of Output Data

When conducting a coarse-grained MD simulation, we are interested in the dispersed structures obtained by changing several parameters. It is necessary for the relaxation simulation to allow sufficient movement of the filler. In this analysis, the results are visualized to determine whether the filler has aggregated. The analysis using radial distribution function is also applicable for the dispersed structure. Further analysis has no particular requirements that need to be noted in this section.

17.2.4 Analysis of Results: Dispersion Structure

In this section, the dispersion structure is analyzed according to changes in (1) the interaction between the filler and polymer and (2) the molecular weight. Here the dispersion structures are examined in the parameter study of (1) and (2). The results are obtained in relaxation simulations using the NPT ensemble written in Sect. 17.2.2. In these simulations, it is important to check how the system relaxes. In this case, a longtime simulation is performed for 10,000 τ. Figure 17.2 shows the volume change in the NPT simulation taken from the UDF file "Poly40-Filler64-ichRX0_out.udf." Figure 17.2 shows that the volume converges in a short time. If the filler distribution structure changes dramatically after this time, it is conceivable that the volume may change in the following simulation. Here, analysis is performed using the stabilized structure obtained in a shorter-time simulation.

Using the relaxed structure obtained by NPT simulation, further NVT simulations are conducted to obtain the stabilized dispersed structure. It is recommended that this simulation is performed for a longer time than the previous NPT simulation. Although the cell size is optimized in the previous NPT simulation, both the filler-dispersed structure and the polymer chains near the filler are relaxed in the NVT simulation. A longer time relaxation simulation is thus recommended, and here, a 25,000 τ calculation is performed.

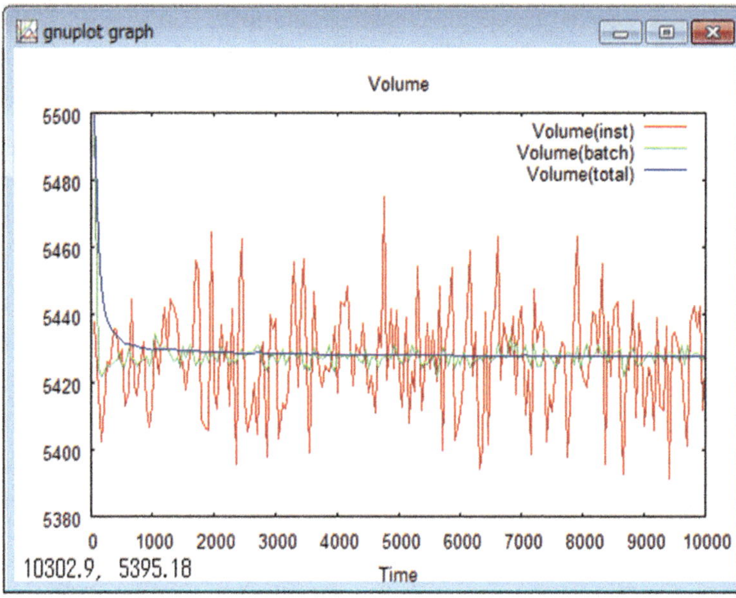

Fig. 17.2 Change in volume during the NPT calculation

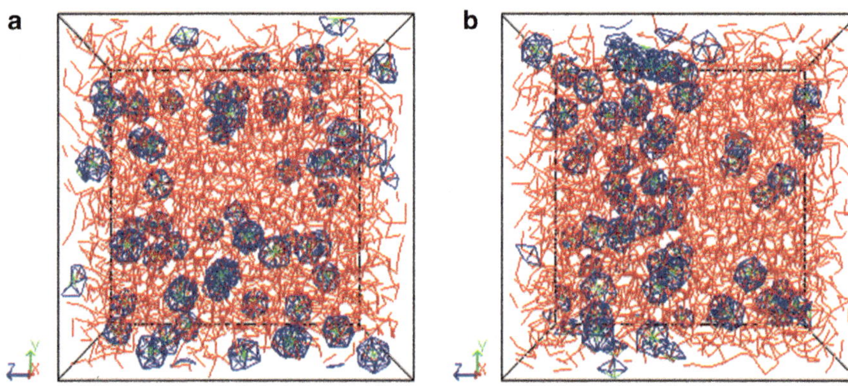

Fig. 17.3 Snapshots obtained by calculation for (**a**) $\varepsilon = 1.0$ and (**b**) $\varepsilon = 2.0$

Figure 17.3 shows snapshots of the structure obtained in NPT simulations. In these simulations, the coefficient ε of the Lennard-Jones potential is 1.0 and 2.0 for a chain length of 40. The value of ε controls the strength of the repulsive force; the force becomes more repulsive as ε increases. The filler-dispersed structure is obtained in the case of $\varepsilon = 1.0$, and a structure with partially aggregated fillers is obtained in the case of $\varepsilon = 2.0$. With an increase in the repulsive force acting between the filler and polymer, the aggregated structure of fillers becomes more stable than the completely dispersed structure.

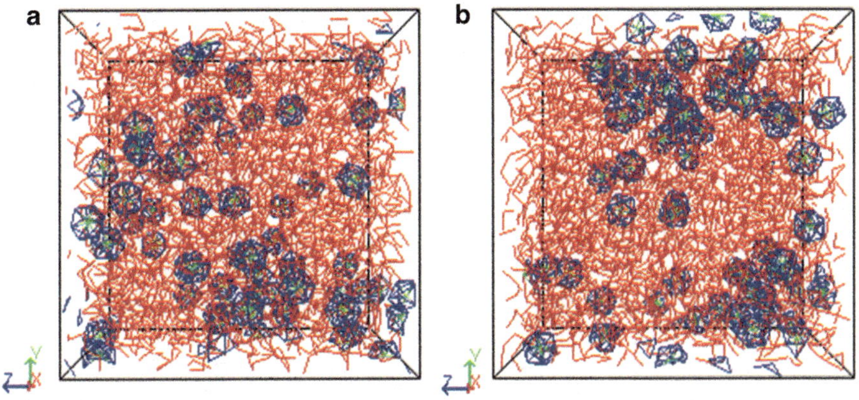

Fig. 17.4 Snapshots obtained by calculation; results for $\varepsilon = 2.0$ and chain lengths of 100 (**a**) and 200 (**b**)

To ascertain the aggregated structure with coefficient $\varepsilon = 2.0$, simulations for a chain length of 100 and 200 are performed. Figure 17.4 shows snapshots of these simulations and reveals partially aggregated structures in all cases of $\varepsilon = 2.0$. The series of simulations produced inhomogeneous structures even for a small system size and short-time simulation (total time of 25,000 τ). Note that the aggregated structure varied momentarily in the high-temperature simulation.

17.3 Elongational Properties

Section 17.2 discussed the filler dispersion structures in a polymer melt. We now conduct an elongation simulation of the cross-linked polymer system to discuss the physical properties of the elongational material. The added filler contributes to stress in the elongation, and this feature can also be observed in coarse-grained MD simulation. Here stress during elongation is estimated by conducting simulations while changing the coefficient of the strength of interaction ε. Specifically, the example that the filler and polymer are attracted to each other is described, and the effect of the attractive interaction is examined.

17.3.1 Calculation Model

This section describes the creation of the calculation model and preparation of the initial structure. First, the cross-linked network structure based on the bead–spring model must be prepared. Section 17.2 already presented the structure of melt polymers with filler. A cross-link reaction is applied to this structure to produce the initial structure for the expansion simulation. As a structure prior to cross-linking, the filler-dispersed structure with a polymer length of 200 is taken as a sample. Using this structure, the cross-link reaction is applied. In this reaction model, a random reaction is applied, where the cross-link bond is made between beads randomly; details are given in Chap. 15.

In the following, the relaxation simulation is performed while changing the filler–polymer interaction. This is one of the most important processes in adopting a filler–polymer interface. It is necessary to allow an adequate period of time for the relaxation calculation. In particular, for the relaxation of a polymer near the interface of a filler or solid substrate, it is expected that the relaxation of an interfacial system takes longer than that of a normal bulk polymer melt. If the attractive interaction is added between substrate and polymers, the mobility of the polymers near the substrate is reduced, and it takes longer for those polymers to relax. This feature can be demonstrated expressly by increasing the glass transition temperature of the polymer near the attractive substrate. For these reasons, it is recommended to conduct the relaxation simulation of the interfacial system for a longer time than the relaxation simulation of the bulk system.

Using the structure prepared in the previous procedure, an expansion simulation is performed. In performing the expansion simulation using COGNAC, there are several methods available. Here, the *Cell_Deformation* method is used. *Cell_Deformation* is a method in which affine deformation is applied to the unit cell, and as a result, expansion or compression can be realized in coarse-grained MD simulation. Using this method, an expansion simulation is performed to obtain the tensile stress in the process and to estimate the stress–strain curve.

17.3.2 Tips in Creating Input Data

This section describes the specific process of creating the input file. First, the generation of the cross-linked molecular structure is addressed. Details of the random cross-linking method are described in Chap. 15. Here the cross-link is made by the cross-link reaction in MD simulation, and in the case of chain length of 100 and 200, 98 and 198 bonds of cross-link are generated and those structures are obtained as the UDF files of "Poly200-Filler64-ichRX2CR100rlx_in.udf" and "Poly200-Filler64-ichRX2CR200rlx_in.udf," respectively. After the following relaxation simulations, the initial structure for the expansion simulation can be prepared.

Section 17.2 applied a repulsive interaction between the filler and polymer. Here, the interaction is changed to attraction. In the case of an attractive interaction, the existence of trapped polymers near the filler is expected. More precisely, bound rubber near the carbon black is a typical example of trapped polymers. In the bound rubber layer, the polymer molecules are restricted in their motion by the attractive interaction with carbon black, as has already been confirmed by nuclear magnetic resonance measurements. It is also noted that bound rubber may contribute to the stress. To realize the effect of bound polymers, a model with an attractive interaction between filler and polymer is introduced. Using this model and changing the strength of the attractive interaction, the results of a stress–strain curve can be examined.

Changing the cutoff distance from 1.12246 to 2.0 to realize attractive interaction between the filler and polymer, and changing the coefficient ε as 1.0 or 2.0 in the UDF files "Poly200-Filler64-ichRX2CR100rlx_out.udf" and "Poly200-Filler64-ichRX2CR200rlx_out.udf," further relaxation simulation is enforced. Although ε controls the strength of the attractive interaction, the dependence of the strength of interaction can be analyzed in the series of simulations. The additional relaxation simulation is performed for 25,000 τ to ensure that sufficient relaxation occurs. In this way, four files having the initial structures for expansion simulation are generated, namely, "Poly200-Filler64-ichRX2CR100rlxA1Exp_in.udf," "Poly200-Filler64-ichRX2CR100rlxA2Exp_in.udf," "Poly200-Filler64-ichRX2CR200rlxA1Exp_in.udf," and "Poly200-Filler64-ichRX2CR200rlxA2Exp_in.udf," where A1 and A2 refer to $\varepsilon = 1.0$ and $\varepsilon = 2.0$, respectively.

Uniaxial extension simulations are performed using these structures. As mentioned in Sect. 17.3.1, extension simulation was performed using the *Cell_Deformation* method in COGNAC. Three methods for the extension simulation can be selected in *Cell_Deformation*, namely, *Deformation Rate*, *Simple Elongation*, and *Oscillation*. Detailed descriptions of the methods are given in Sect. 5 of the COGNAC manual. Here *Simple Elongation* is selected. In the *Simple Elongation* method, three parameters—*Elongation_Rate*, *Poisson_Ratio*, and *Axis*—must be set. Axis should be set to the z direction. The other two parameters are set by users according to their own needs. Specifically, the *Poisson_Ratio*, under the normal conditions of rubber, is set as 0.5, which condition is the fixed condition of the total volume. However, *Poisson_Ratio* can be changed to a smaller value. For example, if users set the ratio as 0.0, fracture or cavitation can easily be observed by increasing the total volume. These settings depend on the needs of the users; users are thus responsible for the selection of appropriate settings. Users need to take care in setting *Elongation_Rate*. This is because its value is closely related to the total simulation time. To reduce the total simulation time, beginners sometimes set a larger value of *Elongation_Rate*, and a fast elongation simulation will be performed. As a result, even in the case of rubber, it is possible for the material to expand at speeds that are impossible in reality. To avoid this problem, it is recommended that the expansion simulation is performed using smaller values of *Elongation_Rate*. Here, examples are shown in which *Elongation_Rate* is set as 0.01 σ/τ. If *Elongation_Rate* takes a value smaller than the value of the gyration radius of the polymer divided by the relaxation time, as a discussion of the order level, the expansion is accomplished as the expansion of a rubber-state material. This indicator expects the extension under the condition of time when the polymer having the same molecular weight of cross-linking points can be relaxed. Naturally, this indicator strongly depends on the molecular weight, number of cross-links, and temperature, and the indicator thus must be used with care.

17.3.3 Analysis of Output Data

In this section, we carry out the analysis to obtain the stress–strain curve. To plot the stress–strain curve, the value of stress must be measured. In the case of uniaxial expansion in the z direction, it is recommended to obtain the stress in the following way. It is recommended to choose values of *$Statistics_Data.Stress.Total.Batch_Average* for stresses in each direction, and using each stress value of σ_{xx}, σ_{yy}, and σ_{zz}, total stress of σ is obtained as

$$\sigma = \sigma_{zz} - \frac{1}{2}\left(\sigma_{xx} + \sigma_{yy}\right).$$

Using this equation, the inner stress derived from the second term can be deleted, and zero stress at zero strain is calculated. In this estimation of the stress–strain curve, *Plot_SS_Curve in Action* can be used.

If we perform the simulation for a smaller system, we sometimes obtain statistically inadequate results owing to the limited number of samples. Therefore, in the expansion simulation of a small system, it is recommended that three expansion simulations, one each in the x, y, and z directions, are performed, and the final stress value is obtained as the averaged value for the three results. In the expansion simulation using COGNAC, the direction of expansion must be set to the z direction owing to the specifications of COGNAC. To perform the expansion simulation in the directions of x and y, the structure of each particle is changed, and exchanges between z and x positions or between z and y positions are performed using the Python scripts "XZexchange.py" and "YZexchange.py," respectively. Using these scripts, three simulations of the expansion along x, y, and z directions are performed, and the average results of stress are obtained to draw the stress–strain curve.

17.3.4 Analysis of Results: Analysis of Stress–Strain Curves

Figure 17.5 shows the stress–strain curves obtained in expansion simulations. Results are shown for $\varepsilon = 1.0$ and $\varepsilon = 2.0$. The values of stress are average values for three results. It is seen that stress for $\varepsilon = 2.0$ is a little larger than that for $\varepsilon = 1.0$. Initially, there is hardening of the polymer layer near the filler like that for the bound rubber layer near the carbon black filler in the case of the strong attractive interaction between polymer and filler. It is believed that this hardening affects the overall stress. If the difference between the two curves in Fig. 17.5 is significant, it appears that the difference in stress is derived from the difference in the number of fillers.

Figure 17.6 shows snapshots from expansion simulations corresponding to the stress–strain curves in Fig. 17.5. In the expansion process, it is found that the

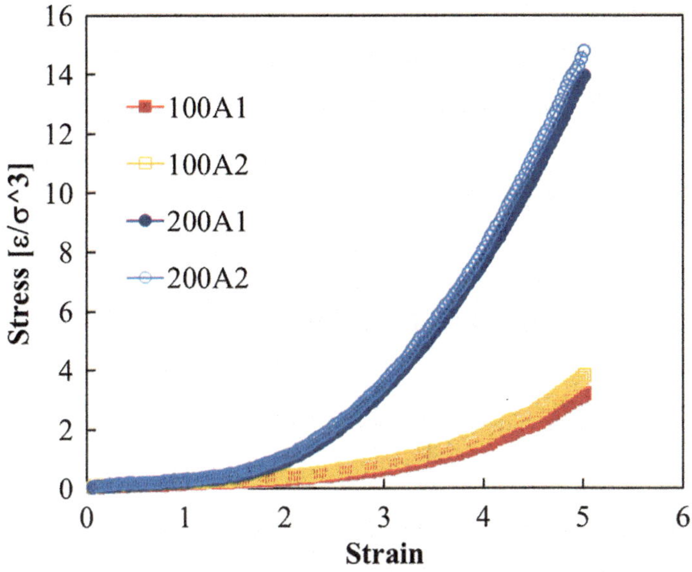

Fig. 17.5 Stress–strain curves; 100 and 200 indicate the numbers of cross-links, while A1 and A2 indicate $\varepsilon = 1.0$ and 2.0, respectively

distribution of filler changes. In the case of 500 % strain, the density of fillers differs greatly at each position, although the distribution of filler in the initial structure is almost homogeneous. Meanwhile, in terms of the dynamics of polymers, expanded bonds in the stretched direction can be found in the case of 500 % strain and 200 cross-linkers. In the case of large strain, the energetic elasticity is derived from these stretched bonds, although, in the smaller strain case, the entropic elasticity can be derived from the dynamics of each polymer chain.

Figure 17.5 plotted the average values of stress for expansions in *x*, *y*, and *z* directions. Figure 17.7 shows the average curve and three original curves for the expansions in *x*, *y*, and *z* directions for 100 cross-linkers using $\varepsilon = 1.0$. The curve for the *x* direction is almost the same as the average curve by chance, while the curves for *y* and *z* directions are lower and higher than the average curve, respectively. Although these simulations are performed using the same initial structure, the obtained stress is different owing to the distribution of the filler or polymer network. Furthermore, this simulation is performed using a small system, and deviation thus occurs easily. Therefore, much more care is needed when conducting a simulation for a smaller system. There is sometimes an appreciable difference in the stress. From this point of view, the difference between the two curves shown in Fig. 17.5 is small, and care may be needed in examining this difference.

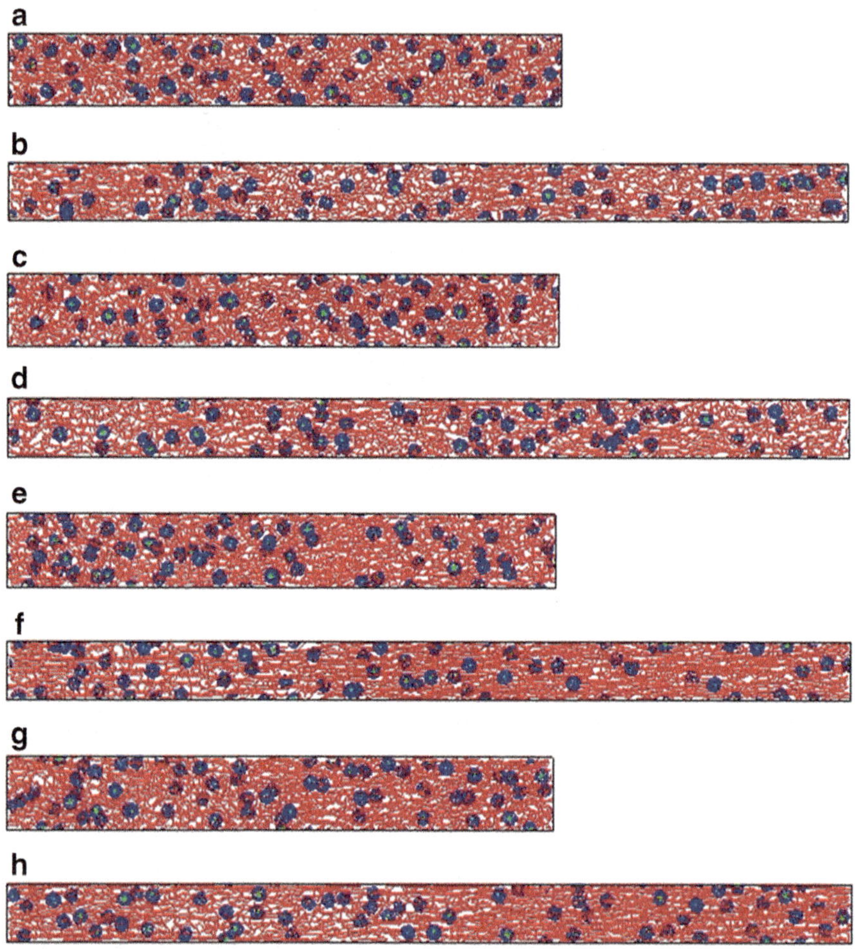

Fig. 17.6 Snapshots from elongation simulations. Simulation conditions of these figures are listed as follows, (**a**) number of cross-links = 100, $\varepsilon = 1.0$, strain = 300 %, (**b**) number of cross-links = 100, $\varepsilon = 1.0$, strain = 500 %, (**c**) number of cross-links = 100, $\varepsilon = 2.0$, strain = 300 %, (**d**) number of cross-links = 100, $\varepsilon = 2.0$, strain = 500 %, (**e**) number of cross-links = 200, $\varepsilon = 1.0$, strain = 300 %, (**f**) number of cross-links = 200, $\varepsilon = 1.0$, strain = 500 %, (**g**) number of cross-links = 200, $\varepsilon = 2.0$, strain = 300 %, (**h**) number of cross-links = 200, $\varepsilon = 2.0$, strain = 500 %

17.4 Download Instruction

Sample files are available from the OCTA website. See the instruction on http://octa.jp/download_bookdata.html. Password for the sample (zip file) of this chapter is “JaaGDUEh.”

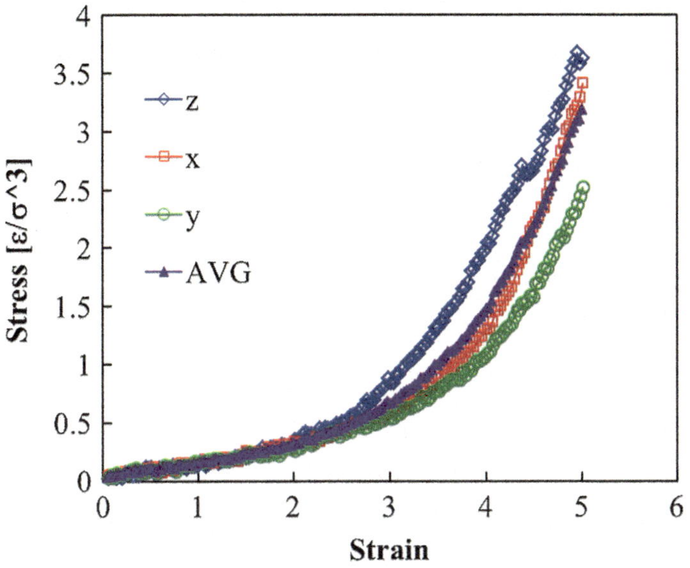

Fig. 17.7 Original stress–strain curve for 100 cross-linkers using $\varepsilon = 1.0$ in the expansion in x, y, and z directions

References

1. F.W. Starr, J.F. Douglas, S.C. Glotzer, J. Chem. Phys. **119**, 1777 (2003)
2. K. Akutagawa, Nihon Gomu Kyokaishi **82**, 227 (2009)
3. H. Morita, Nihon Gomu Kyokaishi **86**, 222 (2013)

Chapter 18
Structures of the Surface and Interface

Shigeru Yao

18.1 Experimental Methods for the Estimation of Surface and Interface Structures of Polymers and the Simulation

If one substance influences another substance, i.e., if it has a function, then an interface must necessarily exist between them. Adhesion, adsorption, surface structure (smooth, coarse, porosity), flexibility, and solidity are physical properties that influence the expression of functions. These properties are defined by the molecular structure of polymers in the vicinity of the surface, unless the surface is modified by physical techniques such as plasma treatment.

Methods for analyzing the surface properties of materials can be divided into two types: noncontact, such as with the use of X-ray or light, and contact, where the shape of the surface is actually measured by using a probe.

Among noncontact analytical methods, attenuated total reflection-infrared (ATR-IR) spectroscopy can obtain information on the chemical structure and molecular orientation over a relatively wide area. From X-ray diffraction or scattering at a very shallow angle (grazing incidence), information such as the crystal, long-period, and phase structures in the vicinity of the surface can be obtained.

Among the contact methods, atomic force microscopy (AFM) can be used to determine the extremely high-resolution structure of a substrate surface without the need for any special pretreatment, i.e., the actual surface condition can be analyzed. Various analytical methods have been proposed based on the analytical principle of AFM. Scanning force microscopy (SFM) can be used to evaluate the

S. Yao (✉)
Department of Chemical Engineering, Fukuoka University, Fukuoka, Japan
e-mail: shyao@fukuoka-u.ac.jp

Japan Association for Chemical Innovation, *Computer Simulation of Polymeric Materials*, DOI 10.1007/978-981-10-0815-3_18

physical properties, such as the viscoelastic property of the surface. Chemical force microscopy (CFM) can be used to evaluate the distribution of functional groups on the surface.

With these advances in methods for analyzing the polymer surface, we have gained a better understanding of the conformation of polymer chains near the surface [1].

On the other hand, the results obtained from polymer simulations using coarse-grained molecular dynamics and/or mean field theory have been reported to replicate these experimental findings [2, 3]. Thus, the use of a polymer simulation is a suitable method for calculating a desired surface state, which means that material design is possible with the use of a polymer simulation. In addition, with the use of a polymer simulation, it is possible to obtain information on the conformation of each polymer chain. Thus, the polymer simulation can be used to clarify the specific site that contributes to the expression of a particular function.

In addition, a polymer simulation can simulate the orientation and aggregation state of polymer chains not only at the surface but also at any depth. Information for chains at depth cannot be obtained experimentally. Especially, at the interface region, experimental observations can only be obtained by making a cut surface. In contrast, the conformational state of a polymer chain and the interaction forces at any depth can be estimated by a polymer simulation. Thus, a polymer simulation can be used to clarify the characteristics of function and for material design, such as for polymer alloys, polymer blends, and filler systems.

This section addresses the effect of the chain end, which is the simplest issue regarding the polymer surface. In addition, the surface and interior state of a mixed system composed of low- and high-molecular-weight polymers will be introduced by using SUSHI of OCTA. These calculations can be performed very easily by using examples in OCTA and provide important information to understand the basic movement of polymer chains [4].

18.2 The Distribution of Polymer Ends at the Surface of a Thin Film

A linear polymer has only two chain ends, which have almost the same chemical structure as other sites. However, the number of degrees of freedom of possible conformations is very high, and thus, it has been suggested that the chain ends are near the surface. This means that the surface characteristics strongly depend on the properties of the chain end. Thus, it is very important to understand the behavior of a polymer end for the material design of polymers.

It is very easy to construct an input file in SUSHI to calculate the behavior of a polymer end. We only need to modify a suitable sample file for a block copolymer as follows:

1. Add “axis condition [1]” to the “boundary condition” and change both [0] and [1] to “WALL” (the original value is “periodic”).

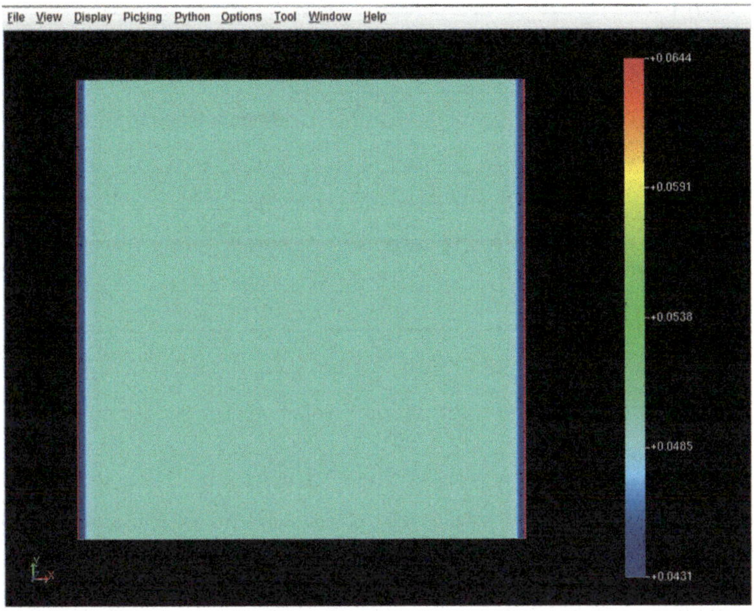

Fig. 18.1 Volume fraction of end segments. The surfaces are set at both the *right* and *left* sides. The cell size is 128. On both sides, the last layer is very concentrated. After this thin layer, a dilute layer occupies a very wide region

2. Add "block[2]" to "polymer" of "components" and make an ABA-type block copolymer. If we construct a 1:48:1 ABA-type block copolymer, the "A monomer" becomes an end site.
3. Set the "chi parameter" between A and B to 0.0.

Figure 18.1 shows the calculated results by using a modified input file.

This figure shows the distribution of end segments. In the near-surface region on both the right and left, the volume fraction of end segments is very high compared to the corresponding average value. Just below the upper region, there suddenly appears a thin-layer region in which the volume fraction of end segments is very low. As we move deeper, the volume fraction gradually increases to the corresponding average value.

Figure 18.2 shows the simulated results for the 1-D volume fraction for a polymer length of 50 (the lengths of both end segments were set to 1) and a cell size of 128. The vertical axis represents the ratio of end segments to other segments. Consistent with the results in Fig. 18.1, the volume fraction of end segments is high the surface; however, it decreases rapidly and shows a minimum at the fourth cell.

These results are for a single type of polymer chain. Next, we considered the case of a mixture of polymer chains with different lengths and evaluated how the end segments were present.

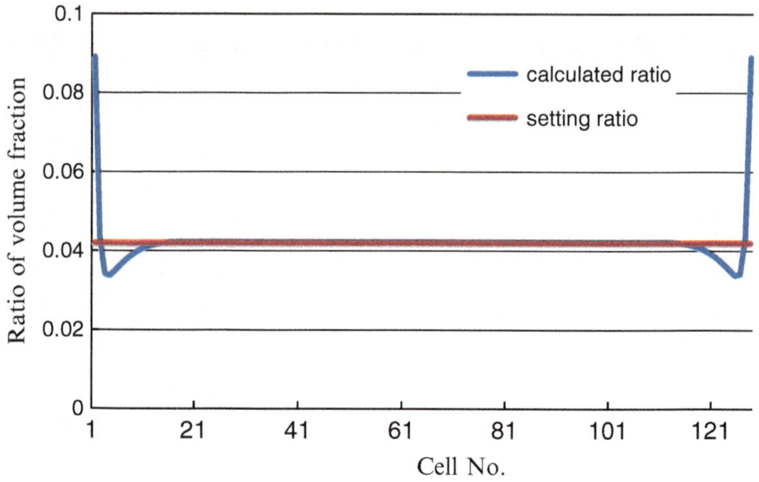

Fig. 18.2 Setting and calculated volume ratio of end segments (Reproduced from [4] with permission from Nihon Reoroji Gakkaishi (J. Soc. Rheology, Japan))

Figure 18.3 shows possible structures very near the surface for polymer chains of different lengths.

Figure 18.3 (a) shows the case where the end segments of both short and long chains exist near the surface, (b) shows the case where short chains form a thin layer near the surface and the end segments of the long chains exist in a deeper layer, and (c) shows the case where the surface layer is composed entirely of short chains and the long chains do not show any orientation.

For the calculation, we just apply the following changes to the input file:

1. "monomer[2]" and "monomer[3]" are added to the monomer field and set to "C" and "D", respectively.
2. "polymer [1]" is added to "polymers" of "components". Type is set to "BLOCK" and "block [0]", "block [1]", and "block [2]" are set to "C", "D", and "C", respectively. (In this calculation, we used a ratio of 1:58:1.)
3. For the "polymer volume fraction" of "volume fraction", "polymer volume fraction[1]" is added. The volume fraction of [0] and [1] are set to a suitable number. (In this case, we set both to 0.5.)
4. For the "chi parameter", from [0] to [5] were all set to 0.

Figure 18.4 shows the simulation results for a system composed of a mixture of polymers of length 20 and 60. The vertical axis is the ratio of the number of end segments to other segments. As shown in Fig. 18.4a, the volume ratio of the long polymer decreases rapidly near the surface and the volume ratio of the short polymer increases near the surface. These results support the general assumptions of Fig. 18.3. On the other hand, the end-segment volume ratio of the long polymer still increases near the surface. This suggests that Fig. 18.3a reflects the simulation results.

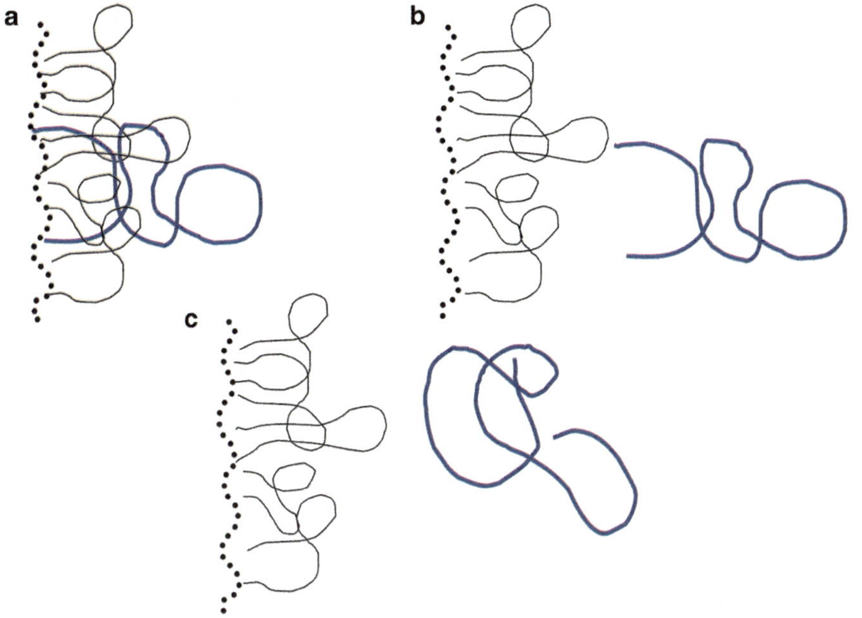

Fig. 18.3 Schematic polymer chain conformations at surface in the case of blend of short and long polymer (Reproduced from [4] with permission from Nihon Reoroji Gakkaishi (J. Soc. Rheology, Japan))

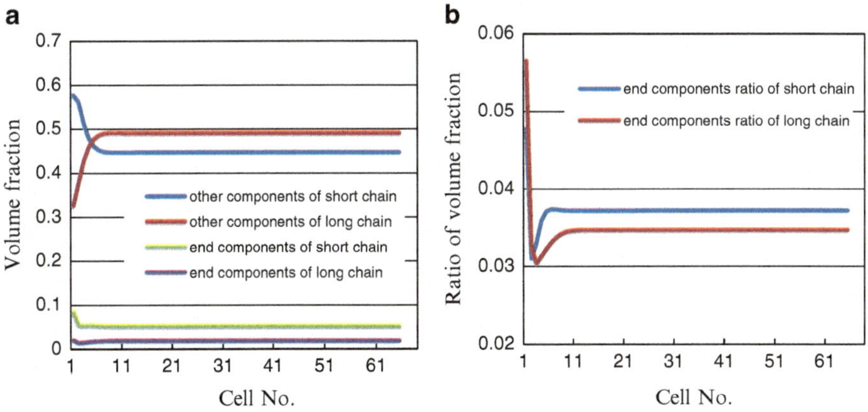

Fig. 18.4 Film thickness dependence of volume ratio (**a**) and ratio of end segments to other segments (**b**) (Reproduced from [4] with permission from Nihon Reoroji Gakkaishi (J. Soc. Rheology, Japan))

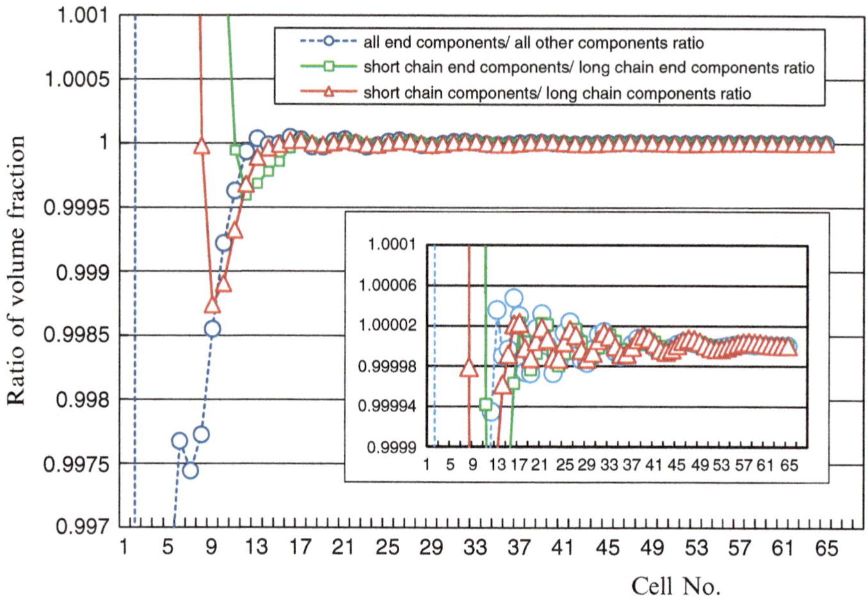

Fig. 18.5 Film thickness dependence of ratio of various volume fractions. ○, all end-segment volume fraction/all component volume fraction; □, end-segment volume fraction of short polymer/end-segment volume fraction of long polymer; Δ, volume fraction of short polymer/volume fraction of long polymer. Inner figure shows more details (Reproduced from [4] with permission from Nihon Reoroji Gakkaishi (J. Soc. Rheology, Japan))

Figure 18.4b shows the ratio of end segments to other segments. This suggests that there is a higher proportion of end segments compared to other segments for a long polymer. In addition, the layer next to the surface with a low volume fraction is much thicker than that in the case of a short polymer.

Figure 18.5 shows the thickness dependence of the ratio of all end segments to other segments (○), the end-segment ratio for a short polymer and a long polymer (□), and the all-segment ratio for a short polymer and long polymer (Δ). In this case, when we used the entire volume fraction of a simulated system, the settled ratio became 1.

In this figure, the thickness of the layer with a concentration of end segments is only about one layer, and after ten layers, the volume fraction of end segments oscillates as we move to the center of the film. In addition, the ratio between short and long polymers oscillates as we move to the center of the film. This simulation result reflects the possibility of a convergence error in the calculation. However, if there is a large difference in molecular weight, the ratio of short component to long component may become heterogeneous in the near-surface layer.

18.3 Summary

We have introduced a method for simulating the surface structure of a homopolymer system by using OCTA. The input file of the simulation was used and the sample input file of OCTA was changed; in all cases the chi parameter was set to 0.

Even with such a simple simulation, we can obtain some interesting results on the influence of the end segments of a polymer at the surface. Especially, in the case of a mixture of short and long polymers, the components are present heterogeneously rather than homogeneously. This suggests that if there is a large difference in molecular weight, phase separation may occur.

By using the input file introduced in this chapter, we can easily simulate the end-segment distributions of different polymer species by properly setting the chi parameter. We can also simulate the bleed-out phenomena of a system that includes a good solvent and/or a poor solvent. If we know the equilibrium state beforehand, we can perform reasonable material design.

18.4 Download Instruction

Sample files are available from the OCTA website. See the instruction on http://octa.jp/download_bookdata.html. Password for the sample (zip file) of this chapter is “HxeEUG9U”.

References

1. J. Takahara, Denso Tech. Rev. **12**, 3 (2007). for a review article
2. T. Aoyagi, J. Takimoto, M. Doi, J. Chem. Phys. **115**, 552 (2001)
3. H. Morita, T. Ikehara, T. Nishi, M. Doi, Polym. J. **36**, 265 (2004)
4. S. Yao, T. Shoji, Nihon Reoroji Gakkaishi (J. Soc. Rheol, Japan) **32**, 117 (2004)

Chapter 19
Glass Transition at the Surface and Interface

Hiroshi Morita

19.1 Introduction

The glass transition temperature (*Tg*) is an important physical property of polymeric materials, and *Tg* itself or a parameter shifted from *Tg* is sometimes used as the material index of temperature. Values of *Tg* for a thin film, surface, and interface are also discussed as indices of local physical properties. For example, as the thickness of a thin film increases, *Tg* of the thin film increases in some cases [1] and decreases in other cases [2]. These results may be explained by the strength of the attractive interaction between polymer and substrate. Meanwhile, *Tg* of a free-standing film is lower than that of bulk material [3]. The properties of a film thus depend on the set conditions affecting of the film. It is therefore important to understand local characteristics such as those of the surface and interface in studying the overall properties of a film.

Tanaka and coworkers [4–6] experimentally studied the *Tg*surface of polystyrene film employing friction force microscopy, which is one of the atomic force microscopy techniques used to observe the physical properties of a surface. Their results revealed that surface *Tg* is lower than bulk *Tg*. Meanwhile, *Tg* near the substrate has also been studied using the lifetime measurement of fluorescence induced by evanescent light and *Tg* near the substrate and was found to be higher than that of the bulk. A simulation comparable to the cited experiments can be performed using coarse-grained molecular dynamics (CGMD) method. In CGMD simulation, the dynamics of the polymer chain can be observed, which is not possible in experiments. Additionally, local regions of the surface or interface can be analyzed. This chapter describes the simulation of *Tg* of a polymeric material

H. Morita (✉)
National Institute of Advanced Industrial Science and Technology, Ibaraki, Japan
e-mail: h.morita@aist.go.jp

Japan Association for Chemical Innovation, *Computer Simulation of Polymeric Materials*, DOI 10.1007/978-981-10-0815-3_19

following [7, 8]. To realize the previous results of [7, 8], the User Definable Format (UDF) files and analysis method are described. Owing to the size limitation of the UDF file, the system described in this section is much smaller than that in [7, 8]. Note that the results presented in this chapter may not be statistically acceptable.

19.2 Calculation Model

To simulate values of Tg for a thin film, surface, and interface, the model of thin film by Kremer–Grest (bead–spring) chain is constructed; this is a coarse-grained (CG) model of a polymer chain. In the present study, how to construct the film made of CG polymer chains is one of the most important and the difficulty of this study is in this point. Experimental results show that the end of the polymer chain must segregate at the surface and interface in the model film. To make an acceptable film structure, the initial structure of the thin film is constructed employing the node density biased Monte Carlo (NDBMC) method, which was proposed by Aoyagi et al. The NDBMC method is detailed in Chap. 4 and the present chapter only gives an overview of the method.

First, a one-dimensional self-consistent field (SCF) simulation along the height axis is performed and the segment density $\phi(i,\mathbf{r})$ is obtained, where i and $\mathbf{r}$ are the segment number from one end and height position, respectively. The SCF simulation can reproduce segregation of the end segment at the surface and interface. Using the segment density distribution, a Monte Carlo simulation is performed to place each bead of a polymer chain and make the initial structure of the film. Employing the NDBMC method, each bead can be placed along the probability of each segment and the distribution of beads almost exactly follows the density distribution obtained by SCF simulation, and segregations of end beads at the surface and interface can be realized.

As noted in Sect. 19.1, the system size is statistically unsatisfactory. In a production run, simulations must be run for a system that is at least five times as large with a large number of time steps.

19.3 Tips in Creating Input Data

The specific procedure of the calculation is described. First, a one-dimensional SCF simulation along the height direction is performed using SUSHI, where the blended system of a solvent and polymer is treated. At the boundary of the substrate, the *WALL* boundary condition is used. At other boundary, the *NEUMANN* boundary condition is used. The polymer phase is put on the substrate and the film phase is formed at the side of the *WALL* boundary. The χ parameter for solvent and polymer, which decides the interfacial force, is needed. In this simulation, the length of the polymer is set as 100, and the χ parameter for solvent and polymer is set as 1.5. The input and output files for SUSHI simulations are "Film1D-01_uin.udf" and

"Film1D-01_uot.udf," respectively. In the SUSHI simulation, the density profile of $\phi(i,\mathbf{r})$ must be outputted and its option is set in the input file "Film1D-01_uin.udf."

Using the result of the density profile $\phi(i,\mathbf{r})$, the NDBMC simulation is run, and the film structure comprising bead–spring chains is obtained. The detailed settings of each parameter are given in the UDF file "Film-MD-01_in.udf." In this NDBMC simulation, the CG chain having a length of 100 used in the SCF method is mapped to the CG chain having a length of 100 used in the CGMD method. In the CGMD method, solvent particles are not introduced to the system, and the simulation is performed under vacuum conditions. If the attractive force between polymers is not active, the polymers in the film spread into the vacuum phase. To keep the film structure, the attractive interaction between polymers is introduced by setting the cutoff distance of the interaction potential between the beads of polymer as 2.0 σ. It is noted that the distance of 2.0 σ is shorter than the distance between the beads of second-nearest neighbors in the closed packed structure, and it becomes a weakly attractive force. An attractive interaction is also needed between the polymer and substrate to avoid detachment of the film from the substrate, and *Interactions. External_Interaction* is also set. Figure 19.1 shows the initial structure in the CGMD simulation. In the simulation, the boundary conditions for upper and lower boundaries are set to *REFLECTIVE1*, which is the corresponding condition of the SCF simulation. At other boundaries, the periodic boundary condition *PERIODIC* is used. The structure shown in Fig. 19.1 is obtained after a relaxation simulation under

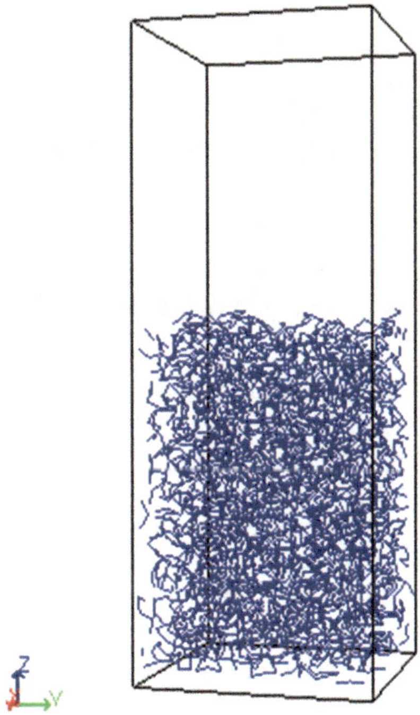

Fig. 19.1 Snapshot of the initial structure of a thin film

the condition of T = 1.0 with at least one million time steps. If the system is enlarged or the polymer is changed to larger ones, further relaxation simulation is needed to relax the polymer chains. The relaxed state can be checked by the change in R^2 or Rg^2. This relaxation simulation is performed using "Film-MD-02relax_in.udf."

To analyze *Tg* obtained from CGMD simulation, the relaxed structure at each temperature and dynamics simulation results are needed. To obtain these data, a relaxation simulation is performed at each temperature. One relaxation simulation is performed under the condition $T = T_1$, and the equilibrium structure at $T = T_1$ is obtained. The next relaxation simulation is performed using $T = T_1 - \delta T$, and the equilibrium structure at the next temperature is obtained. Through a series of simulations at decreasing temperature, the equilibrium structure at each temperature is obtained. *Tg* of the bulk condition has already been estimated as about $Tg = 0.4$. In the case of a temperature around $T = 0.4$, δT is set to 0.05, while in other temperature ranges, δT is set to 0.1. The series of input UDF files from $T = 0.8$ to $T = 0.1$ are "Film-MD-02T08_out.udf" to "Film-MD-02T01_out.udf."

Details of the analysis method are given later. *Tg* of thin film and local values of *Tg* at the surface and interface are measured according to the temperature dependences of the thickness and mean square displacement (MSD), respectively. To estimate *Tg* values using these methods, the equilibrium structure and simulation data for a greater number of time steps are needed. In this section, a simulation is conducted for three million time steps to obtain the equilibrium structure, and a simulation is conducted for two million time steps to obtain long-time simulation data. In these simulations, as described before, a small system is used owing to the size limitation of the UDF file, and data for much longer times are needed to achieve *Tg* value results with higher statistical significance. The analysis of *Tg* for a thin film requires the results of film thickness versus temperature. In the analysis of local *Tg*, the MSD results must be estimated. To analyze the film thickness and MSD results, the Python script "Thickness-MSDAnaly.py" is prepared.

19.4 Analysis of Results: Glass Transition Temperatures of the Thin Film, Surface, and Interface

Several methods of measuring *Tg* have been proposed. For example, difference scanning calorimetry is applied for the analysis of *Tg* of a powder sample and ellipsometry is applied for the analysis of *Tg* of a film sample to measure the thickness change versus temperature. In this chapter, *Tg* of a thin film is measured using the temperature dependence of the film thickness in the CGMD simulation. Several estimation methods of *Tg* based on CGMD have been proposed by Baljon et al. [9], who pointed out that the temperature dependences of density, specific heat, and MSD can be used in an analysis to measure *Tg*. It has been confirmed that consistent *Tg* values can be measured from these property data.

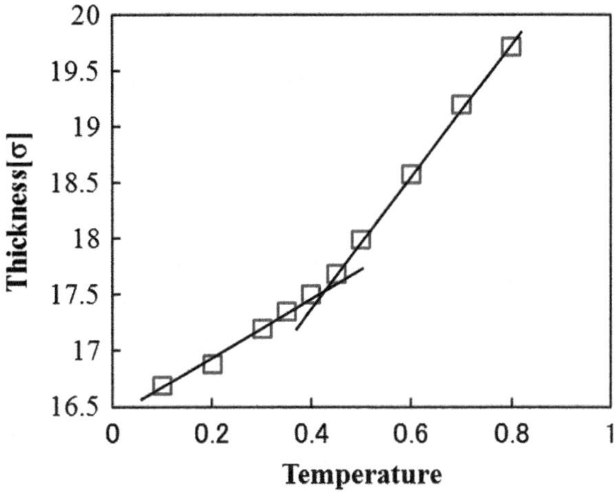

Fig. 19.2 Estimation of the glass transition temperature of a thin film

Here, in the measurement of *Tg* values of the thin film and local *Tg* values of the surface and interface, changes in film thickness and MSD with time are used. In this analysis, to obtain steady results, *Tg* values must be analyzed using the statistically satisfactory simulation data. As statistically stable data, simulation results for two million time steps after the relaxation simulation are used, and the local *Tg* values of the surface, middle region of the film (bulk), and interface (near the substrate) are estimated using the beads that remain in the 3-σ-thick regions of the surface, middle thickness region, and interface, respectively. The local physical properties must be estimated from data in a small region. However, if the region is limited, there are fewer data, and it is thus not possible to use statistically satisfactory data. To balance the needs for locality and statistical accuracy, the thickness of 3 σ is selected. In [7, 8], the horizontal area is taken to be much larger than that in this work. We therefore need to take care in treating the simulation results for the smaller system.

Figure 19.2 shows the analysis results obtained with "Thickness-MSDAnaly.py" and plots the thickness of the film at each temperature. Vertical and horizontal axes show the thickness of the film and temperature, respectively. The curve shows the transition point, which is defined as *Tg*; *Tg* is estimated as 0.419. Figure 19.3 shows analysis results of MSD of the surface, middle thickness region, and interface. Vertical and horizontal axes indicate the MSD and temperature, respectively. The interval time is an important parameter in MSD analysis, and the present analysis uses a much shorter time within the time scale of Brownian motion inside a reputation tube (300 $\tau < \tau_e$). In the same way, transition points are obtained for each curve, and the transition temperatures are defined as *Tg* values of the surface, bulk, and interface; *Tg* (surface) = 0.363, *Tg* (bulk) = 0.391, and *Tg* (interface) = 0.429, and hence, *Tg* (surface) < *Tg* (bulk) < *Tg* (interface), which is the same order obtained in previous studies [7, 8]. In [7, 8], *Tg* values of the surface, bulk, and interface are 0.35, 0.41, and 0.46, respectively, and there are thus shifts of about 0.03.

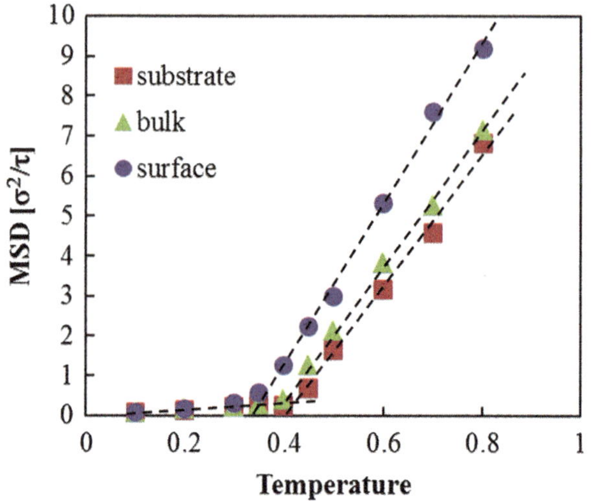

Fig. 19.3 Results for the glass transition temperatures of the surface, bulk, and near-substrate region

As seen in the previous paragraph, a unit of temperature is used as the simulation unit. The temperature unit can be converted to kelvins. In [7], the conversion equation for a thin film of polystyrene is

$$Tg\,[\mathrm{K}] = \beta \mathrm{T\ g}^{\mathrm{sim}}\,(\beta = 927.74)\,.$$

Using this equation, Tg values of the thin film, surface, bulk, and interface are estimated as 388, 324, 380, and 427 K, respectively (Figs. 19.2 and 19.3).

19.5 Download Instruction

Sample files are available from the OCTA website. See the instruction on http://octa.jp/download_bookdata.html. Password for the sample (zip file) of this chapter is "8qWHvzNc".

References

1. W.E. Wallace, J.H. Van Zanten, W.L. Wu, Phys. Rev. E **52**, R3329 (1995)
2. J.L. Keddie, R.A.L. Jones, R.A. Cory, Faraday Discuss **98**, 219–230 (1994)
3. J.A. Forrest, K. Dalnoki-Veress, J.R. Dutcher, Phys. Rev. E **56**, 5705 (1997)
4. K. Tanaka, A. Takahara, T. Kajiyama, Macromolecules **33**, 7588 (2000)
5. T. Kajiyama, K. Tanaka, A. Takahara, Polymer **39**, 4665 (1998)
6. T. Kajiyama, K. Tanaka, N. Satomi, A. Takahara, Sci. Technol. Adv. Mat. **1**, 31 (2000)
7. H. Morita, K. Tanaka, T. Kajiyama, T. Nishi, M. Doi, Macromolecules **39**, 6233 (2006)
8. K. Tanaka, Y. Tateishi, Y. Okada, T. Nagamura, M. Doi, H. Morita, J. Phys. Chem. B **113**, 4571 (2009)
9. A.R.C. Baljon, M.H.M. Van Weert, R.B. DeGraaff, R. Khare, Macromolecules **38**, 2391 (2005)

Chapter 20
Evaporation from Polymer Solution

Hiroshi Morita

20.1 Introduction

Polymer thin films have recently been used in electronic and other devices as highly functional materials. For example, the polymer resist used in lithography is a film coating with thickness ranging from 10 nm to 1 μm. In addition to the structural control of thin films, such as multilayered films, the process of coating an organic photovoltaic cell has also been studied to control the bulk heterojunction and thus produce a highly efficient cell. It is important to clarify the mechanism of the structural formation of a thin film in the coating process [1].

Important aspects of the coating process include the evaporation of the solvent and the dynamics of the solvent and polymer chains. It is difficult to observe experimentally the coating process in situ including the solvent evaporation even if using a cutting-edge observation technique. As another method of clarifying the evaporation process, coarse-grained simulations of evaporation are conducted.

There have been several simulation studies of solvent evaporation in the film coating process. For example, Okuzono and coworkers studied the evaporation process by solving the differential equation on a regular mesh employing a lubrication approximation [2, 3]. Yamamoto [4] and Morita et al. [5] performed evaporation simulations based on the dissipative particle dynamics (DPD) method. They modeled a thin film comprising a solvent and polymer and represented the evaporation by the diffusion of the solvent into the gas phase. Note that the length scales of the simulations of the differential equation and DPD greatly differ.

This chapter describes the model simulation of solvent evaporation using the DPD method following [5] using OCTA. The DPD method uses a coarse-grained polymer model in which the dynamics of the polymer chain are described as those

H. Morita (✉)
National Institute of Advanced Industrial Science and Technology, Ibaraki, Japan
e-mail: h.morita@aist.go.jp

Japan Association for Chemical Innovation, *Computer Simulation of Polymeric Materials*, DOI 10.1007/978-981-10-0815-3_20

of a Rause or phantom chain. Using the DPD method, a timescale longer than that of the coarse-grained molecular dynamics method can be treated. The DPD method thus has the advantage that it can simulate the phase separation using a particle-based polymer chain model. In the process of solvent evaporation, the initial structure is composed of two phases, namely, the air phase and the mixed phase comprising polymer and solvent. After evaporation of the solvent, the structure undergoes a phase change to polymer and air. Given that this is one example of phase transitions, the DPD method can be applied to solvent evaporation.

20.2 Calculation Model

As described above, the initial structure has two phases: the air phase and the mixed (polymer–solvent) phase. To make this initial structure, we apply the node density biased Monte Carlo (NDBMC) method, which was developed by Aoyagi et al. and has already been described in Chap. 4. In particular, the density distributions of the solvent, polymer, and air are obtained using the self-consistent field (SCF) simulation using SUSHI. Using one-dimensional SCF simulation, the one-dimensional density profiles of the two phases (i.e., the air phase and the mixed phase of solvent and polymer) are calculated. In the NDBMC simulation, each DPD particle is placed in a three-dimensional simulation box along the weight probability of the density distribution by SCF simulation. After placing all DPD particles, the relaxation simulation is performed and the initial structure is obtained.

Once the initial structure is obtained, the simulation condition of the interaction parameter is changed so that it is suitable for evaporation and the dynamics simulation is performed to realize the dynamical process of solvent evaporation. In the dynamics simulation, solvent particles diffuse into the air phase and the number of solvent particles in the film decreases and a polymer film forms. In the simulation of this process, the saturation of the solvent in the air phase is a critical problem. A function of the COGNAC simulator is used to avoid this problem. This function changes the particle species from solvent to air. In detail, the COGNAC simulator replaces the solvent particle traveling near the upper boundary with an air particle. Near the upper boundary, all solvent particles are changed to air particles and the number of solvent particles becomes zero at the boundary. The number of solvent particles must be zero at an infinitely distant point in the real system. In our model, however, the zero-density point of the solvent, which is set near the upper boundary through the distance from the surface to the upper boundary, is not infinite. Through this modeling, saturation of the solvent in the air phase does not occur. This modeling is applied to the simulation of solvent evaporation.

20.3 Tips in Creating Input Data

A one-dimensional SCF simulation is performed to obtain one-dimensional density profiles of the solvent, polymer, and air along the height direction. The boundary conditions of this SCF simulation using SUSHI are the same as the conditions given in Chap. 19. The difference in the SCF simulation between this chapter and Chap. 19 is the setting of the χ parameter. In the case of the initial structure, the polymer interacts attractively and repulsively with the solvent and air, respectively. χ parameters are set to reproduce this situation. The input file of the SUSHI simulation is "Film1D-Evp_uin.udf", and the obtained result file is "Film1D-Evp_uot.udf."

Using the density profiles along the height direction, the NDBMC simulation is conducted to produce a phase-separated structure made from DPD particles. The settings of the NDBMC simulation are written in the User Definable Format (UDF) file "Film-Evp-01_in.udf". This chapter gives an overview of the settings. The boundary conditions of height and lateral directions are chosen as *REFLECTIVE1* and *PERIODIC*, respectively. In some cases of NDBMC simulation, a few air particles are placed in the film phase, the contents of which must only be solvents and polymers. To extract these air particles from the film, the Python script "C_pos_conv.py" is prepared. Using this script, all air particles in the film are moved to the air phase. After using this script, the relaxation simulation is performed. The input file of the relaxation simulation is "Film-Evp-02relax_in.udf." The relaxed structure is used as the initial structure in the evaporation simulation.

In the evaporation simulation, the interaction parameters are changed. While the solvent does not attractively interact with the air particles in the NDBMC simulation, it does so in the evaporation simulation, where solvent particles penetrate the air phase. Given that the threshold value a_{SV} (a parameter for solvent and air) of the phase separation is 25.0, a value of a_{SV} smaller than 25.0 must be set to avoid the phase separation of solvent and air in the air phase.

Two-hundred forty polymer chains having 10 particles, 3600 particles of solvent, and 6000 particles of air are placed in the simulation box, where the system size is $10 \times 10 \times 40$. In this initial structure, the concentration of polymer in the film is rather high at 40 %. Decreasing the concentration of polymer in the film leads to a thinner polymer film. Thus, the total volume of polymer decides the final thickness of the film.

20.4 Analysis of Output Data

In the evaporation simulation, it is important to analyze the structure of the film and the distributions of the polymer and solvent. The analysis of the one-dimensional density profile is of particular importance. This analysis can be easily done using an Action program in COGNAC. Snapshots are also important to the structural analysis

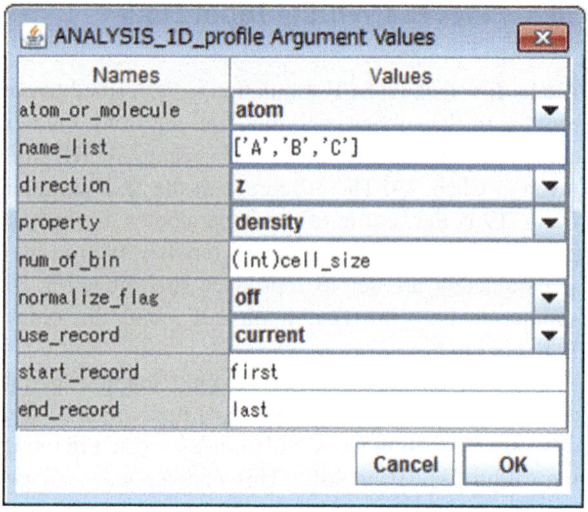

Fig. 20.1 Small window that opens for the Action program ANALYSIS_1D_profile

of a transient film. Because there are many air particles inside the simulation box in this simulation, there is a greater calculation cost in showing air particles. When the calculation result is drawn, it is thus recommended to draw only the species of polymer and solvent.

Here the analysis of the one-dimensional density profiles is explained. This analysis is done using the Action program **ANALYSIS_1D_profile**. Figure 20.1 shows the window in the working process of the Action program **ANALYSIS_1D_profile**. On the line of *name_list*, the value is set to "['A', 'B', 'C']." On the line of *direction*, "z" is chosen. After these lines are set and the "OK" button is pressed, the result of the one-dimensional density profile is displayed in the gnuplot window. On the line of *normalize_flag*, the default condition is "on," and the normalized density is plotted. If we want to plot the raw density value, the *normalize_flag* is set to "off."

20.5 Analysis of Results: Structural Analysis of Solvent Evaporation

The parameter values of a_{XY} for evaporation simulations are given in Table 20.1. The simulation involves A, B, and C particles, which represent polymer, solvent, and air particles, respectively. Two simulations are performed with input files "Film-Evp-03Evp_in.udf" and "Film-Evp-03Evp2_in.udf."

a_{XY} is a repulsive parameter and the repulsion increases as the parameter increases. Here, a_{AC} is set to a large value of 60.0 because the polymer segregates from air. After evaporation of the solvent, the large value is effective in obtaining a separated structure of the polymer and air phases. The difference between the two

Table 20.1 Parameter values of a_{XY}

X	Y	Film-Evp-03Evp_in.udf	Film-Evp-03Evp2_in.udf
A	A	25.0	25.0
A	B	25.0	25.0
A	C	60.0	60.0
B	B	25.0	25.0
B	C	25.0	10.0
C	C	25.0	25.0

input files is the setting of a_{BC}. a_{BC} is the interaction between the solvent and air and it affects the evaporation. To understand the mechanism of the evaporation model, the principle of this model is described.

In this simulation, the difference in the chemical potential $\Delta\mu$ is the driving force of the solvent evaporation:

$$\Delta\mu = \mu_{\text{polymer-Solv}} - \mu_{\text{air-Solv}},$$

where the first and second terms are the chemical potential of the mixed phases of polymer and solvent and that of air and solvent, respectively. For the first term of the chemical potential of the mixed phase of polymer and solvent, the interaction parameter of a_{AB} (for the interaction between polymer and solvent) is effective. Meanwhile, for the latter term, the interaction parameter of a_{BC} (for the interaction between solvent and air) is effective. If the value of a_{BC} decreases, the solvent can interact with air more attractively and $\mu_{\text{air-Solv}}$ decreases and $\Delta\mu$ increases. According to this discussion, in the case of a smaller value of a_{BC}, it is expected that evaporation will accelerate. Thus, there is faster evaporation for "Film-Evp-03Evp2_in.udf."

Figure 20.2 shows snapshots of the simulation results. In the lower region inside the simulation box, there is a concentration of the blue lines that indicate the polymers. The green points indicate solvent particles and the red points indicate air particles that were originally solvent particles. In these plots, the original air particles are not drawn, but air particles were placed in the upper region. From the series of snapshots, the diffusion of solvent particles into the air phase and the evaporation of solvent can be obtained. In the early stage of evaporation, the result for "Film-Evp-03Evp2_in.udf" indicates faster evaporation. In Fig. 20.2 at time 200 τ (a) and time 50 τ (b), there is a concentrated polymer layer at the surface of the film. This concentrated layer is seen as a skin layer in the simulation and can be analyzed according to the density profile along the height direction.

The depth density profile along the height direction can be estimated using the Action program **ANALYSIS_1D_profile**; the results are shown in Fig. 20.3. The figure presents results for (a) "Film-Evp-03Evp_in.udf "at times 0, 100, and 2000τ and (b) "Film-Evp-03Evp2_in.udf" at times 0, 50, and 2,000τ. In the figure, only the gnuplot window of **ANALYSIS_1D_profile** is shown. The horizontal and vertical axes indicate the position of the height direction and the raw density value,

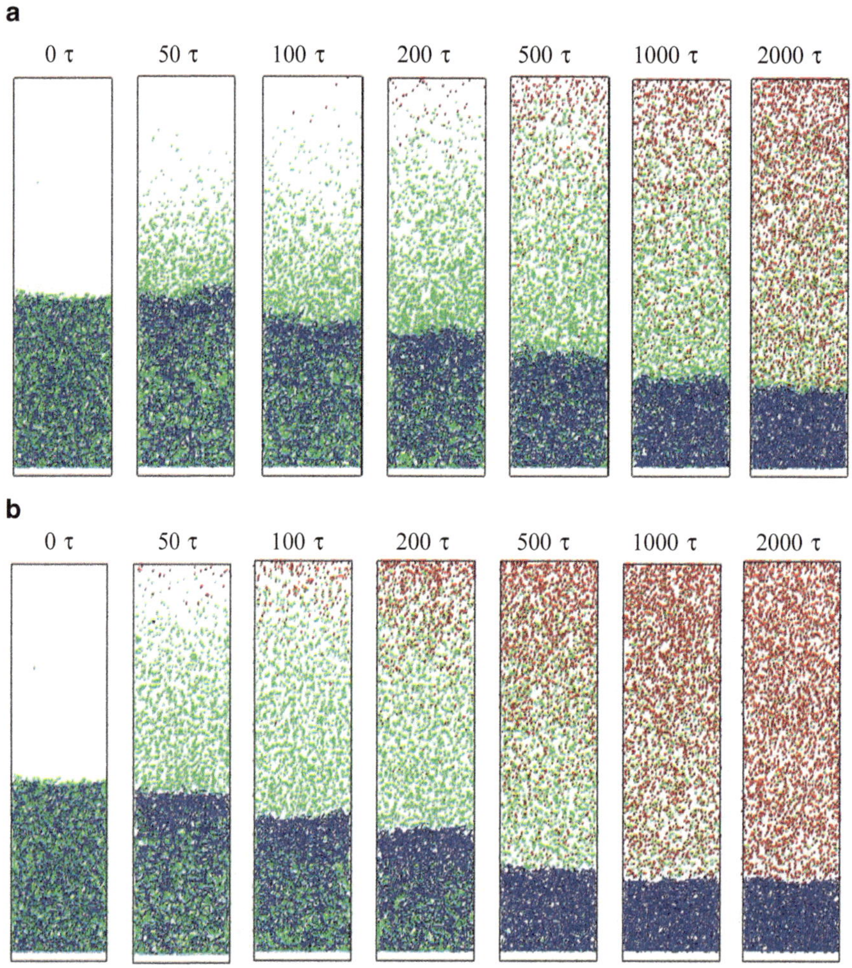

Fig. 20.2 Snapshots of evaporation simulation. (**a**) and (**b**) show the results for "Evp-03Evp_in.udf" and "Film-Evp-03Evp2_in.udf," respectively. In both plots, snapshots are shown for times 0, 50, 100, 200, 500, 1000, and 2000 τ

respectively. On the horizontal axis, the zero position indicates the interface of the substrate (bottom position in Fig. 20.2) and the right region indicates the air phase. In each plot, the line of phi A indicates the density profile of the polymer. The concentrated layer of the polymer can be found at time 100 τ in Fig. 20.2 (a) and time 50 τ in Fig. 20.2 (b). This is the skin layer in the simulation. However, this result must be treated carefully. These DPD simulations treat the model simulation in a nanoscale, and the results of the skin layer are also found as a nano-skin layer. A comparison based on experiments and further validations are needed. Meanwhile, the exchange of solvent with air particles is treated in the upper region of the

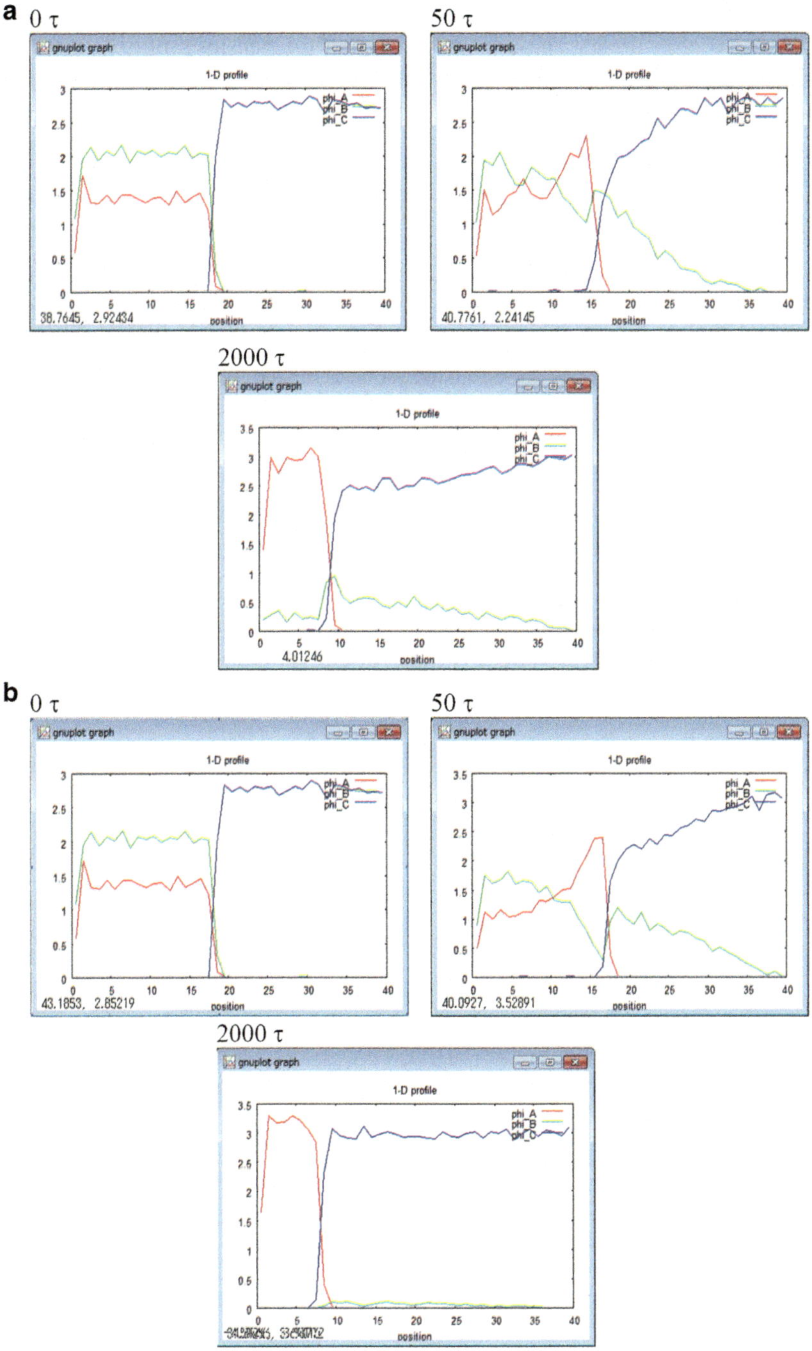

Fig. 20.3 Analysis of the density profile along the height axis. (**a**) Density profiles at times 0, 100, and 2000 τ for "Film-Evp-03Evp_in.udf." (**b**) Density profiles at times 0, 50, and 2000 τ for "Film-Evp-03Evp2_in.udf"

simulation box using a function of COGNAC. In Figs. 20.2 and 20.3, no saturation of the solvent in the air phase is observed, and the density profile of the solvent indicated as the line of "phi B" decreases as the height position increases. The results show that this function works well.

20.6 Download Instruction

Sample files are available from the OCTA website. See the instruction on http://octa.jp/download_bookdata.html. Password for the sample(zip file) of this chapter is "H4gX2GFp".

References

1. Y. Harasaki, *Coating Engineering* (Asakura Shoten, Tokyo, 1971) (in Japanese)
2. T. Okuzono, K. Ozawa, M. Doi, Phys. Rev. Lett. **97**, 136103 (2006)
3. T. Okuzono, M. Doi, Phys. Rev. E **77**, 030501 (2008)
4. S. Yamamoto, Nihon Reoroji Gakkaishi **32**, 295 (2004)
5. H. Morita, T. Ozawa, N. Kobayashi, H. Fukunaga, M. Doi, Nihon Reoroji Gakkaishi **36**, 93 (2008)

Chapter 21
Crystallization in Thin Films of N-Alkanes

Takashi Yamamoto

21.1 Introduction

Long-chain molecules such as n-alkanes and their derivatives are materials of popular interest in polymer science, colloid science, biology, and pharmacology [1]. Many long-chain molecules have hydrophilic heads and hydrophobic tails and have various, interesting morphologies such as micelles, vesicles, and flat bilayers in aqueous solutions. In the present chapter, however, we restrict our discussions to thin films of simple n-alkanes on solid substrates; extension of the present simulations to more complicated molecules is simple and straightforward.

Organic thin films on solid substrates have great relevance to many industrial challenges such as lubrication and anticorrosion, as well as to future technologies of molecular electronics, in which crystallization or self-organization is of central importance [2]. The crystallization of n-alkane thin films on metals, silicon oxide, and graphite has long been studied. In general, the chain molecules tend to lie parallel to the substrate when strong attraction is exerted by the substrate of metals and graphite. In contrast, the chain molecules tend to aggregate with their chain axis approximately normal to the substrate when the substrate attraction is weaker in SiO_2 (see Fig. 21.1).

However, the true origin of such molecular organization on a substrate is not well understood. The organized structure is not necessarily determined by the equilibrium thermodynamics, and the kinetics of crystallization are important. Crystallization in the n-alkane thin film is considered to be affected by various parameters such as the film thickness, crystallization temperature, and chain length and interaction with the substrate. Understanding the roles of these parameters

T. Yamamoto (✉)
Graduate School of Science and Engineering, Yamaguchi University, Yamaguchi, Japan
e-mail: yamamoto@mms.sci.yamaguchi-u.ac.jp

Japan Association for Chemical Innovation, *Computer Simulation of Polymeric Materials*, DOI 10.1007/978-981-10-0815-3_21

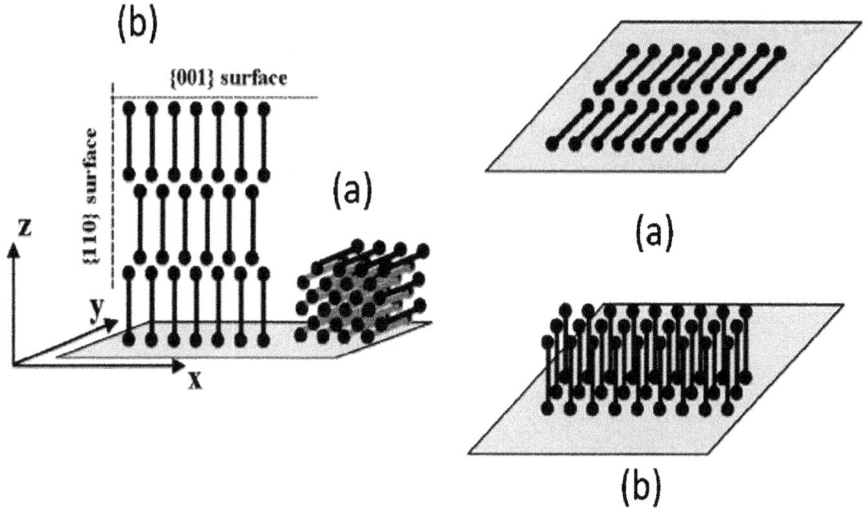

Fig. 21.1 Typical molecular orientations of chain molecules either parallel (**a**) or perpendicular (**b**) to the substrate

is important in devising thin films having desirable properties. However, the complicated effects of many parameters are not easy to grasp, and therefore, molecular simulations are promising.

Although a few simulation works have been reported [3], more systematic investigations are needed. In this chapter, we present simulations to give a brief introduction to this area of research. All are carried out using the coarse-grained molecular dynamics (MD) program (COGNAC).

21.2 Molecular Models and Simulation Conditions

We here consider a short n-alkane $C_{11}H_{24}$ (undecane C_{11}) and a slightly longer n-alkane $C_{19}H_{40}$ (nonadecane C_{19}). To reproduce the crystal structure precisely, especially the space group symmetry, we must consider a full-atomistic model of n-alkanes with explicit hydrogen atoms. However, we here give only a semiquantitative account of the crystallization mechanism by adopting a simple united atom model in which a carbon and hydrogens in CH_2 and CH_3 are combined into a united atom. There are several united atom models that have been proposed, from relatively old and conventional models to more recent and accurate models. We here adopt the conventional united atom model proposed by Rigby and Roe, which has been used in various studies of n-alkanes and polyethylene [4] and is readily available in COGNAC.

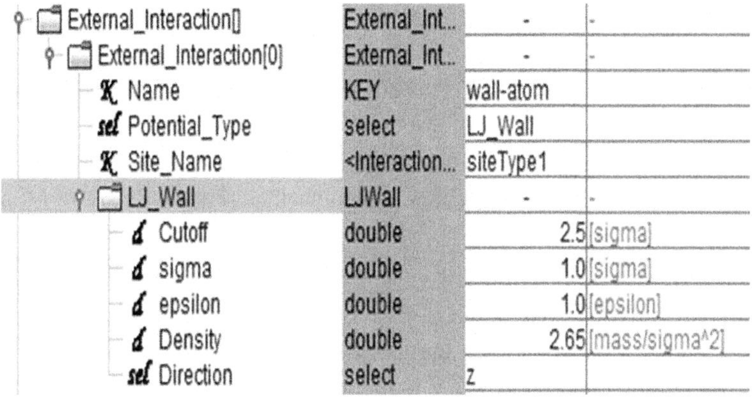

Fig. 21.2 Parameter setting for the LJ flat wall

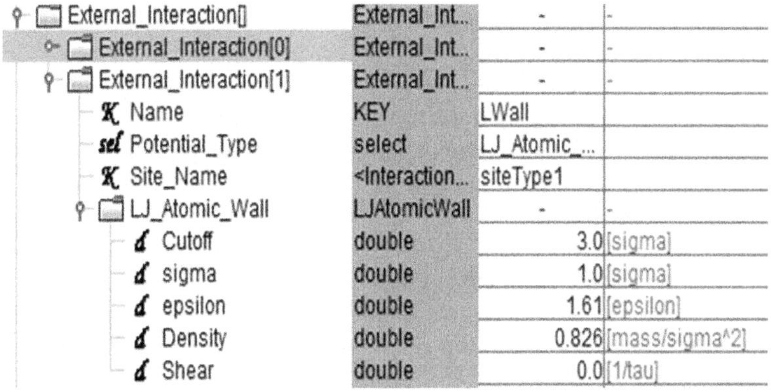

Fig. 21.3 Parameter setting for the LJ atomic wall

Modeling of the substrate is important but not always simple [5]. The substrate models range from very realistic models that consider detailed atomistic structures to much simpler models that consider only attractive and repulsive forces normal to the substrate surface. The simplest model is the Lennard–Jones-type flat wall (LJ flat wall) model (Fig. 21.2), whose interaction potential is only described as a function of the distance z between the chain atoms and the substrate surface.

When we must consider the detailed atomistic structure of the substrate, we can build a substrate of arbitrary lattice symmetry into the model. However, COGNAC has a convenient LJ atomic wall. Although it can handle only a simple 2D cubic lattice, it is useful for initial trial simulations of atomic walls. Figure 21.3 shows how to give related parameter values for a square lattice substrate with given surface density and interaction strength.

21.3 Results of Simulations

21.3.1 Thin-Film Crystallization of C_{11} on the Flat Wall

Let us first consider the short n-alkane C_{11} on the LJ flat wall. We here assume that the interaction potential between the wall and chain atoms has the form

$$U_{\mathrm{ext}}(z) = 4\pi \varepsilon_w \rho_w \left(\frac{1}{5} \left({}^{\sigma_w}\!/_{z} \right)^{10} - \frac{1}{2} \left({}^{\sigma_w}\!/_{z} \right)^{4} \right), \tag{21.1}$$

where z is the distance between the chain atoms and substrate. The minimum of the potential is located at $z = \sigma_w$ with the depth $-\frac{12\pi\varepsilon_w\rho_w}{10} = -3.77\varepsilon_w\rho_w$. Because the potential depth is given by the product $\varepsilon_w\rho_w$, the partitioning to ε_w and ρ_w is arbitrary. By setting the parameter as $\varepsilon_w = \varepsilon$, where ε is the LJ parameter for the united atom CH_2, we here control the depth of the wall potential by ρ_w. In our previous work, the depth of the standard substrate was considered -10ε [3]. For comparison of the present simulations with our previous simulations, we consider that $3.77\ \varepsilon_w\rho_w = 10\varepsilon$ for the standard substrate, i.e., $\rho_w = 2.65$. In this chapter, we also consider $\rho_w = 1.33$ for the weakly attractive substrate.

Here we briefly explain the simulation procedures adopted for the thin film of C_{11}. We first considered an MD cell that was periodic in the X–Y direction and bound by repulsive upper and lower walls. Within the cell having dimensions A = 20σ, B = 10σ, and C = 10σ, 128 chains of C_{11} were placed at random (Fig. 21.4a).

After waiting for the chains to aggregate near the wall at low temperature (180 K) and by assuming a very attractive wall ($\rho_w = 10$), we doubled the MD cell size C along the Z-direction (Fig. 21.4c) and then reset the attractive potential ρ_w to the original potential, $\rho_w = 2.65$ or $\rho_w = 1.33$. In this way, we obtained starting thin films that were adsorbed onto the wall. After relaxation at a higher temperature, which enables the film to reach a relaxed melt state (Fig. 21.5a), we investigated the crystallization of the film at 240 K versus the strength of the attractive force (ρ_w) of the substrate.

Figure 21.5b, c show the final configurations of the crystallized films on the substrates of standard attraction ($\rho_w = 2.65$) and weak attraction ($\rho_w = 1.33$), respectively.

On the attractive substrate ($\rho_w = 2.65$), all chains lie parallel to the substrate and form a two-dimensional crystalline film. The chains form stacked lamellae similar to usual n-alkane crystals (Fig. 21.5b). Because of the short simulation time, the system is not fully equilibrated and misoriented chains remain. Meanwhile, when the chain molecules are crystalized on the weakly attractive substrate ($\rho_w = 1.33$), the chains show a definite tendency to stand up on the substrate, forming a monolayer with perpendicular orientation (Fig. 21.5c). The chain molecules are densely packed to form a hexagonal monolayer with a slight tilting of the chains against the substrate normal. These observations are consistent with those of our previous report [3], which made additional studies of systems with a much larger number of chains than the current report.

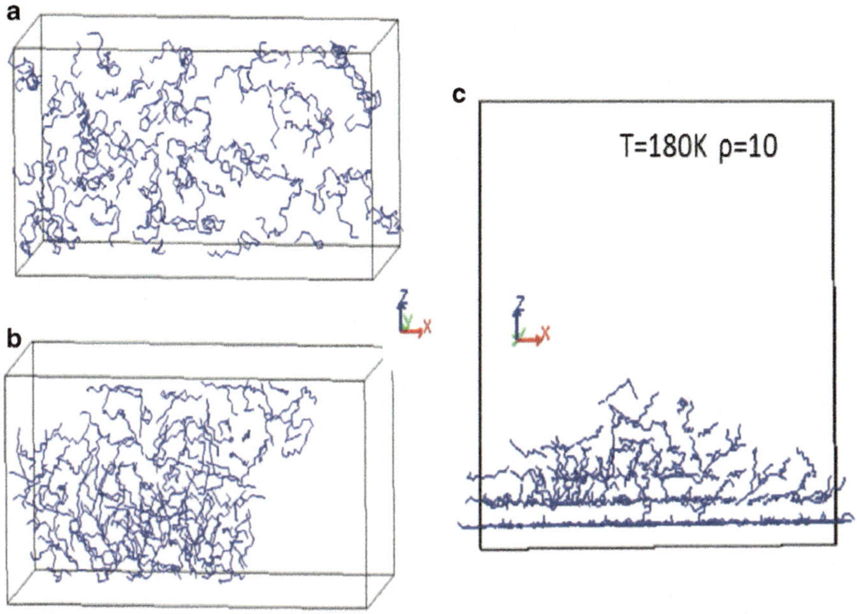

Fig. 21.4 Preparation of a thin film of C_{11} melt on the flat wall (undecane128_initial_in.udf, undecane128_initial_out.udf)

21.3.2 *Crystallization of C_{19} on Atomic Walls: The Effect of Commensuration on the Structure of the Film*

As explained above, thin-film crystallization is greatly affected by the mean strength of the substrate attraction. However, various other factors also contribute. For example, the matching or epitaxy between the substrate lattice and the n-alkane crystal is important. This section explains simple examples of the epitaxial crystallization of C_{19} on the substrate.

In crystals of n-alkanes of medium-chain length, the interchain separations in the present model are around 0.44 nm. We here consider two kinds of atomic wall having a two-dimensional square lattice with lattice constants a = b = 0.42 nm = 1.1σ and a = b = 0.38 nm = 1.0σ, where σ is the LJ parameter for CH_2 and taken as 0.38 nm. The lattice constants are nearly commensurate for the substrate and incommensurate for the n-alkane crystals. We here use the LJ atomic wall potential in COGNAC and assume a similar LJ potential for the interactions between chain atoms and substrate atoms.

For the incommensurate substrate ($a = 0.38$ nm), the atomic density on the substrate is $\rho_w = 1/a^2 = 1.0 \quad [1/\sigma^2]$, while $\rho_w = 1/a^2 = 0.82 \quad [1/\sigma^2]$ for the commensurate substrate ($a = 0.42$ nm). To make the expected average attraction $\varepsilon_w \rho_w$ of the substrate similar to that of the weakly attractive substrate considered

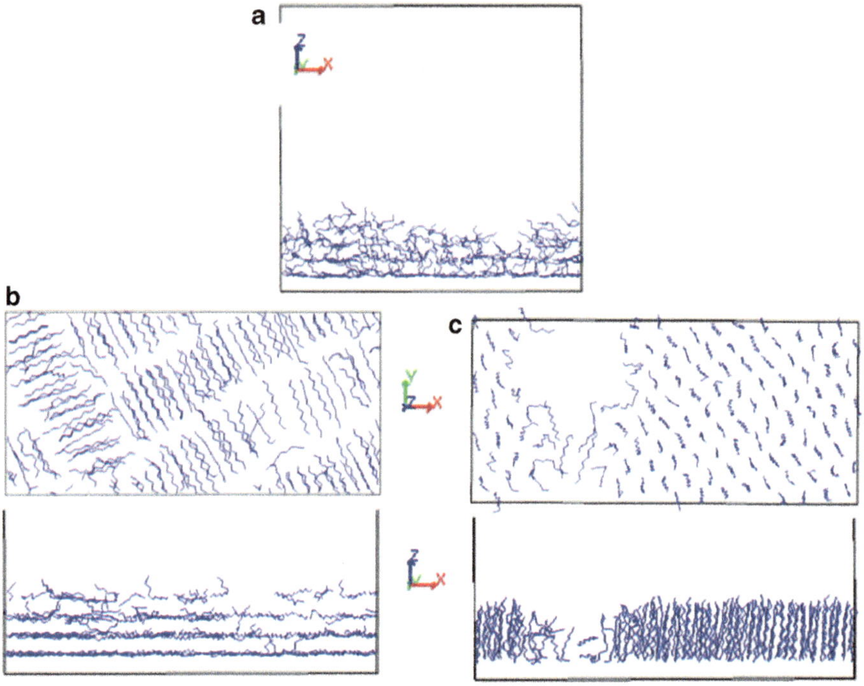

Fig. 21.5 Crystallization and chain orientation in the thin film of C_{11} versus strength of the attraction by the substrate; the case of an attractive substrate (**b**) and that of a weakly attractive substrate (**c**), both the top view along the Z-axis (*above*) and the side view along the Y-axis (*below*) (undecane128_Xtalz1_in.udf, undecane128_Xtalz1_out.udf, undecane128_Xtalz05_in.udf, undecane128_Xtalz05_out.udf)

in the previous section, we take $\varepsilon_w \rho_w = 1.33$. This means that $\varepsilon_w = 1.33\varepsilon$ for the former and $\varepsilon_w = 1.62\varepsilon$ for the latter. Because each MD cell is connected by the periodic boundary condition in the X and the Y directions, the dimensions of the MD cell A and B must be integer multiples of the substrate lattice spacing. We took $A = B = 20\sigma$ for the incommensurate system ($a = b = 1.0\sigma$) and $A = B = 22\sigma$ for the commensurate system ($a = b = 1.1\sigma$). The height C of the MD cell is taken to be a little larger than $C = 30\sigma$ used before. Within the MD cell, we placed 256 chains of C_{19}, about twice the number used in the previous section.

After equilibrating the melt film (Fig. 21.6), we cooled the system to 280 K and observed crystallization.

In the case of the incommensurate substrate, many of the chains were found to crystallize by standing up from the substrate while others lay parallel to the substrate (Fig. 21.7); the crystalline film was a mixture of parallel and perpendicular crystallites.

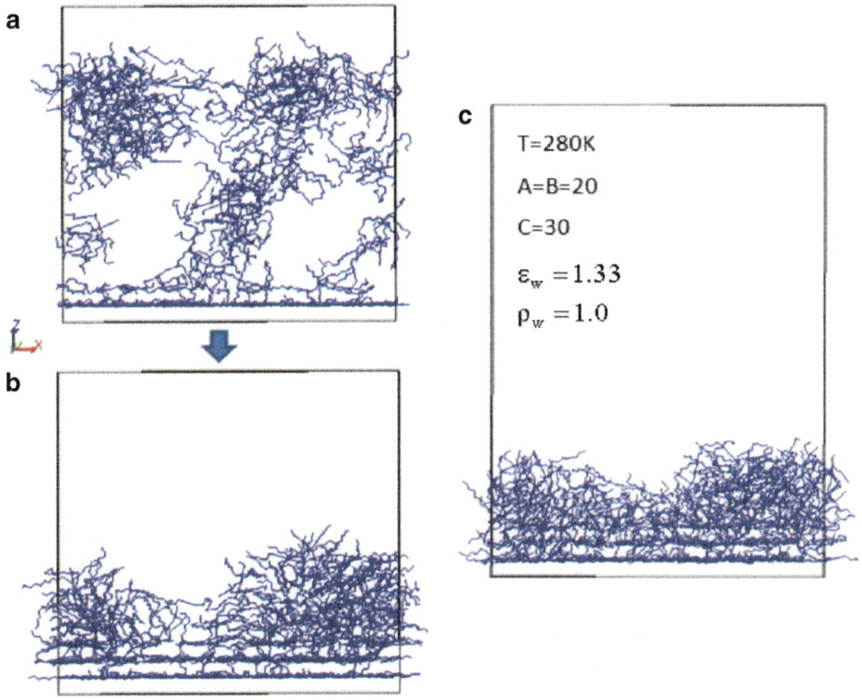

Fig. 21.6 Preparation of a thin film of C_{19} melt on the atomic wall (nonadecane256_initial_in.udf, nonadecane256_initial_out.udf, nonadecane256_AtomicWall_A_Relax_in.udf, nonadecane256_AtomicWall_A_Relax_out.udf)

Meanwhile, on the commensurate substrate, the chains had a marked tendency to lie parallel to the substrate (Fig. 21.8). It is also noted that the chains lay along the X-axis of the substrate crystal, showing good epitaxy.

21.4 Conclusions and Remarks

By assuming simple geometry for the solid substrate and LJ interactions between the substrate and chain atoms, we presented simulations of n-alkane crystallization on a substrate. Two typical chain orientations, perpendicular and parallel to the substrate, were reproduced depending on the chain length, strength of the substrate attraction, and the commensurability between the substrate lattice and the crystal lattice of the n-alkane.

The crystallization we observed here occurred within a very short time. The degree of supercooling must be much larger than that under the usual crystallization condition. It is known from experiments that the chain orientation in crystalline films

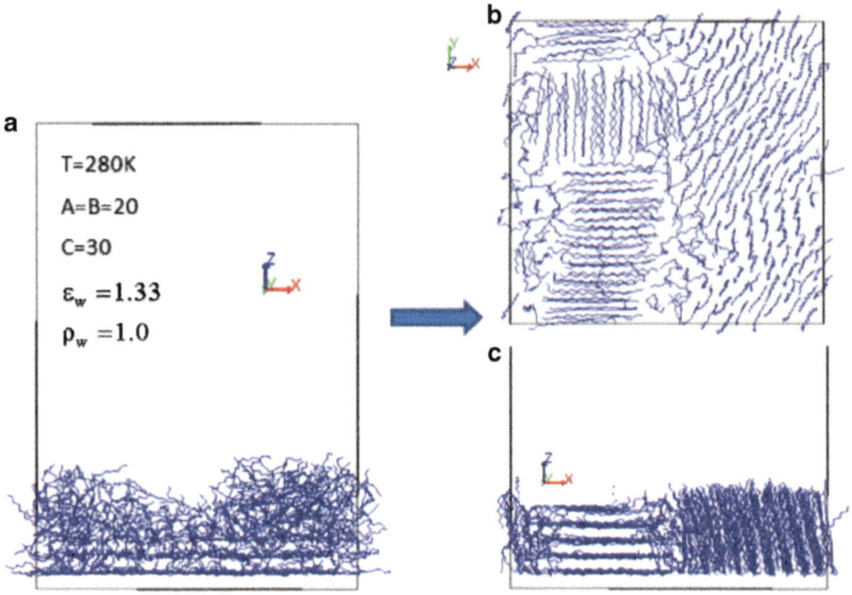

Fig. 21.7 Crystallization and chain orientation in the thin film of C_{19} on the incommensurate substrate; *top view* (**b**) and *side view* (**c**) (nonadecane256_AtomicWall_A_Xtallz_in.udf, nonadecane256_AtomicWall_A_Xtallz_out.udf)

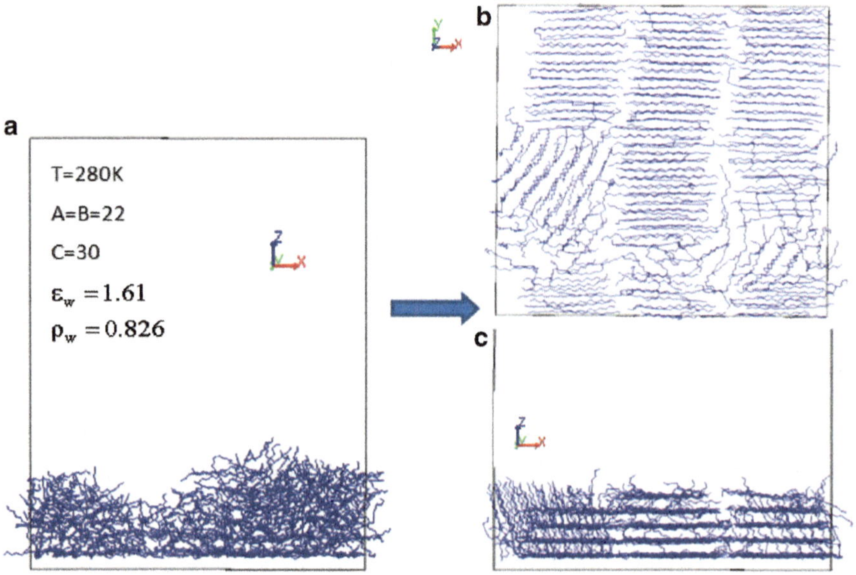

Fig. 21.8 Crystallization and chain orientation in the thin film of C_{19} on the commensurate substrate; *top view* (**b**) and *side view* (**c**) (nonadecane256_AtomicWall_B_Xtallz_in.udf, nonadecane256_AtomicWall_B_Xtallz_out.udf)

is also affected by other factors, such as the crystallization rate and film thickness, where crystallization kinetics cannot be ignored. The control of morphology is important in organic thin-film technologies. We need a much deeper understanding of fundamental molecular processes, for which computer modeling will make great contributions.

21.5 Download Instruction

Sample files are available from the OCTA website. See the instruction on http://octa.jp/download_bookdata.html. Password for the sample(zip file) of this chapter is "9FNwJPYS".

References

1. M. Small, *The Physical Chemistry of Lipids* (Plenum, New York, 1986)
2. A. Ulman, *An Introduction to Ultrathin Organic Films* (Academic, San Diego, 1991)
3. T. Yamamoto, K. Nozaki, A. Yamaguchi, N. Urakami, J. Chem. Phys. **127**, 154704 (2007)
4. D. Rigby, R. Roe, J. Chem. Phys. **87**, 7285 (1987)
5. W.A. Steele, Surf. Sci. **36**, 317 (1973)

Chapter 22
Improvement of Adhesive Properties Through the Segregation of Oligomers and an Investigation of the Mechanism Using SUSHI Simulation

Hiroshi Sasaki

22.1 Introduction

To design novel high-performance materials using chemical compounds, it is important to address the mesoscale region, up to an order of 10^6 in the length dimension, between the nanoscale region of molecular structures and the macroscale region in which actual material properties are evaluated. From a coarse-grained point of view, a polymer chain can be described as a "very thin and long thread twisting together and rolled into a ball." Indeed, this description aptly encompasses the distinctive features of a polymer chain. One such feature is "segregation," which is the bleeding out of small molecules or oligomers, which is defined as low molecular weight (Mn < 10,000) polymer, from the polymer bulk and localization at the interface between the polymer layer and substrate. Segregation seems to be a low cost and effective way of improving the macroscale material performance through the formation of a mesoscale structure.

In an adhesion application in which more than two different substrates are joined, it is well known that the interface makes a large contribution to the adhesive performance. By controlling the molecular weight and/or solubility of properly functionalized oligomers, one can concentrate them to an interface through segregation and possibly improve macroscale properties, such as adhesion.

Recently developed personal information devices, such as smartphones and tablet personal computers, widely use touch panels. The touch panel component is assembled employing adhesives, including pressure-sensitive adhesives (PSAs), with superior optical properties; these adhesives are called optically clear adhesives (OCAs). The adhesive layer should join parts together firmly while maintaining

H. Sasaki (✉)
Toagosei Co. Ltd., Nagoya, Aichi, Japan
e-mail: hiroshi_sasaki@mail.toagosei.co.jp

Japan Association for Chemical Innovation, *Computer Simulation of Polymeric Materials*, DOI 10.1007/978-981-10-0815-3_22

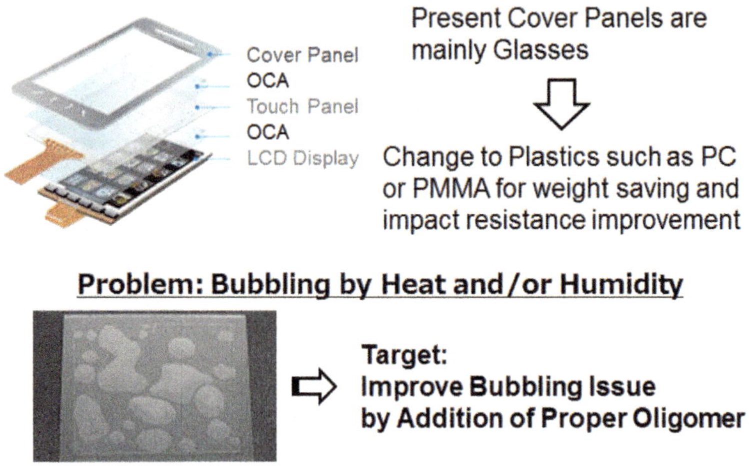

Fig. 22.1 Touch panel devices and bubbling issue of plastic panels

high optical performance. As the OCA performance is directly linked to usability, manufacturers wish to improve the properties of OCAs.

Recent research on cover panels has investigated weight reduction, improvement of the impact resistance and aesthetic design, and the substitution of glass with plastics, such as polycarbonate (PC) and polymethylmethacrylate. Plastic cover panels, however, present a problem in that bubbles often form in an accelerated evaluation under high temperature and humidity. The prevention of bubbling during this process is important for further development of touch screens. Schematic structure of touch panel devices and bubbling issue of plastic panels are summarized in Fig. 22.1. In our previous development of an OCA for plastic panels, we found that the addition of selected oligomers effectively suppresses bubble generation and that the suppression mechanism can be explained by the segregation of oligomers into the interface of the substrates.

This chapter explains the development process of novel oligomers for bubbling suppression and presents a SUSHI simulation investigation of the suppression mechanism of oligomer addition.

22.2 Development Flow

22.2.1 Estimation of the Bubble Generation Mechanism

Using a PC panel adhered to polyethylene terephthalate (PET) film with PSA, we investigated the mechanism of bubble generation.

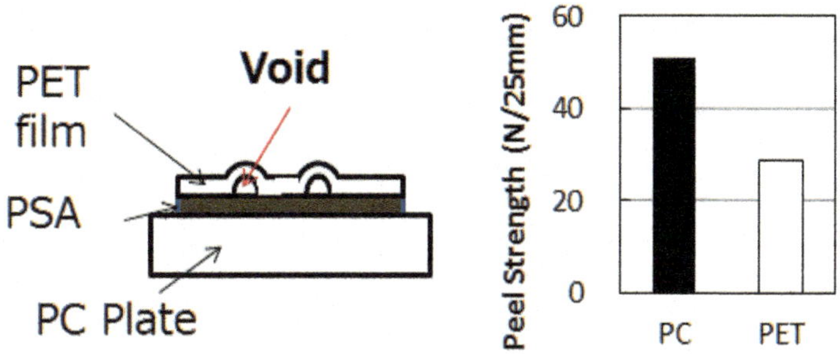

Fig. 22.2 Evaluation of bubble generation and adhesion strength for different substrates

Microscopic observation of bubbled samples from a vertical direction view and perpendicular cut view revealed that bubbles tended to form around the interface between the PSA layer and PET film. The peeling strength of the PSA for PET was much lower than that of the PSA for PC. Schematic model of bubble generation and adhesion strength for different substrates are shown in Fig. 22.2. Additionally, generation of out gas from the PC panel by heating was confirmed by gas chromatography.

Considering the above results, a model of the bubble generation mechanism is constructed as follows:

- First, out gas is generated from the PC panel by applying heat.
- The generated gas diffuses through the PSA layer.
- Finally, the diffused gas generates bubbles in the weakest region, which is the PSA/PET interface.

This model suggests that improving the relatively inferior adhesion strength—in this case that of the PET interface—would be effective for bubble suppression.

22.2.2 *Investigation of Oligomers*

Employing a copolymerization technique, the properties of the polymer can be changed and controlled by selecting proper monomer combinations. In this research, the "solubility parameter" (SP) and "glass transition temperature" (T_g) are considered controlling parameters of oligomer properties. Using these two parameters, a variety of oligomers were synthesized through the copolymerization of different acrylic monomers.

Three oligomers with different SP values (T_g approximately 80 °C, which is the atmospheric temperature in the heat test) were formulated with PSA, and bubble generation through the application of heat (85 °C, 85 %RH, 24 h) was evaluated by visual inspection. The cloud point of each oligomer formulation was OL-1: 4 wt%, OL-2: 9 wt%, and OL-3: 16 wt%.

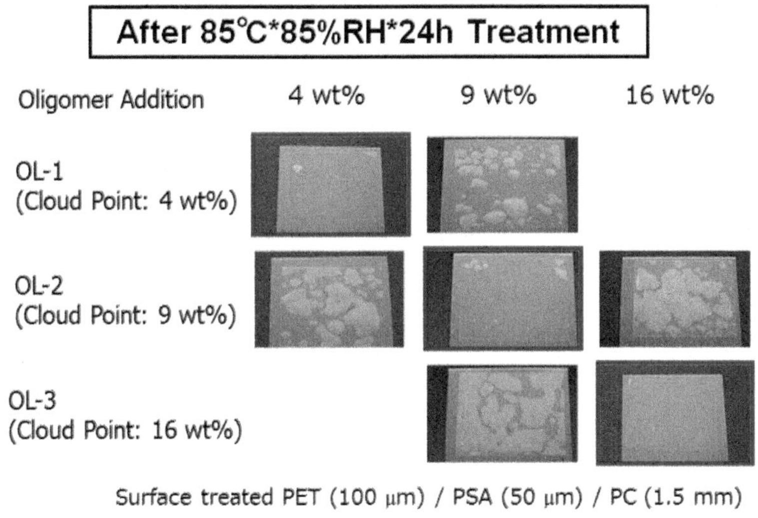

Fig. 22.3 Bubbling suppression (85 °C ×85 %RH ×24 h)

Bubble generation was effectively suppressed around the cloud point addition of oligomers, but a lesser or excess amount of oligomer addition had no effect on bubble suppression as shown in Fig. 22.3. According to the bubble generation model, the adhesion strength of the PET/PSA interface is likely enhanced around the cloud point.

22.2.3 Segregation of Oligomers into the Interface

The solubility of oligomers and polymers can be discussed using a phase diagram, which can be theoretically treated using Flory–Huggins theory. χ parameters, which represent the interaction strength of each segment in Flory–Huggins theory, can be estimated from SP values. Under experimental conditions, because of thermal fluctuation, the clouding of mixture samples can occur just inside the binodal line, which is called the metastable binodal region. For the three previously described oligomer formulations, an experimental binodal line can be drawn using the oligomer amount for the cloud points as shown below (Fig. 22.4).

In the region near the interface, where the length is in the order of R_g (radius of gyration) of the polymer chain, the conformational entropy of polymers decrease. On the other hand, the conformational entropy loss for oligomers is much smaller than long-chain polymer. As a result of lower loss of conformational entropy, oligomers tend to concentrate in this area instead of the polymer chain. This phenomenon likely relates to an effective increase in the χ parameter near the

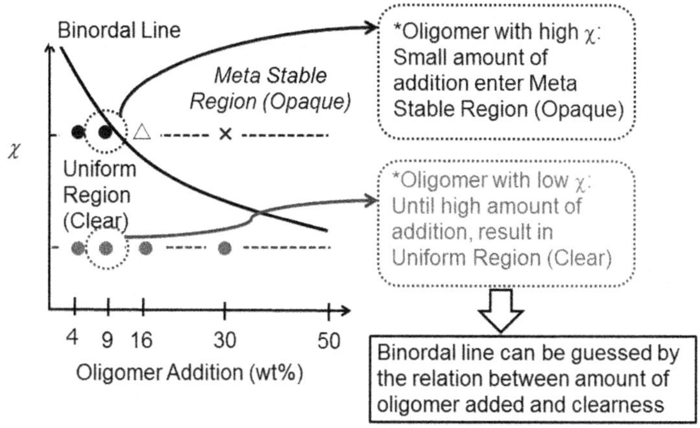

Fig. 22.4 Phase diagram of oligomers/polymers

interface. Segregation of the oligomer having a lower molecular weight can be explained by this mechanism.

Considering the interface effect shown above (i.e., with the addition of the oligomer around the cloud point), the entire formulated system can be clearly and uniformly mixed, and the amount of segregated oligomer near the interface might increase. Previously presented results of bubbling suppression by the addition of the cloud-point amount of oligomers (Fig. 22.3) can be explained by the enhanced adhesion strength of the PET/PSA interface due to the thick segregation of oligomer with proper T_g at the interface.

The segregation content of oligomers was examined for the same formulations evaluated in Fig. 22.3 using X-ray photoelectron spectroscopy (XPS). XPS measurements are summarized in Fig. 22.5 in conjunction with peel strength data of PET/PSA for each adhesion sample. For all oligomer formulations, the addition of the cloud-point amount of oligomer resulted in almost 100 % surface concentration of oligomer, and the peel strength also reached a maximum at that point. The addition of excess oligomer resulted in an almost saturated concentration but reduced the peel strength.

22.3 Investigation of the Bubbling Suppression Mechanism Using SUSHI Simulation

To establish a proper model that can explain the "experimental results" presented so far, one-dimensional self-consistent field (SCF) calculations were conducted using the SUSHI simulator. The local concentration of each component (polymer or oligomer) near the interface can be calculated in this simulation, and each profile

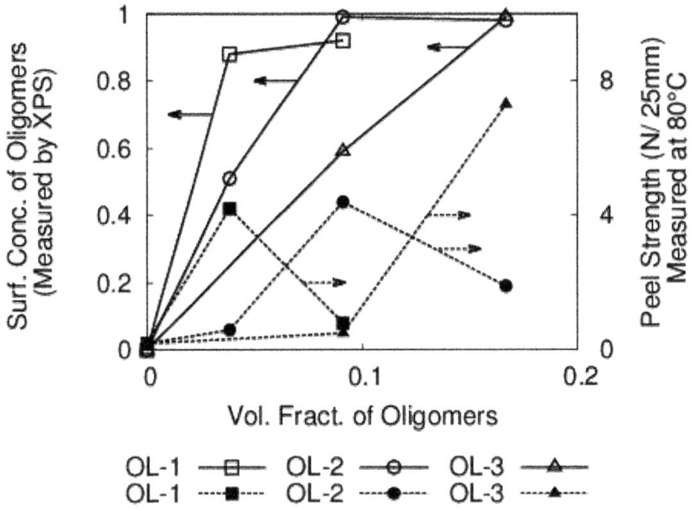

Fig. 22.5 Oligomer amounts added with surface concentrations by XPS and peel strength at 80 °C

can be checked visually. Additionally, using the calculated profiles of components, the surface concentration measured by XPS is estimated.

22.3.1 Setting of the Interaction Parameter χ in Accordance with Experimental Results

As previously noted, the phase separation of polymer blends will occur with a formulation content in the binodal region, because of the rather high thermal fluctuation effect. Therefore, the phase separation experimentally determined by the "cloud point measurement" should correspond to the binodal formulation. Meanwhile, in SCF calculations based on density functional theory, no phase separation is seen in the metastable binodal region, which is below the spinodal formulation ratio, without extra thermal fluctuation. When employing a grand canonical ensemble simulation and introducing a wall into the simulation system, it is known that there will be phase separation in the binodal regime caused by the segregated component near the wall acting as a nuclei.

Noting the points discussed above, it seems proper to set the simulation χ parameter according to a phase diagram calculated using Flory–Huggins theory, so that the experimentally determined phase separation at the cloud point corresponds to the binodal formulation ratio of the diagram. The phase diagram shown in Fig. 22.6 was drawn for the formulation of an oligomer that consists of six A segments ($N_A = 6$) with a polymer that consists of 600 B segments ($N_B = 600$).

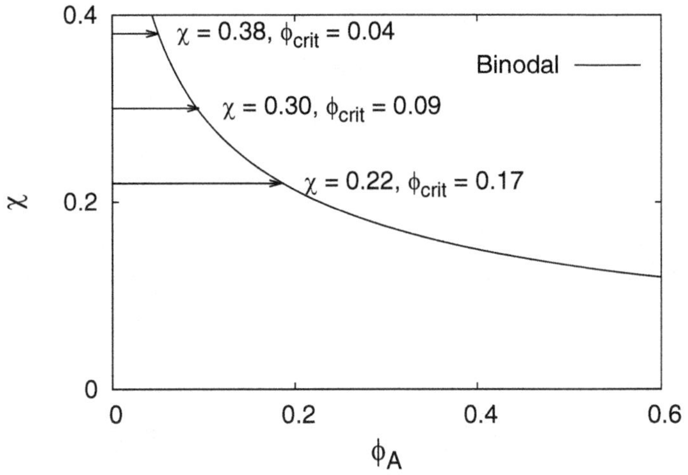

Fig. 22.6 Phase diagram for formulation of $N_A = 6$ oligomer with polymer ($N_B = 600$) and correspondence to experimental results

Table 22.1 SUSHI conditions

Polymer and oligomer	The oligomer consists of six *A* segments and the polymer of 600 *B* segments ($N_A = 6, N_B = 600$)
Simulation system	One dimension, Mesh = 0/20/80 (dx = 0.25)
Boundary condition	"WALL" and "NEUMANN"
Interaction parameter (χ_{AB})	$\chi_{AB} = 0.22, 0.30, 0.38$, corresponding to experimental data (Fig. 22.6)
Ensemble	Grand canonical ensemble

The segment numbers were determined according to the experimental molecular weights. The same formulation was used for SUSHI simulations discussed lately.

22.3.2 Simulation Conditions for SUSHI

To visualize the oligomer profile near the interface, one-dimensional SCF calculations were conducted using the SUSHI simulator. The conditions selected for SUSHI simulations are listed in Table 22.1.

Selecting a grand canonical ensemble, a formulation imbalance, such as an excess or shortage of a component due to segregation in the system, can be compensated by using an outer reservoir. This situation corresponds to focusing on the area near the wall of a large flask in an experiment, with the flask bulk acting as the reservoir.

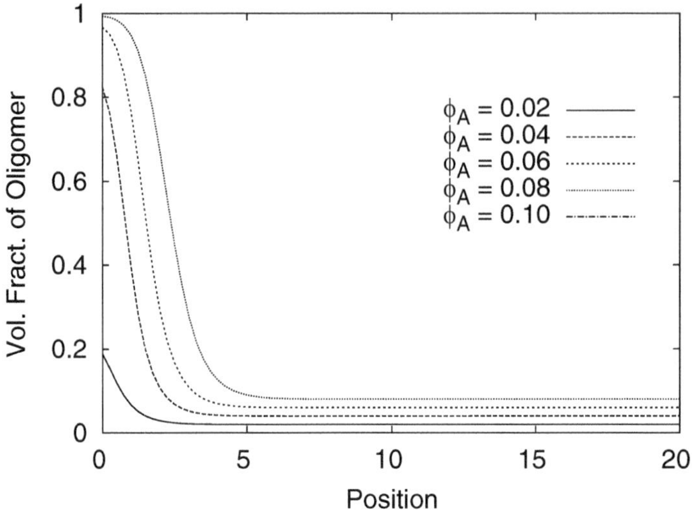

Fig. 22.7 Oligomer concentration profile for $\chi_{AB} = 0.3$ formulations near Interface (Correspond to OL-2 (uniform until 9 %))

22.3.3 Simulation Results for OL-2 ($\chi_{AB} = 0.3$)

The simulated results for $\chi_{AB} = 0.3$ formulations (simulating OL-2: uniform until 9 % addition) with a wall (position = 0 in Fig. 22.7) are summarized and shown in Fig. 22.7.

A greater amount of oligomer addition until the binodal line resulted in a thicker layer of the oligomer-rich region near the wall (interface). After entering the metastable regime ($\phi_A \geq 0.1$), local phase separation occurred and the simulation box was filled with oligomer, which is not recognizable in Fig. 22.7 because of overlapping to the top line of the graph.

22.3.4 Consistency with XPS Results

XPS is a surface-sensitive quantitative spectroscopic technique that measures the elemental composition that exists within a material. XPS spectra are obtained by irradiating a material with a beam of X-rays while simultaneously measuring the kinetic energy and number of electrons that escape from the top 0 to 10 nm of the material being analyzed. The mean free path (λ) of the photo-emitted electron, which implies the length over which the number of electrons will be reduced by a factor of $1/e$ because of collision with atoms in the material, is known to be about 3 nm in normal organic materials.

A measurement in the depth direction can be made by changing the angle of slant of the sample and electron detector. The direction normal to the sample surface is set to $\theta = \pi/2$, and the increase in depth is expressed by shortening the mean free path in the form of $\lambda \sin\theta$. According to this idea, the volume fraction of each component $\phi(d)$ at a distance d from the interface (i.e., the wall in the simulation) can be expressed by the following Eq. (22.1), using l_f as a fitting parameter in combination with the simulation length d:

$$\begin{cases} \phi_A^{XPS}(sim) = \dfrac{\sum\limits_d \phi_A(d) \cdot \exp\left(-\dfrac{l_f d}{\lambda \sin\theta}\right)}{\sum\limits_d \exp\left(-\dfrac{l_f d}{\lambda \sin\theta}\right)} \\ \phi_B^{XPS}(sim) = \dfrac{\sum\limits_d \phi_B(d) \cdot \exp\left(-\dfrac{l_f d}{\lambda \sin\theta}\right)}{\sum\limits_d \exp\left(-\dfrac{l_f d}{\lambda \sin\theta}\right)} \end{cases} \tag{22.1}$$

Employing experimental XPS results for the angle of slant of $\theta = 45°$, the experimental data were fitted with the simulation volume fraction of $\phi(d)$ using l_f as a fitting parameter. Good correlation was seen for $l_f = 4$, and it was found that a simulation dimensionless length of 1 corresponded to an experimental length of 4 nm. Using this fitting parameter, XPS measurement results in conjunction with data calculated from simulation results for $\chi = 0.22, 0.30, 0.38$ are summarized and shown in Fig. 22.8.

According to this fitting investigation, the segregated oligomer thickness for $\chi_{AB} = 0.3$ (shown in Fig. 22.7) was estimated to be about $10 \sim 20$ nm near the cloud point.

22.3.5 *Local T_g Near the Interface*

It is known that T_g for polymer mixtures can be estimated using the Fox equation and the weight fraction of each component. Assuming that the specific gravities of oligomer and polymer are almost the same, this equation can be rewritten using the volume fraction given below:

$$T_g \simeq \frac{T_{gA} \cdot T_{gB}}{\phi_A \cdot T_{gB} + \phi_B \cdot T_{gA}} \tag{22.2}$$

Employing this relation, local T_g near the interface is calculated for $\chi_{AB} = 0.3$ and shown in Fig. 22.9.

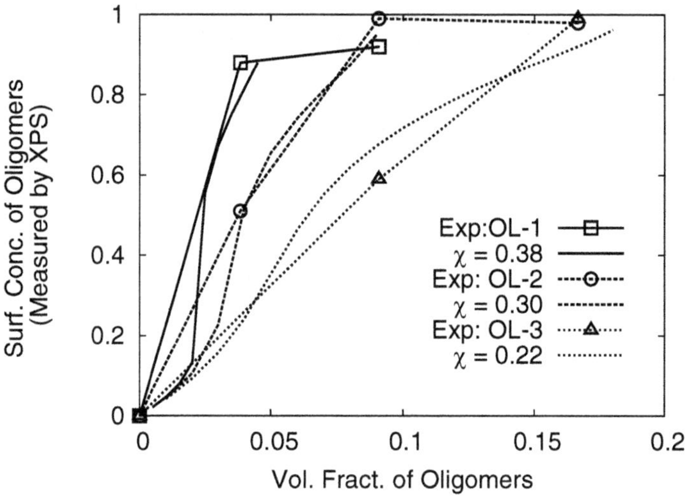

Fig. 22.8 XPS experimental results and calculated data using fitting parameter $l_f = 4.0$

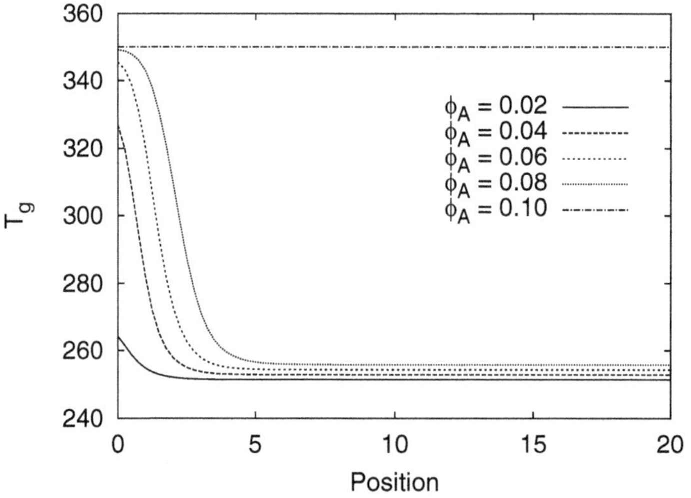

Fig. 22.9 Local T_g calculated for $\chi_{AB} = 0.3$

22.3.6 Summary of the Simulation

Comparing XPS measurements data with the interfacial oligomer segregation profile obtained in the SUSHI simulation, the thickness of the segregation layer was estimated to increase until $10 \sim 20$ nm near the addition of the cloud-point amount of oligomer. Furthermore, calculation of local T_g revealed that oligomer addition nearly up to the cloud point (binodal line) effectively enhanced local T_g near the interface.

Considering the above simulation results, the maximum experimental peel strength near the cloud-point addition of oligomer seemed to be established by a high-T_g layer with proper thickness near the interface. An excess amount of oligomer beyond the cloud point resulted in phase separation, and the peel strength was lowered by the fragile oligomer layer.

22.4 Conclusions

This chapter introduced, as an example for material design employing a SUSHI simulation, an improvement in the adhesion strength using the segregation of oligomers at an interface. An investigation based on the comparison of the oligomer concentration profile obtained by a simple one-dimensional SCF calculation with experimental XPS data provided a simple model for enhanced adhesion.

The combination of the simulation approach with experimental data provides greater clarity for polymer behavior at the mesoscale. Through this combinational approach, we can better understand such interfacial properties, which are relatively micro and local phenomena, and advances in adhesion applications are expected.

Chapter 23
Adsorption of Polyelectrolytes

Hiroki Kubo

23.1 Introduction

The adsorbed structure that forms when a cationic copolymer is adsorbed onto a negatively charged flat surface from a solution is simulated employing SUSHI's static self-consistent field (SCF) method. This chapter describes how to determine the main parameters, the modeling method, and calculation results for the example in which SUSHI conducts a supplementary examination of a report published by van de Steeg et al. [1]. In addition, the SUSHI manual provides details of a theoretical approach for incorporating the effect of an electrostatic field into the SCF method.

23.2 Method and Model Used by van de Steeg et al.

Van de Steeg et al. used a method in which Scheutjens-Fleer theory (employing a discrete-type SCF based on a lattice model) [2] is expanded to polyelectrolytes. In this method, electrostatic fields are handled basically in the same way as in SUSHI; however, the difference is that SUSHI employs an SCF described by a continuum-type model. Accordingly, when a supplementary examination is conducted using SUSHI, improvements must be devised by taking into consideration this difference; these improvements will be explained in this chapter.

H. Kubo (✉)
Analytical Science Research Center, Kao Corporation, Chūō, Japan
e-mail: kubo.hiroki@kao.co.jp

Japan Association for Chemical Innovation, *Computer Simulation of Polymeric Materials*, DOI 10.1007/978-981-10-0815-3_23

When conducting an adsorption simulation, van de Steeg et al. used a homopolymer in which electric charges are uniformly allocated to every monomer unit without taking into consideration every monomer arrangement of cationic/nonionic monomers in the cationic copolymer. Additionally, the homopolymer had a degree of polymerization (number of segments) of 100. Furthermore, the difference in the copolymerization ratio of the cationic monomers is represented by changing the electric charge (segment charge) per monomer. (Hereafter, the whole constituent unit, including a monomer unit, a hydrated ion, and water, is referred to as a "segment.") A segment is equal to a lattice unit in a lattice model, and all constituents of a polymer, such as the monomer unit, hydrated counterion, added salt ion, and water, are represented as one segment, the diameter of which is 0.6 nm.

The calculation is performed for a negatively charged flat plate using a model such as those described above and changing the polymer's segment charge, adsorption interaction parameter χ_s for the wall and a polymer segment, and the concentration of the added salt in the solution to compare the concentration profiles of the adsorbed layers and adsorbed amounts.

23.3 System Modeling Using SUSHI

Modeling is done to reproduce the model of van de Steeg et al. as faithfully as possible. As for the charged surface, a charged segment having an electric charge of -1 is grafted to a wall to achieve an appropriate graft density, thereby creating a surface having a surface electric charge density of -0.01 C/m^2 as in van de Steeg's model. Additionally, a homopolymer having 100 segments is used to handle the cationic copolymer. The other components are a counterion of the charged segment that is grafted, a counterion of the cationic polymer, positive and negative ions of added salt, and water. The SCF calculation is performed in one dimension with a concentration distribution only in the direction perpendicular to the wall (which is the same condition used by van de Steeg et al.). Figure 23.1 shows a conceptual drawing of the system.

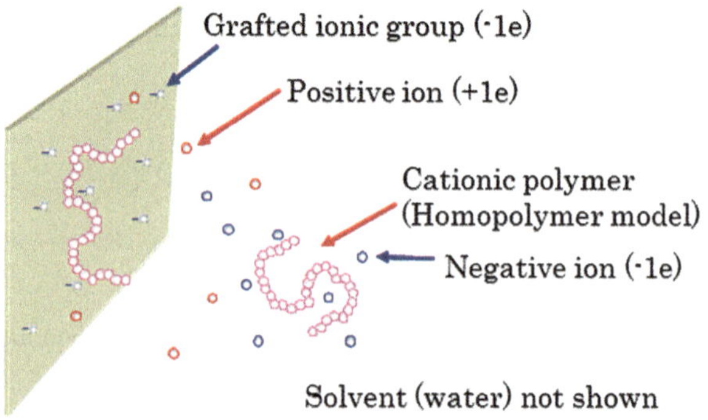

Fig. 23.1 Conceptual drawing of the system (although the actual calculation is performed using a one-dimensional SCF)

23.4 Determination of SUSHI Parameters

SUSHI is a continuum-type SCF but is similar to a discrete-type SCF; various parameters are thus set to have the same or nearest possible value as that of van de Steeg et al. The mesh size (=1) and effective bond length of segments (=1), which are made dimensionless in SUSHI, are regarded as corresponding to the segment size (lattice size) used by van de Steeg et al. (i.e., 0.6 nm in a system having dimensions).

Moreover, in allowing the simulation of an electrolyte system to correspond to a system having dimensions, the main question is how to determine the dielectric constant of the system when the Poisson equation is made dimensionless. Generally, the Poisson equation is written as $\nabla[\varepsilon \nabla \mathrm{U}] = -\rho$ using a nabla operator ∇ [1/m], a dielectric constant ε [C^2/Jm], an electrostatic potential U [J/C], and electric charge density ρ [C/m^3]. Meanwhile, each parameter in SUSHI is made dimensionless using a specific basic unit. The following basic units are used for conversion between a system with and without dimensions:

- Length (segment size) $l_0 = 0.6$ [nm]
- Energy $E_0 = k_B T = 1.381 \times 10^{-23}\,[\mathrm{J/K}] \times 298\,[\mathrm{K}] = 4.11 \times 10^{-21}\,[\mathrm{J}]$
- Elementary charge $e_0 = 1.60 \times 10^{-19}\,[\mathrm{C}]$
- Dielectric constant in a vacuum $\varepsilon_0 = 8.85 \times 10^{-12}\left[\mathrm{C}^2/\mathrm{Jm}\right]$

Using these basic units, the dimensionless Poisson equation is obtained as

$$\left(E_0\,\varepsilon_0\,\mathrm{l}_0/e_0{}^2\right)\,\nabla^{\sim}\varepsilon^{\sim}\nabla^{\sim}\mathrm{U}^{\sim} = {}^{-}\rho^{\sim}.$$

Here, the superscript "~" attached to each parameter indicates that the parameter has been made dimensionless, and the coefficient on the left side of the equation $\left(E_0\,\varepsilon_0\,\mathrm{l}_0/e_0{}^2 = 8.51 \times 10^{-4}\right)$ is a setting value for the "dielectric constant of the system."

The other main parameters are set as follows:

1. Boundary condition: [0] WALL/[1] NEUMANN is set.
2. Monomer: A five-component system consisting of a polymer segment (+), a counterion (−), a surface ionic group (−), a counterion (+) of the surface ionic group, and water or a seven-component system additionally including an added salt (+) and an added salt (−) is set.
3. Components: A homopolymer in which 100 charged segments are linearly linked is used as the polymer, and the number of segments is set to 1 for all other components.

   For ionic groups that are to be grafted to the wall, type:GENERAL is selected to set junction pairs[0], and "first" is set to zero and "second" is set to 1 for the graft condition.
4. Volume fraction: The whole component "canonical" is set.

   - Polymer: A range from 0.3 to 2.0 % is input so that the bulk concentration at equilibrium (at the end of calculation) is at least 0.1 %.

- Counterion: The same amount as that of the polymer segment is set.
- Surface ionic group: The wall's surface electric charge density is set to be -0.01 C/m^2.[1]
- Counterion of the surface ionic group: The same amount as that of the surface ionic group is set.
- Added salt ions: The mole concentration is converted to a volume fraction using the following conversion formula used by van de Steeg et al., and the volume fraction is input for both positive and negative ions:

$$\Phi = CN_{\mathrm{A}}V = 0.130\mathrm{C}\,[\mathrm{mol/l}]\,,$$

  where Φ is the volume fraction, C is the molar concentration of ions, N_{A} is Avogadro's number, and V is the volume of a single segment, namely, 0.6^3 nm^3.
- Water: The residual amount obtained by subtracting the aforementioned amount of each component from 1 (where the total sum of the volume fractions is set to 1).

First, the surface electric charge density of -0.01 C/m^2 is converted into the number of ionic groups per unit area using an elementary charge, 1.60×10^{-19} C, in one surface ionic group having an electric charge of -1 to obtain $(0.01/(1.60 \times 10^{-19}))/10^{18}$ nm^2 $= 0.0625$ ionic groups/nm^2. Because a surface ionic group is assumed to be a segment having a diameter of 0.6 nm, when the area per surface ionic group is assumed to be 0.6×0.6 nm^2, the area ratio of the surface ionic groups in the wall surface becomes 0.0625 ionic groups $\times$ 0.6×0.6 nm^2/nm^2 $= 0.0225$. This means that the segment volume fraction of the surface ionic groups is 0.0225 in the lattice of the first layer (0–0.6 nm) adjacent to the wall. In addition, the specific concentration (volume fraction) of the surface ionic groups is input considering the system size. When the system size is set to 50, for example, the charging concentration of the surface ionic groups becomes $0.0225/50 = 4.5 \times 10^{-4}$ (and, as a matter of course, all charged surface ionic groups are grafted to the wall).

5. Chi parameters (χ parameters): In the same manner as for the χ parameters of van de Steeg et al., the χ parameters for a polymer segment and ion and for a polymer segment and water are set to 0.5, and all other χ parameters for other combinations are set to zero.
6. Surface chi parameters (χs):

   - χs parameters for a polymer segment and the wall are set in the range of $\chi\mathrm{s} = 0$ to -0.6 (where the definition of χs differs between SUSHI and van de

[1]The setting for the amount of surface ionic groups corresponding to a surface electric charge density of -0.01 C/m^2 is determined by specifically considering that the input parameters are determined by substituting a lattice model of a discrete-type SCF in the following manner.

Steeg's method; specifically, the way of assigning a positive or negative sign is reversed, with SUSHI assigning χs a negative value).

- χs = −10 is set for a surface grafted ionic group and the wall (although this setting may be unnecessary because of grafting).
- All other χs parameters for other combinations are set to zero.

7. Graft conditions: Graft conditions are set to create surface ionic groups.
8. Electrostatic parameters of segment: The electric charge density (charge) and the relative dielectric constant (dielectric constant) for each segment are input as follows.
   <charge>
   - Polymer segment: Charge = 0.2 is set in the case of representing a polymer in which, for example, 20 mol% of a cationic monomer having an electric charge of +1 is copolymerized.
   - Counterion of the polymer: A value having the sign opposite that of the polymer segment is input (where charge = −0.2).
   - Water: Zero charge is set.
   - Others: Charges are set for the surface ionic group (−1), counterion thereof (−1), and added salt ions (+1, −1).

   <dielectric constant>
   - Polymer segment and surface ionic group: A value of 20 is set.
   - Counterion and added salt ions: A value of 80 is set because of the assumption of hydrated ions.
   - Water: A value of 80 is set.

23.5 Calculation and Analysis

1. SCF convergence: There are often difficulties in converging SCF calculations of electrolytes. To ensure early convergence, it is first important that the values for constV and constW be set as small as possible initially in order to find a condition that allows progression, albeit slowly, toward convergence. Thereafter, the calculation is performed by inputting somewhat larger values of constV and constW such that the slope of residual error reduction versus the number of steps becomes large. In some cases, it may be necessary to form and thoroughly test a matrix of constV and constW values. Meanwhile, the residual error sometimes decreases to a certain value and thereafter fluctuates. In such a case, a stop file (input.stp) is put into a holder in which the calculation is performed to complete the calculation. Because halfway results are written out in an output UDF file to complete the calculation, the output UDF file can be used as a restart file when the calculation is performed again after changing constV and constW, thereby shortening the calculation time.

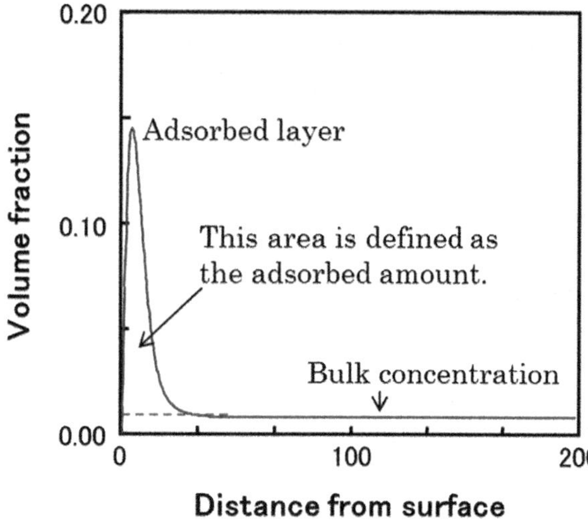

Fig. 23.2 Concentration distribution of the adsorbed polymer

Trial and error (over 1–2 h) is required until the optimal condition for constV and constW is fixed; however, once fixed, calculation converges to an SCF error of 1E−6 in 2–3 h. In addition, the system size is changed from 200 to 50 halfway through the series of studies because the time required for convergence becomes longer as the segment charge increases and as the concentration of the added salt increases. When the system size is changed, it has already been confirmed that the concentration profile on the bulk side (solution side) does not have a slope but is flat.

2. Calculation of the adsorbed amount (surface excess): The adsorbed amount (strictly speaking, the surface excess) is determined from the distribution of the concentration in the adsorbed layer obtained after SCF convergence. Figure 23.2 shows an example of the concentration distribution of the adsorbed polymer in a system having a system size of 200. The segment volume fraction becomes flat (without a slope) in a region apart from the surface, and the volume fraction in this region is defined as the bulk concentration.

   Meanwhile, the region near the wall face has a concentration higher than the bulk concentration because of the existence of the adsorbed layer. The area of the portion where the concentration is higher than the bulk concentration, the area surrounded by the solid and dotted lines in Fig. 23.2, is defined as the adsorbed amount.

23.6 Comparison of Calculation Results

The results of calculation performed under various conditions are described in comparison with the results of van de Steeg et al.

23.6.1 Effect of Segment Charge

Figure 23.3 shows the concentration distributions in the adsorbed layer calculated for four different values of the segment charge of the polymer within a range from +0.005 to +0.2 under conditions that $\chi s = 0$ and no salt is added. As shown in the figure, concentration distributions similar to the results of van de Steeg et al. are obtained.

Furthermore, the adsorbed amount is determined from the concentration distribution, and Fig. 23.4 shows graphs obtained by plotting the adsorbed amount for each segment charge (and also the results obtained by setting χs to −0.6). The adsorbed amount calculated by SUSHI is somewhat less than the adsorbed amount obtained

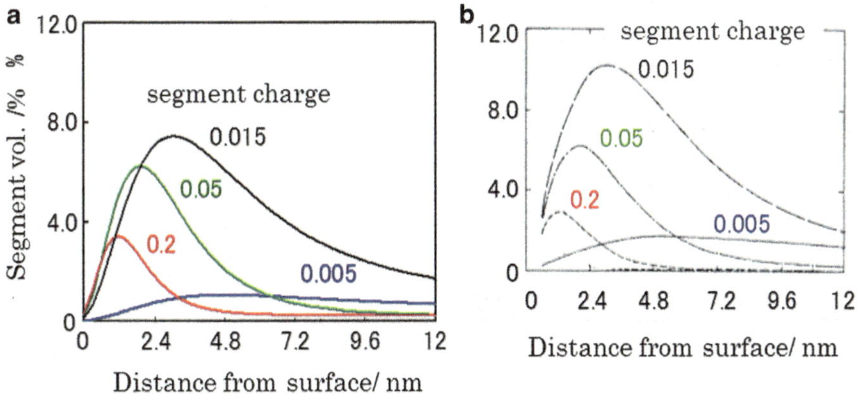

Fig. 23.3 Concentration distribution of the adsorbed layer: results of SUSHI (**a**) and van de Steeg et al. [1] (**b**)

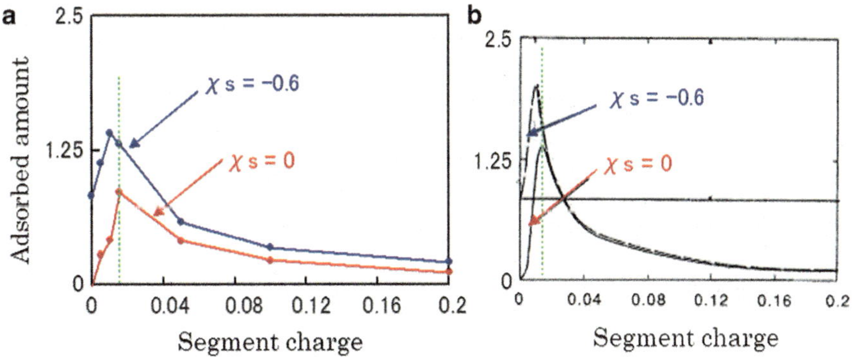

Fig. 23.4 Relation between adsorbed amount (dimensionless) and segment charge: results of SUSHI (**a**) and van de Steeg et al. [1] (**b**)

by van de Steeg et al.; however, the value of the segment charge where the adsorbed amount reaches a maximum for each of $\chi s = 0$ and -0.6 agrees with the results of van de Steeg et al.

23.6.2 Effect of the Concentration of Added Salt

Figure 23.5 shows the relation between the adsorbed amount and concentration of added salt obtained by performing the calculation for three different values of the segment charge within a range from 0 to 0.25 under the condition that $\chi s = -0.6$. The tendency of the adsorbed amount to increase or decrease in association with an increase in the concentration of added salt strongly depends on the segment charge value, and the tendency agrees with the results of van de Steeg et al. Figure 23.6

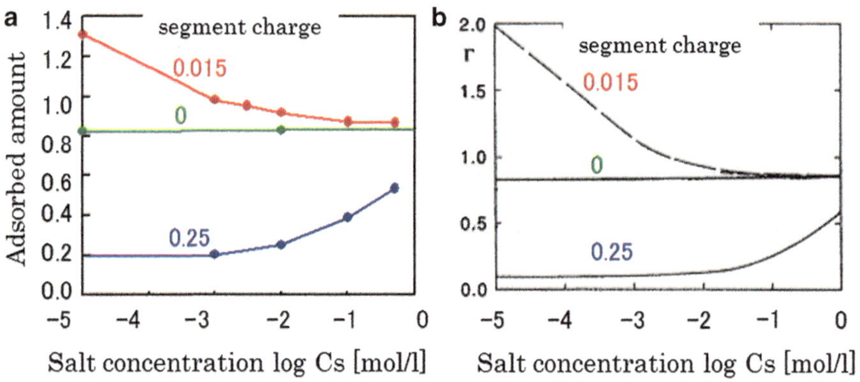

Fig. 23.5 Relation between adsorbed amount (dimensionless) and concentration of added salt at $\chi s = -0.6$; results of SUSHI (**a**) and van de Steeg et al. [1] (**b**)

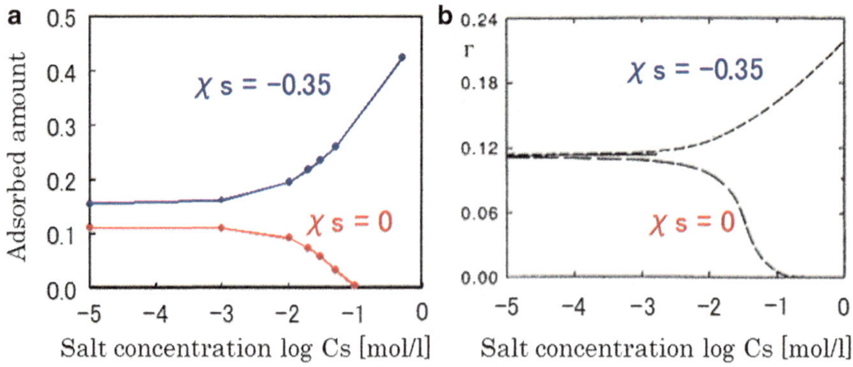

Fig. 23.6 Relation between adsorbed amount (dimensionless) and concentration of added salt at a segment charge of 0.2; results of SUSHI (**a**) and van de Steeg et al. [1] (**b**)

shows the relation between the adsorbed amount and concentration of added salt obtained by performing the calculation for the two cases $\chi s = 0$ and -0.35 under the condition that the segment charge is 0.2. The tendency of the adsorbed amount to increase or decrease in association with an increase in the concentration of added salt strongly depends on the value of χs, and the tendency agrees with the results of van de Steeg et al.

23.7 Concluding Remarks

Using OCTA SUSHI, polyelectrolyte adsorption onto a charged surface was simulated. The unit of length, the coefficient of Poisson's equation, and the graft density of charged segments were determined by regarding the SUSHI SCF as a discrete-type SCF, particularly with respect to modeling. As a result, SUSHI reproduced the results of van de Steeg et al. closely, though not exactly. In this chapter, we introduced an example of a homopolymer in which electric charges are equally allocated to all segments; however, SUSHI can easily handle not only polymer chains in which monomer arrangements of ionic and nonionic monomers differ but also polymer chains having different branches, such as comb-shaped and star-shaped polymer chains, as well as a wide range of simulations other than the one introduced in this chapter.

23.8 Download Instruction

Sample files are available from the OCTA website. See the instruction on http://octa.jp/download_bookdata.html. Password for the sample (zip file) of this chapter is "ubP8H2go".

References

1. H.G.M. van de Steeg, M.A. Cohen Stuart, A.D. Keizer, B.H. Bijsterbosch, Langmuir **8**, 2538 (1992)
2. G.J. Fleer, M.A. Cohen Stuart, J.M.H.M. Scheutjens, T. Cosgrove, B. Vincent, *Polymers at Interfaces* (Chapman & Hall, London, 1993)

Chapter 24
Adsorbed Structures and Surface Forces

Hiroki Kubo

24.1 Introduction

Polymer dispersants are often used in pigment dispersion systems (e.g., those for paint, ink, and cosmetics). They prevent aggregation through the action of a polymer adsorbed onto the surfaces of fine particles. (See [1] regarding the attraction between fine particles, which is the cause of aggregation.) The mechanism by which polymers prevent aggregation is thought to be as follows. Fine particles exhibit Brownian motion within the dispersion medium, repeatedly and frequently colliding throughout the volume of the medium. When fine particles approach each other, the conformational spaces of the polymer chains adsorbed on their respective surfaces overlap, and the osmotic pressure of the solvent that accompanies the local increase in polymer concentration resulting from these overlaps causes repulsion. Furthermore, if the fine particles become closer still, the conformational spaces of the adsorbed polymer chains become deformed, and a strong repulsion originating from entropic elasticity is generated. In nonaqueous solvent systems, for which the contribution of electrostatic interactions is not important, it is considered that polymer dispersants prevent aggregation by generating these two types of repulsion. To design superior dispersants, it is therefore important to understand the relationships among three factors: the molecular structure of the dispersant, the adsorbed structure, and the repulsion.

If the fine particles are sufficiently large with respect to the polymer, then the surface of the particles can be regarded to be a flat with no curvature. The OCTA-SUSHI one-dimensional static self-consistent field (SCF) calculation is an easy calculation with simple parameter settings for the concentration distribution (equilibrium structure) of the polymer adsorbed onto the surface (wall). Further-

H. Kubo (✉)
Analytical Science Research Center, Kao Corporation, Chūō, Japan
e-mail: kubo.hiroki@kao.co.jp

Japan Association for Chemical Innovation, *Computer Simulation of Polymeric Materials*, DOI 10.1007/978-981-10-0815-3_24

more, because the excess free energy output following SCF convergence can be used to obtain the surface force curve with respect to the wall separation, an analysis that compares the concentration distribution of the adsorbed polymer chains against surface forces can be performed. (Refer to the SUSHI manual for the theoretical background behind excess free energy.) The relationships among the aforementioned three factors can thus be clarified. OCTA includes Surface Simulator, which is specifically designed for performing surface calculations using SUSHI. The use of Surface Simulator allows several independent SCF calculations at different wall separations to be performed in a single RUN by creating only a single input file. Furthermore, this convenient simulation automatically outputs a list of values for the wall separation versus excess free energy that are needed to plot the force curve. However, because Surface Simulator has several restrictions in terms of the calculation condition settings, this chapter describes how SUSHI itself is used to perform the surface calculations and thus serves as a reference for those wishing to perform calculations using a variety of conditions.

24.2 Background and Purpose of the Example Analysis

The analysis targets a polymer dispersant with a comb block structure. The main chain block can be adsorbed onto the surface, whereas the side chain block cannot. In general, adsorption of such dispersants tends to be pictured as a structure in which the main chain block adsorbs flat against the surface, while the side chain block floats in the solvent in a manner resembling seaweed. Therefore, in many cases, theory and simulation are discussed only in terms of a brush model in which the side chain block is grafted directly onto the surface. However, in cases where the main chain block has some level of affinity with the solvent, it has been pointed out that the main chain block may float (i.e., form a loop or tail conformation); thus, the brush model is not always suitable. However, although concentration distributions are displayed for each of the main chain and side chain blocks, if adsorption simulation is performed using SUSHI, then no direct information on the adsorbed structure adopted by the main chain (loop, train, or tail) can be obtained. Against this backdrop, instead of setting parameter χs for the interaction between the wall and polymer segment, we analyze the relationship between surface forces by creating adsorption-type models for the main chain that grafts its end onto the wall.

24.3 Method for Modeling the Adsorbed Structure

If only the main chain section is being considered, then it may be assumed that the adsorbed structure in a comb block chain may be expressed as a loop, train, tail or a combination thereof. This will be explained using the example of the comb

Fig. 24.1 Adsorbed structure model of the comb chain

chain adsorbed structure shown in Fig. 24.1. This adsorbed chain consists of three units in which the main chain blocks form a loop, a train, and a tail, with the main chain sections having chain lengths of *l*, *m*, and *n*. In this method, the three units are treated as unlinked independent chains (i.e., as a mixture of three types of chains). Hence, the loop unit is formed by grafting only its ends onto the wall. The train unit is formed by grafting all main chain segments onto the wall. The tail unit is created by grafting only one end of the unit onto the wall. Although the χs parameter is not set, by changing the number of points of grafting onto the wall, the degree of affinity to the wall can be indirectly expressed.

Using such a method, and by keeping the chain length of the entire main chain $(1 + m + n)$ constant, various adsorbed structures can be created by changing each of the chain lengths (and furthermore, ensuring that each volume fraction is altered in accordance with the length of each chain).

24.4 Calculation Conditions

24.4.1 Structure of Comb Block Chains

When creating input user definable format (UDF) files, the loops, trains, and tails are treated as independent chains. Before explaining this, the structure of comb block chains will be explained. Comb block chains consist of a main chain block made up of 40 segments, to which 10 side chain blocks made up of 23 segments are linked at regular intervals, while the solvent is expressed as one segment. Figure 24.2 shows the models for the comb block chain and solvent. To further simplify the model, all segments (i.e., those of the comb block chain's main and side chains, together with that of the solvent), are treated as being of the same type; that is, all χ parameters are zero, and the solvent is considered to be a good solvent with respect to both the main chain and side chain blocks.

24.4.2 Adsorbed Structure Models

Two adsorbed structures are shown in Fig. 24.3. One is a train model, which has a main chain segment—to which side chains are linked—grafted onto the wall.

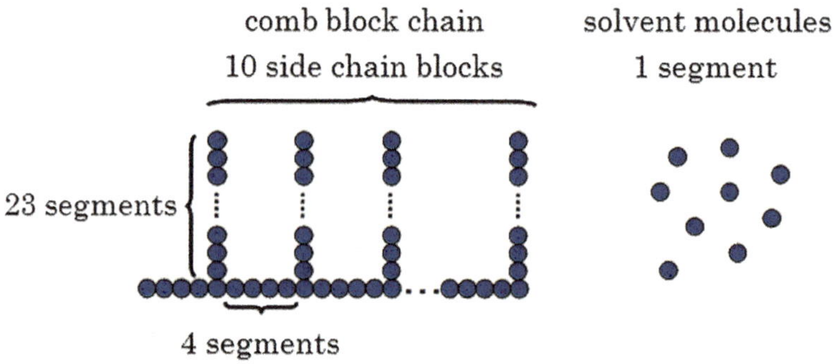

Fig. 24.2 Coarse-grained model

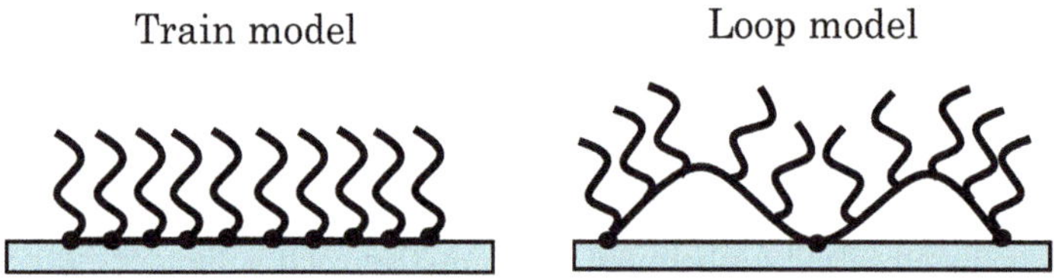

Fig. 24.3 Two adsorbed structures used in force curve calculations

The other is a loop model consisting of two loops of equal chain length, with the main chain grafted onto the wall at only three points, namely, the two ends and center. When creating an input UDF file for the loop model, as described in the previous section, only a loop chain of half the chain length of the main chain (with only the ends of the main chain grafted) is input, and the volume fraction of the chain is input so as to be equal to that for the train model. To create the graft chain, the "polymer_ID" to be grafted and the "segment junction_ID" are input into "graft_conditions" in the input UDF file.

24.4.3 Adsorbed (Grafted) Amount

For the aforementioned two adsorbed structures, a comparison is made of cases in which the dimensionless adsorbed amount (i.e., the quantity of chains grafted onto the wall) is 2.5 and 5.0, and the effect of adsorbed amount is investigated. Here, the adsorbed amount indicates the thickness (i.e., the distance between the two walls) when all solvent in the system has been eliminated. The reason for this is that the thickness when the chain segments are placed in this manner (i.e., without solvent) is proportional to the adsorbed amount. Furthermore, in the current example, because all the included polymer chains are grafted onto the wall, the product of "chain volume fraction" and "wall separation" in the input UDF file equals the adsorbed amount.

24.4.4 System Conditions

The chain is grafted onto the wall of one side only. The bulk phase is filled only with solvent. A large number of input UDF files are prepared, with the wall separation being varied in increments of one between a minimum value (corresponding to the adsorbed amount) and a maximum value of 35; the equilibrium structure of the graft chain (concentration distribution) and excess free energy F_{exess} are then calculated for each wall separation using a one-dimensional static SCF ("canonical" for the chain and "grand canonical" for the solvent). When creating the input UDF files, the quantity of polymer is kept the same for all wall separations; that is, for chain volume fraction Φ and wall separation X shown in Eq. (24.1), Φ is set based on X so that the value of their product does not vary:

$$\Phi X = \text{constant} \, (= \text{adsorbed amount}) \tag{24.1}$$

24.5 Method for Analyzing the Calculation Results (Generation of the Force Curve)

If the standard value of excess free energy $F^0{}_{\text{exess}}$ is set to a value under conditions for which the walls are far apart (i.e., in the region where F_{exess} is constant and independent of X), then, as shown in Eq. (24.2), the value given by subtracting $F^0{}_{\text{exess}}$ from the excess free energy $F_{\text{exess}}(X)$ obtained at wall separation X is the interaction energy $W(X)$ (per unit area) acting on the walls (planes) at that wall separation:

$$W(X) = F_{\text{exess}}(X) - F^0{}_{\text{exess}} \tag{24.2}$$

In addition, using the Derjaguin approximation [1] given by Eq. (24.3), $W(X)$ can be converted to the surface force acting between planar and spherical surfaces $F(X)$:

$$\begin{aligned} &F(X)_{\text{sphere-plane}} = 2\pi R W(X)_{\text{plane-plane}} \\ &(R\text{: radius of curvature of spherical surface}) \end{aligned} \tag{24.3}$$

Because R is a constant, $F(X)_{\text{sphere-plane}}$ and $W(X)_{\text{plane-plane}}$ are proportional. In this guide, the relationship between $W(X)$ and X is therefore referred to as the force curve. As a means of measuring the force curve, if atomic force microscopy (AFM) is used, the force between the substrate (planar surface) and probe (spherical surface) is measured. The measured value corresponds to the left-hand side of Eq. (24.3). Alternatively, $W(X)$ on the right-hand side can be obtained from the SCF calculation via Eq. (24.2). Hence, it is possible to compare AFM and SCF data. (However, suitable dimensionalization of length and energy is necessary when performing a quantitative comparison.)

24.6 Results and Discussion

24.6.1 Concentration Distribution

Figure 24.4 shows the segment concentration distribution for an adsorbed amount of 2.5 under conditions in which the walls are far apart.

The vertical axis is the segment volume fraction Φ, while the horizontal axis is the distance between the walls *X*, with a distance of 1 being equal to the segment size.

The figure depicts the segment concentration distributions for the grafted main chains, side chains, and solvent. Compared with the concentration distribution for the train model (left graph), in which the side chains are fixed at the wall, the concentration distribution for the loop model (right graph) shows a broad distribution of main chains, and an accompanying broad distribution of side chains grafted onto such main chains.

Figure 24.5 shows the change in the concentration distribution for both models due to a reduction in wall separation (compression) for the same adsorbed amount of 2.5. Because the distributions of both main chains and side chains for the loop model extended out to long distances prior to compression (Fig. 24.4), deformation of the concentration distribution due to compression occurs from relatively large distances in the loop model (right graph) compared with the case for the train model (left graph). (This can be seen clearly by comparing the concentration distribution for the two models at $X = 10$.) At $X = 5$, after further compression, there is no large difference between the side chain and solvent distributions in the two models, but the loop model exhibits a broader main chain distribution.

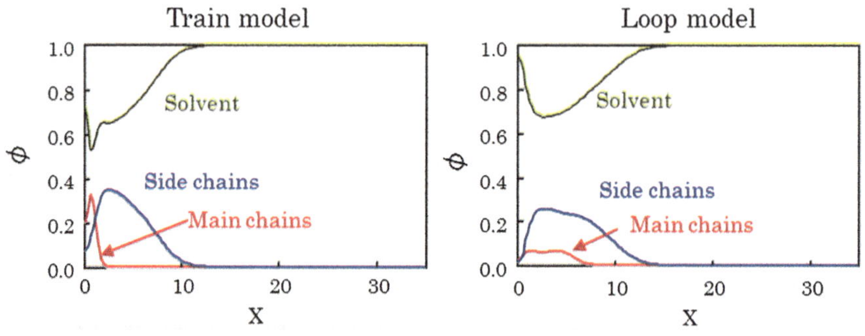

Fig. 24.4 Segment concentration distribution for an adsorbed amount of 2.5 with walls far apart (*left graph*: train model, *right graph*: loop model)

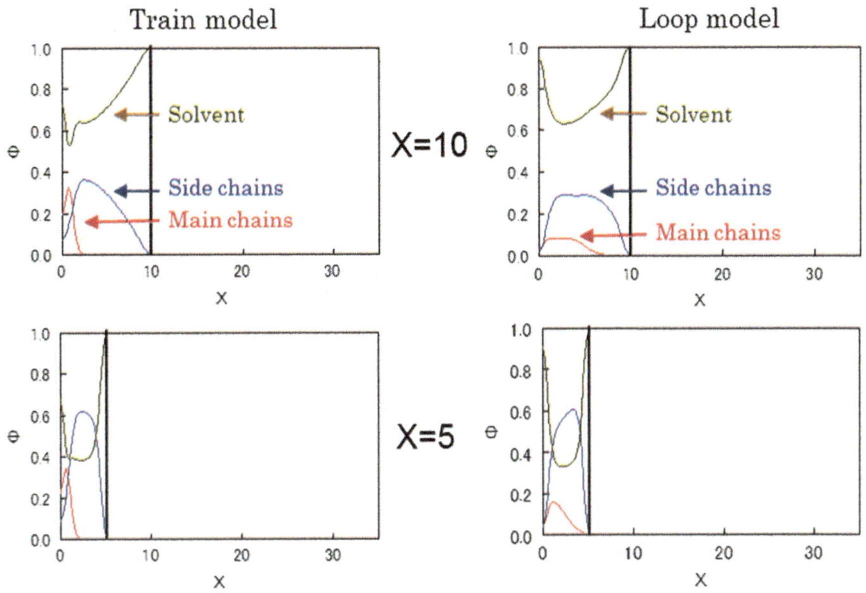

Fig. 24.5 Change in concentration distribution due to a reduction in wall separation (X) for an adsorbed amount of 2.5 (*left graph*: train model, *right graph*: loop model)

24.6.2 Force Curve

For an adsorbed amount of 2.5, the force curves for both models are shown in Fig. 24.6 (linear plot) and Fig. 24.7 (semi-logarithmic plot). In both figures, the vertical axis is $W(X)$, plotted exactly as obtained from Eq. (24.2). As the figures show, repulsion occurs from larger distances in the loop model than in the train model. It may be inferred that this is because the main and side chains in the loop model both spread out to further distances, as shown in Fig. 24.4. Consequently, deformation of the concentration distribution occurs from further away, with the entropy loss caused by this deformation being reflected in the excess free energy value.

24.6.3 Effect of the Adsorbed Amount

The effect of the adsorbed amount on the force curves for the train and loop models is investigated. Comparing the force curves for adsorbed amounts of 2.5 and 5.0 for both models, the results show that, for both models, repulsion occurs from a greater distance for the higher adsorbed amount. This indicates that the larger adsorbed

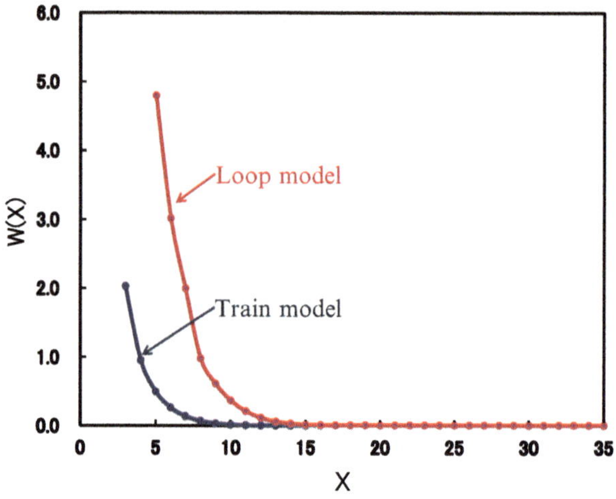

Fig. 24.6 Force curves (linear plot) of both models for an adsorption quantity of 2.5

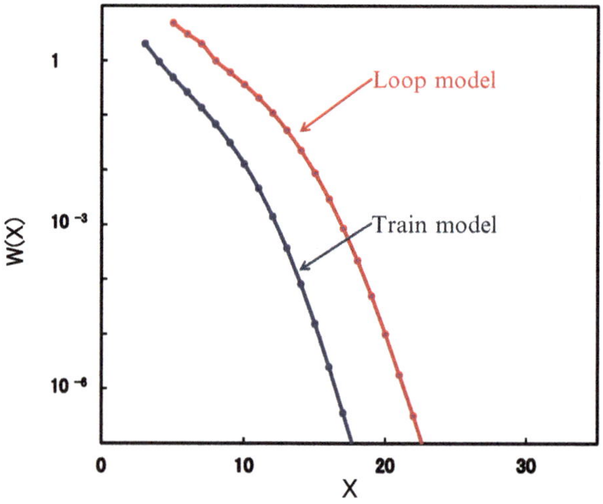

Fig. 24.7 Force curves (semi-logarithmic plot) of both models for an adsorption quantity of 2.5

amount results in a distribution extending further away and a repulsion at a greater distance. In addition, for an adsorbed amount of 5.0, the results of comparison are the same as those described in the previous section for an adsorbed amount of 2.5, with the slopes in both the near and far regions being slightly smaller for the loop model than the train model.

To summarize the foregoing, a comparison of the force curves of the train and loop models shows that, compared with the train model, the loop model exhibits repulsion from a relatively large distance and has a force curve with a slightly smaller slope. These results are the same for a system having a different adsorbed amount.

24.7 Conclusions

Using the functions of OCTA-SUSHI for grafting onto a wall surface, modeling methods that allow various adsorbed structures of polymer dispersants to be expressed were described. Additionally, using the excess free energy of OCTA-SUSHI, the force curve was described. These methods allow the relationship between predetermined adsorbed structures and force curves to be investigated. This chapter described the results obtained using simple models, but these relationships can also be analyzed for more complicated adsorbed structures (i.e., forms in which all three features—the loop, train, and tail—coexist and also forms in which each of these has a chain length distribution).

However, when comparing against force curves obtained through actual measurement, it is necessary to exercise care with regard to several points, of which two in particular are now briefly discussed. First, because these modeling methods employ the grafting function to fix specific segments of the chain (e.g., both ends of a loop chain or one end of a tail chain) onto the wall, the chain and wall adsorption points are assumed to be unchanging, with wall separation having no effect on them. In other words, it is assumed that the process of compression does not cause any rearrangement of adsorption points. Second, the adsorption layer structure obtained from the static SCF calculation represents the equilibrium structure for the specified wall separation. This means that the corresponding force curve—which is obtained by connecting together these calculated results—is, for all wall separations, the curve obtained by bringing the walls together slowly so that the equilibrium structure is maintained at all times. This second point does not apply only to the adsorption model described here; it is a shared feature that must also be considered wherever χs is used to create an adsorption layer.

Finally, although in the cases described here, calculations were performed for chain grafting onto only one of the two walls, adsorption can also be carried out by grafting onto both of the walls or by setting χs. Various other condition settings are also possible with SUSHI, and it is expected that these will be used in further research on dispersants in the future.

24.8 Download Instruction

Sample files are available from the OCTA website. See the instruction on http://octa.jp/download_bookdata.html. Password for the sample(zip file) of this chapter is “Wr9Z2XAo”.

Reference

1. J.N. Israelachvili, *Intermolecular and Surface Forces*, 2nd edn. (Academic, London, 1991)

Chapter 25
Analysis of Relaxation Mechanism of Thread-Like Micelle Solution

Satoru Yamamoto, Taku Ozawa, and Kosuke Ohata

25.1 Introduction

Surfactants in an aqueous solution aggregate to form a variety of assemblies such as spherical and rodlike micelles and bilayer membranes. The stable shape of the aggregation depends on the molecular structure, concentration, and temperature [1]. In particular, surfactant thread-like micellar solutions exhibit pronounced viscoelastic behavior as well as polymer systems with entanglement. For instance, cetyltrimethylammonium bromide (CTAB) containing sodium salicylate (NaSal) or *p*-toluenesulfonic acid sodium (Na*p*TS) form long and stable thread-like micelles in an aqueous solution and exhibit viscoelasticity. This system appears similar to a polymer melt or solution; however, the relaxation mechanism of the thread-like micellar system is completely different from that of a polymer system having broad relaxation spectra. The thread-like micellar system shows a Maxwell-type relaxation with a single relaxation time. This characteristic feature of the viscoelastic behavior of a thread-like micellar system can be explained using a phantom network model [2–4]. However, the actual dynamics at an entanglement point cannot be directly observed.

Computer simulation is a possible means of exploring the thread-like micellar system that is difficult to directly observe in an experiment. There are two approaches adopted in the computer simulation of a micellar system. One is to conduct an atomistic simulation that considers, for example, the molecular dynamics of the bilayer membrane of surfactants in an aqueous solution. The other is to

S. Yamamoto (✉)
Dassault Systemes Biovia K.K., Tokyo, Japan
e-mail: satoru.yamamoto@3ds.com

T. Ozawa • K. Ohata
JSOL Corporation, Tokyo, Japan

Japan Association for Chemical Innovation, *Computer Simulation of Polymeric Materials*, DOI 10.1007/978-981-10-0815-3_25

use coarse-grained simulation techniques to study the aggregation dynamics of surfactants. It is expected that the latter approach is more suitable for studying the dynamics of a thread-like micellar system; however, such research has not yet been reported. There have been several simulation studies on the stability of thread-like micelles [5]; however, thread-like micelles are always supported by a periodic boundary condition of the simulation cell. It is thus uncertain whether the micelle can maintain the thread-like structure without a periodic boundary condition.

In the present study, the simulation condition is examined first to produce a stable thread-like shape. The crossing dynamics of thread-like micelles are then simulated to verify the phantom network model [6].

25.2 Viscoelastic Behavior of the Thread-Like Micellar System

First, we summarize the available experimental knowledge [2–4] of the rheology of CTAB containing salts. The shape of the micelle begins to change from spherical to thread-like with the addition of salts, and thread-like micelles that are long enough to become entangled with one another form in the concentrated regime in which the molar ratio of salt/CTAB is greater than 1 (whereas no change is observed at a much higher ratio). Pronounced viscoelastic behavior emerges at a molar ratio of around 0.3, and the dependence of the viscoelastic spectra on the molecular weight is similar to that for concentrated polymer solutions. This suggests that the growth of the thread-like micelle is due to an increase in the salt concentration and a relaxation mechanism of reptation similar to that of the polymer system. However, the system has a single relaxation time for a molar ratio exceeding 1 and has Maxwell-type behavior. The intensity of relaxation increases in proportion to the square of the surfactant concentration, which is similar to the case in the semi-concentrated regime of a polymer solution. This suggests that entangled thread-like micelles have high flexibility and entropic elasticity like a polymer chain. However, the relaxation mechanism seems to be completely different from that of a polymer chain. For this system, Shikata et al. [2–4] proposed the phantom network model to explain the relaxation behavior at an entanglement point of thread-like micelles (Fig. 25.1). A polymer chain consists of covalently bonded atoms, and therefore, breakage never occurs at an entanglement point. Meanwhile, a thread-like micelle is constructed via intermolecular interactions of surfactants and the structure can recombine, which involves breakdown and reformation. In the model of Shikata et al., thread-like micelles fuse at an entanglement point to form a four-armed branch point, and the micelles then pass through each other like phantoms after a finite lifetime. This crossing event is dominant for the longest relaxation time observed for a molar ratio exceeding 1. The surface of the thread-like micelle is positively charged at a molar ratio below 1, and micelles relax through a reptation mechanism similar to that for a polymer chain. At a molar ratio of 1, the surface charge becomes neutral, and micelles start to pass through each other. As the salt concentration increases, the longest relaxation time gradually decreases.

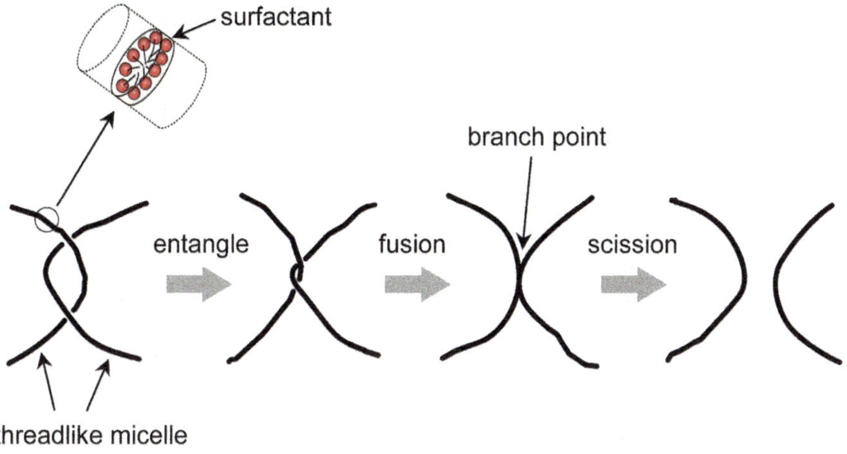

Fig. 25.1 Schematic representation of a phantom crossing process for thread-like micelles (phantom network model) [2–4]

25.3 Dissipative Particle Dynamics (DPD) Model of the Thread-Like Micelle

Several simulations have been conducted for the thread-like micelle; however, the thread-like shape has been supported by a periodic boundary condition in the simulation cell [5]. Therefore, it is not clear whether a micelle can maintain a thread-like shape without a periodic boundary. In the present study, a stable micelle shape was first investigated for a nonperiodic system.

A simple model of surfactant molecules for CTAB is considered as shown in Fig. 25.2. They consist of one hydrophilic head particle (h: head) and two hydrophobic tail particles (t: tail). Water particles (w) are considered to be the same size at the surfactant components.

The set of conservative forces of DPD as a standard condition is listed in Eq. (25.1). The parameters are set at 25 for the diagonal terms and the hydrophilic interaction between head and water particles and 70 for the hydrophobic interaction between tail and water particles and head and tail particles.

$$a_{ij} = \begin{pmatrix} & h & t & w \\ h & 25 & 70 & 25 \\ t & 70 & 25 & 70 \\ w & 25 & 70 & 25 \end{pmatrix} \tag{25.1}$$

With these parameters, the thread-like shape is stabilized under a periodic boundary condition. The number of surfactants is 700 and the dimensions are $20(x) \times 40(y)$

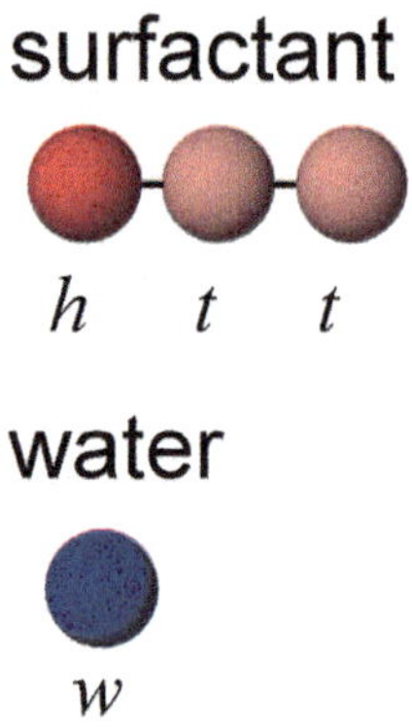

Fig. 25.2 DPD model

$\times$ 20(z). The length scale is non-dimensionalized by particle size. From this initial condition, the DPD simulation is performed without a periodic boundary condition for the y-direction.

The micelle shape, which usually depends on the balance between effective volumes of head and tail components [7], can be changed using the repulsive force between head groups in this simulation. The micelle shape changes to a prolate shape with a standard parameter set. As repulsive forces between the hydrophilic head groups a_{hh} increase, the micelle gradually elongates. It is confirmed that a thread-like micelle is stabilized with a moderate repulsive force of $a_{hh} > 30$.

The validity of parameters is then checked for the mobility of surfactants in the thread-like micelle. The lateral diffusion constant (D_{lat}) of surfactant in the thread-like micelle was measured in experiments. For instance, the reported D_{lat} values for CTAB:NaSal solution are in the range of 9×10^{-6} to 10×10^{-6} cm$^{2/}$s at 25 °C [8]. The D_{lat} value of the surfactant is calculated to be $D_{\text{lat}} = \langle R_y(t)^2 \rangle /2t$ for a thread-like micelle consisting of 350 surfactants, in a box having dimensions $20 \times 20 \times 20$ and a periodic boundary condition. Here, $\langle R_y(t)^2 \rangle$ is the mean-square displacement along the long thread-like micellar axis in the y-direction. D_{lat} decreases in proportion to the repulsive force between hydrophilic head groups. Nondimensional time is used for the DPD calculation; therefore, an actual time scale should be estimated in the comparison of the D_{lat} value of water particles with experimental results [9]. The self-diffusion constant of water particles in a pure water system is estimated to be $D_{\text{self}} = \langle R(t)^2 \rangle \ / \ 6t = 0.26$. From a comparison between the simulated D_{self} value of water and the experimental value [10] of 2.57×10^{-5} cm^2/s, the magnitude of the simulated D_{lat} value of the surfactant ranges from 6.7×10^{-6} cm$^{2/}$s ($a_{hh} = 30$) to 4.9×10^{-6} cm$^{2/}$s ($a_{hh} = 45$). The lateral diffusion constant obtained here compares well with experimental values for CTAB:NaSal solution [8]. It is thus concluded that the mobility of surfactants in the thread-like micelles can simulate corresponding aspects of an actual system.

25.4 Simulation of the Crossing Dynamics

The crossing dynamics of two thread-like micelles are studied in DPD simulation. Two surfactant thread-like micelles, consisting of 350 surfactants each, are set in a simulation box as shown in Fig. 25.3. The dimensions of the simulation box are $L_x = L_y = 20$ and $L_z = 40$. One micelle (micelle 1) is periodic in the y-direction, while the other micelle (micelle 2) is periodic in the x-direction. Although periodic boundary conditions are adopted to simplify the calculation, it is reasonable that the model be considered as the focal part of very long thread-like micelles, because the micelle can be stabilized as thread-like under nonperiodic boundary conditions with the parameters set as discussed before.

From this initial condition, micelle 1 moves in the $+z$-direction and micelle 2 moves in the $-z$-direction such that they come into contact with each other and a crossing event occurs. In the actual simulation, surfactants of micelle 1 existing in the ranges of 0–0.1L_y and 0.9L_y–L_y move in the $+z$-direction, and surfactants of micelle 2 existing in ranges of 0–0.1 L_x and 0.9L_x–L_x move in the $-z$-direction. Here, the vicinity of the micelle edge moves in opposite directions. The magnitude of the velocity for this motion is set at $v = 0.04$, which corresponds to $kT = mv^2/3 = 0.0016$, and is negligible in relation to the thermal motion. A set of 20 runs is performed for each of the values $a_{hh} = 30$, 35, 40, and 45, until time $t = 1{,}000$ ($\Delta t = 0.05$, 20,000 steps). In this study, parameters are set as $kT = m = r_c = 1$, $\rho = 5$, $\sigma = 3$, $C = 100$, and $r_s = 0.73$ (the average distance of the nearest neighbor at a particle density of 5). The spring force between connected beads for a surfactant

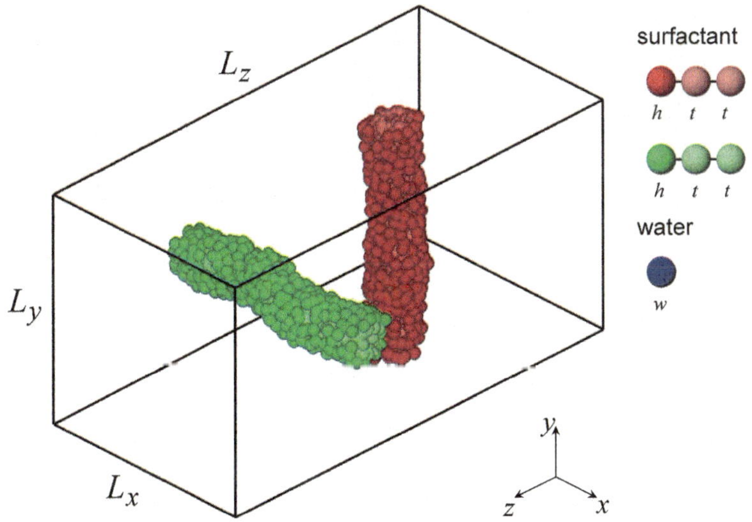

Fig. 25.3 Simulation system for the crossing dynamics at an entanglement point of surfactant thread-like micelles

is considered as a harmonic spring for the equilibrium bond distance. This treatment may guarantee a stable shape of a thread-like micelle. All simulations are performed on Linux PC (Xeon 3 GHz), and the computation time for each calculation is 14 h (for a system of 80,000 particles).

25.5 Results and Discussion

In the simulations with $a_{hh} = 30$, two thread-like micelles fuse to form at an entanglement point in all 20 runs, and a crossing process as shown in Fig. 25.4 is observed in four runs. Two thread-like micelles come into contact with each other ($t = 340$) and fuse ($t = 370$). Two different types of surfactants of micelles 1 and 2 start to mix with each other through the branch point by diffusional motion. The branch point is elongated ($t = 650$) and becomes thinner ($t = 670$). Finally, scission occurs, producing two thread-like micelles ($t = 700$).

The total transient energy of the system is estimated from the conservative force. It is found that the total energy increases by approximately 300 kT when an entanglement point forms, and the energy recovers to the original level by way of a fusion process. The energy then increases again by about 400 kT with elongation at the branch point. After the crossing of two thread-like micelles, the energy recovers again to the original level. A crossing process can be considered as a two-step reaction involving both fusion and scission.

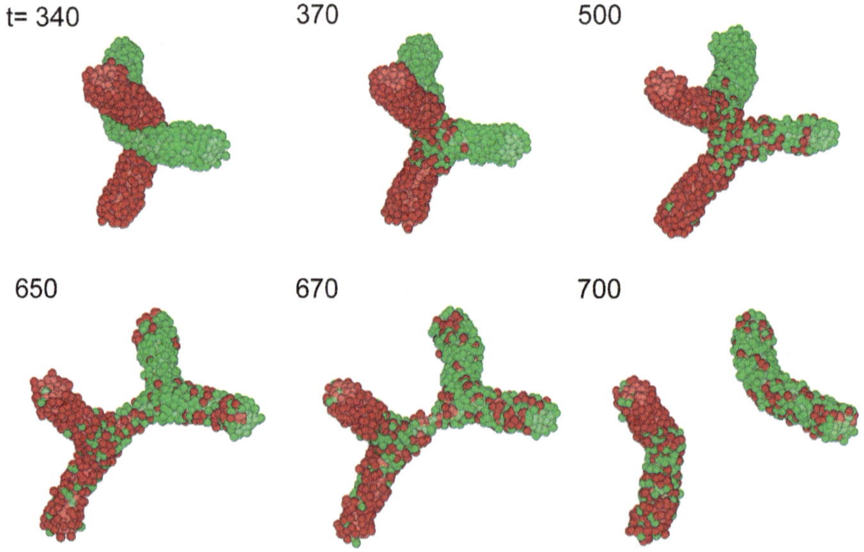

Fig. 25.4 Snapshots of the time evolution for a phantom crossing process ($a_{hh} = 30$)

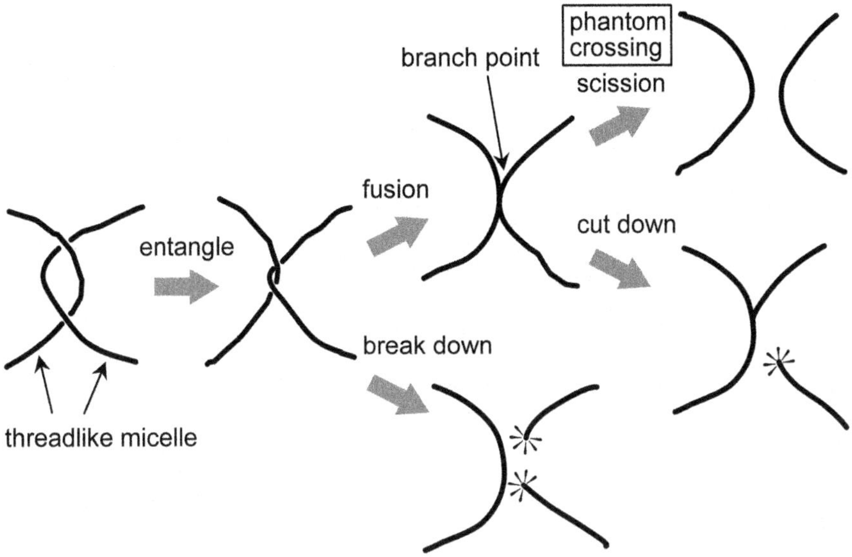

Fig. 25.5 Schematic illustration of the crossing dynamics at an entanglement point of surfactant thread-like micelles

Another simulation result is that one micelle is cut down at the branch point after the fusion of two thread-like micelles at an entanglement point. The energy gap in this reaction has almost the same magnitude as that of the phantom crossing process. There are two types of process after the fusing to form a branch point: the branch point is elongated and scission occurs or one of the micelles is cut down at a branch point. As repulsive forces between the hydrophilic head groups increase, the frequency of observing breakdown before fusion increases.

The crossing dynamics observed in the simulation are schematically summarized in Fig. 25.5. After the fusion process occurs, a phantom crossing is occasionally observed in all cases for approximately one in five runs.

Finally, to explain the evidence that terminals of the thread-like micelles are hardly observed in electron micrographs [11], additional simulations were performed as shown in Fig. 25.6. Considering the thermal motions of the thread-like micelles, there may be random coiling conformation similar to that for a polymer chain. Therefore, on a connected part between two entanglement points in a thread-like micelle, exertion occurs via an entropic stretching force in the same way as for an entangled polymer chain. When a terminal part is generated by the breakage, it may merge into the thread-like micelle part and vanish.

This hypothesis is examined by DPD simulation. From an initial condition of a part of a thread-like micelle including a terminal, the DPD simulation is performed as stretching in the horizontal axis. Using an initial simulation cell with dimensions of $40 \times 40 \times 20$, the density of the particles in the cell is kept constant ($\rho = 5$). As

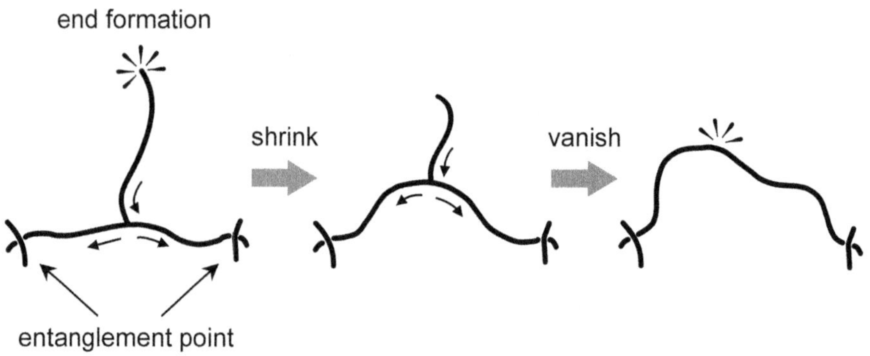

Fig. 25.6 Schematic illustration of the terminal vanishing model

the part of the thread-like micelle is elongated, the terminal part shrinks and finally merges into one thread-like micelle. This demonstration indicates the possibility of the terminal-vanishing model (Fig. 25.6).

25.6 Conclusions

DPD was applied to study the crossing dynamics at an entanglement point of surfactant thread-like micelles. It is confirmed that the thread-like micelle is stabilized with a moderate repulsive force between hydrophilic head groups. When two thread-like micelles are encountered at an entanglement point, they fuse to form a branch point. A phantom crossing is then occasionally observed after elongation of the branch point. A cutting down of one of the thread-like micelles at a branch point was also frequently observed. If the repulsive force between head groups is strengthened, one of the micelles is easily broken down at an entanglement point before fusion. In these three schemes (phantom crossing, cutdown, and breakdown), the breakage occurs somewhere along the thread-like micelle. This breakage may be an essential process in the relaxation mechanism, and a phantom crossing can be seen as a special case of these processes.

25.7 OCTA Example Run

A follow-up calculation was performed using OCTA (specifically, SUSHI and COGNAC). The main points regarding the creation of the initial structure and calculation conditions are explained below.

1. Creation of the initial structure

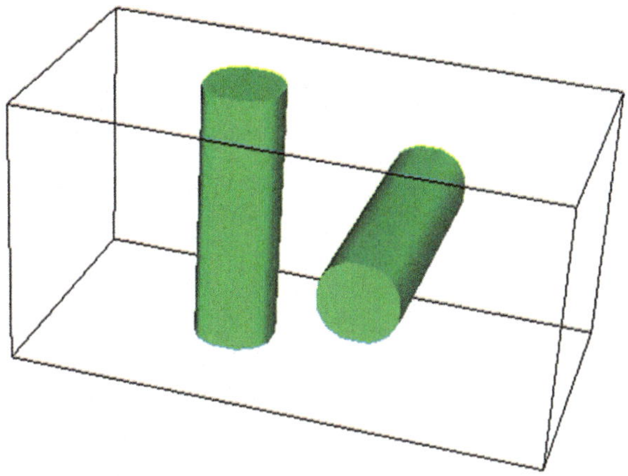

Fig. 25.7 Hypothetical density distribution obtained using SUSHI (input file, "density_field_uin.bdf"; output file, "density_field_uout.bdf")

There are methods of creating a dissipative particle dynamics (DPD) model of thread-like micelles at arbitrary positions in a simulation cell. Here, the zooming function of OCTA is used.

First, as shown in Fig. 25.7, two hypothetical cylinders are created as a phase separation structure, using SUSHI. The components of the SUSHI model do not require realistic surfactant models that consist of a hydrophobic head and hydrophobic tail, and instead, appropriate homopolymers are sufficient to create the structure. The volume fractions, which correspond to the size of the cylindrical domains, do not need to exactly follow the condition of the subsequent simulation using the DPD method. A rough approximation should be enough. The positions of the cylinders can be controlled using SUSHI's domain specification function. After the SCF calculation is conducted using SUSHI, a structure that has two smooth cylindrical domains is formed (input file, "density_field_uin.udf"; output file, "density_field_uout.bdf"; included in the sample data).

Second, the surfactant molecules for DPD simulation are located in the cylinders, and solvent particles are located in the remaining region, using the density-biased Monte Carlo (DBMC) method. The settings for the DBMC function of COGNAC are shown in Fig. 25.8. To identify each of the two thread-like micelles in post-processing, two types of molecules, for example, h1-t1-t1 and h2-t2-t2, are used. However, these molecules are assigned the same pair interaction parameters, so in this simulation, one micelle is considered the same as the other micelle.

At this point, the surfactant molecules are only located in cylindrical domains. From this state, the molecules are adjusted into thread-like micelle structures. With the conservative interactions between DPD particles as described above, under the external force of the density field (user definable field (UDF) path: *Interactions.External_Interaction[].Density_Field*) with the density result obtained previously using SUSHI as described above, the hydrophobic particles naturally align outward without destroying the cylindrical forms (input file, "01_initial_structure.bdf"; output file, "01_initial_structure_out.bdf"; included in the sample data).

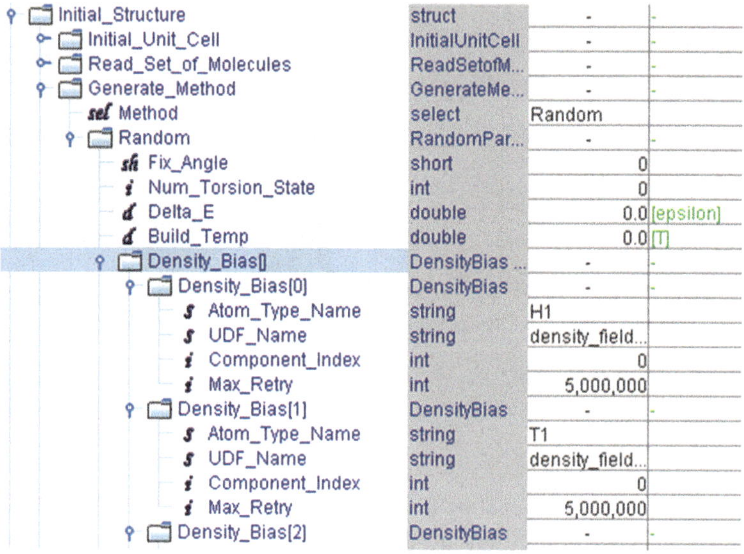

Fig. 25.8 Setting of the density-biased Monte Carlo method for the creation of the initial structure (input file, “01_initial_structure.bdf”; output file, “01_initial_structure_out.bdf”)

Next, after removing the density field from the UDF file by setting *Simulation_Conditions.Calc_Potential_Flags.External* to zero, particles of surfactant molecules near the boundary are constrained using the Python script “set_param_for_relax.py,” included in the sample data. The script registers the particle indices that satisfy the above condition in *Simulation_Conditions.Constraint_Conditions* (i.e., the condition that all the components of velocity are set to zero in Fig. 25.9). According to the COGNAC data format, however, the coordinates of the particles may exist outside of the boundary. In such cases, when executing a Python script such as this one acting on COGNAC files, the particle coordinates in the UDF file are previously treated with the boundary conditions using the Python script “move_molecules_into_cell.py”. Otherwise, it is necessary to determine coordinates under periodic boundary conditions in the Python script.

The relaxation calculation is then performed with the input file obtained through the above procedure (input file, “02_relax.bdf”; output file, “02_relax_out.bdf”; included in the sample data). The two micelles keep a thread-like shape, although the shape fluctuates slightly, and the diameter of the unconstrained regions of the thread-like micelles does not differ dramatically with that of the constraint regions.

2. Crossing of the thread-like micelles

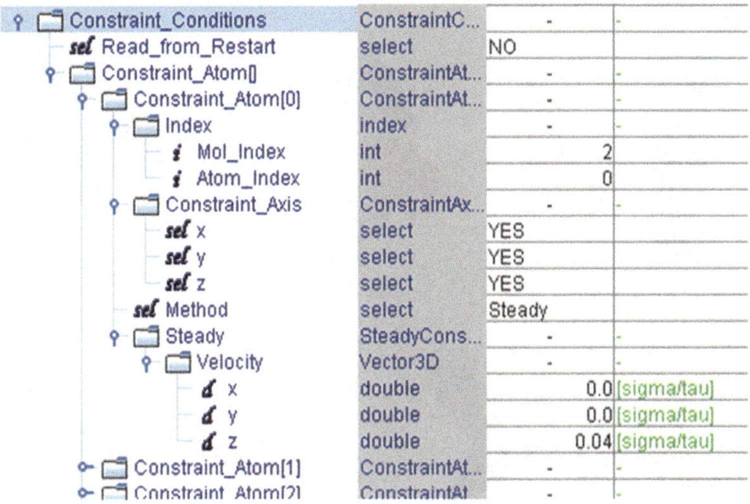

Fig. 25.9 Constraint conditions used to move the thread-like micelles (input file: "03_crossing.bdf")

With the initial structure obtained in the above procedure, a component of the velocity of each of the constrained atoms located near the boundary is entered so that the micelles proceed in the mutually approaching direction (the Python script "set_param_for_crossing.py", included in the sample data; also see Fig. 25.9). Here, the structure relaxed for 3500 τ is used as the initial structure. The use of other structures can produce different results such as the splitting of the thread-like micelles at the branch point.

Figure 25.10 shows the results (input file, "03_crossing.bdf"; output file, "03_crossing_out.bdf"; included in the sample data). For clarity in the snapshots, only the thread-like micelles are shown, and the water particles are not displayed (Python script "show.py", included in the sample data). The result shows that the two thread-like micelles fuse to form a branch point and then cross through each other. The time evolution is, however, slightly different from that in Fig. 25.4 given that the procedure to create the initial structure and the position of the micelles are not the same as those discussed in the previous section.

25.8 Download Instruction

Sample files are available from the OCTA website. See the instruction on http://octa.jp/download_bookdata.html. Password for the sample (zip file) of this chapter is "fNHRLPYY".

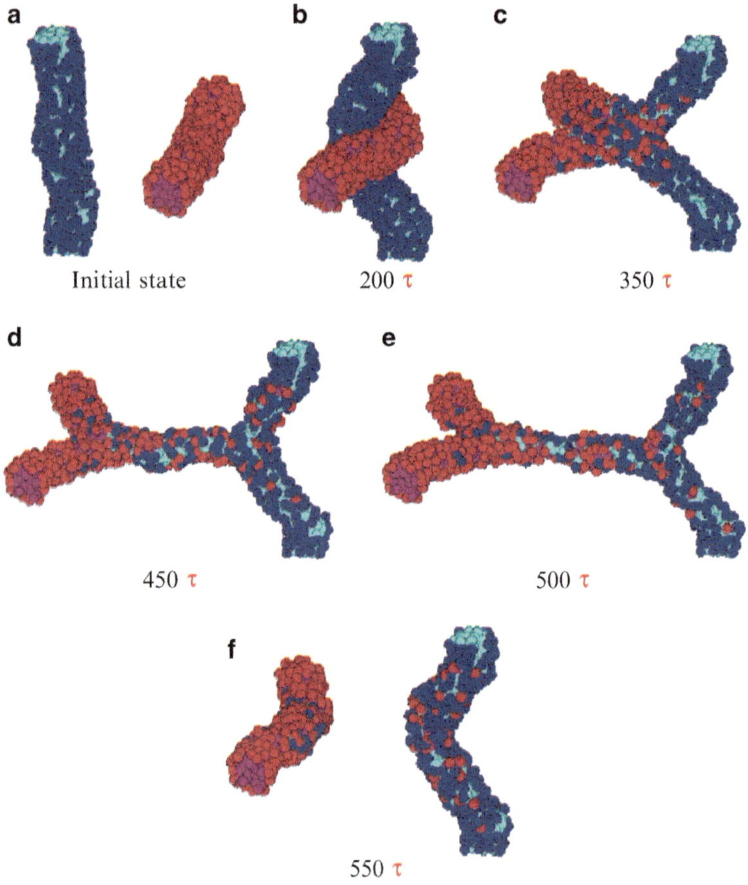

Fig. 25.10 Crossing of the thread-like micelles obtained using COGNAC (input file, "03_crossing.bdf"; output file, "03_crossing_out.bdf")

References

1. Y. Murakami, *Basic and Advanced of Supramolecular Chemistry* (NTS, Tokyo, 1996)
2. T. Shikata, H. Hirata, T. Kotaka, Langmuir **3**, 1081 (1987)
3. T. Shikata, H. Hirata, T. Kotaka, Langmuir **4**, 354 (1988)
4. T. Shikata, H. Hirata, T. Kotaka, Langmuir **5**, 398 (1989)
5. R. Goetz, R. Lipowsky, J. Chem. Phys. **108**, 7397 (1998)
6. S. Yamamoto, S. Hyodo, J. Chem. Phys. **122**, 204907 (2005)
7. J.N. Israelachvili, *Intermolecular Force and Surface Force* (McGraw Hill, New York, 1991)
8. T. Shikata, S. Imai, Y. Morishima, Langmuir **14**, 2020 (1998)
9. R.D. Groot, K.L. Rabone, Biophys. J. **81**, 725 (2001)
10. W.J. Kauzmann, D. Eisenberg, *Structure and Properties of Water* (MisuzuShobo, Tokyo, 1975)
11. T. Shikata, Y. Sakaiguchi, H. Urakami, A. Tamura, H. Hirata, J. Colloid Interface Sci. **119**, 291 (1987)

Chapter 26
Vesicle Formation

Satoru Yamamoto and Taku Ozawa

26.1 Introduction

Amphiphilic molecules such as a surfactant and lipid molecules aggregate in aqueous solution and exhibit a variety of structures, e.g., spherical micelles, rodlike (cylindrical) micelles, and bilayer membranes. The stability of an aggregate depends on the molecular structure, concentration, and temperature [1]. Both a biological membrane and ribosome consist of lipids, and their basic structures are bilayer membranes. Theoretical and experimental studies have investigated the aggregation types and explained the mechanism of spontaneous aggregation according to equilibrium thermodynamics. Because the existence of a biological membrane is essential for functional expression of a living organism, the aggregation of amphiphilic molecules has been vigorously studied. In particular, the ribosome is important to the biosynthesis of protein in a living cell, and there is a need to clarify the formation mechanism and dynamics of a vesicle, which is the type of aggregation of lipids. A vesicle exhibits interesting dynamics such as fusion, fission, and transformation into a multilayer vesicle. For instance, the red blood cell is a biconcave (discocyte) vesicle, and its shape changes into crenated (echinocyte) and knizocyte (leptocyte, stomatocyte, and codocyte) forms depending on aspects of the surrounding environment such as the pH, temperature, and invasion of an ion or foreign substance [2]. Moreover, the pinching off of the microvesicle from the parent vesicle of the echinocyte state has been experimentally observed [3]. It has been pointed out that local asymmetry in the composition of the vesicle due to the

S. Yamamoto (✉)
Dassault Systemes Biovia K.K., Tokyo, Japan
e-mail: satoru.yamamoto@3ds.com

T. Ozawa
JSOL Corporation, Tokyo, Japan

Japan Association for Chemical Innovation, *Computer Simulation of Polymeric Materials*, DOI 10.1007/978-981-10-0815-3_26

invasion of a foreign substance plays an important role in such an extraordinary transformation of shape [3].

Meanwhile, computer simulations have been attempted to investigate the structure and dynamics of lipid bilayers in a solution. There are two types of simulations. One uses a continuous surface model, where a bilayer is treated as a smooth and continuous thin surface and the dynamics are analyzed on a macroscopic scale. In this model, the molecular structures of the amphiphile are not taken into account explicitly. The other uses molecular dynamics, where the dynamics of molecules in the bilayer are investigated. Owing to the scale limitation of the approach, only part of the bilayer membrane is considered (i.e., the shape of the membrane as a whole is not discussed). It is preferable to conduct this investigation by combining approaches at various scales to understand the structure and dynamics of the bilayer membrane. In recent years, mesoscale simulations such as those based on molecular dynamics of coarse-grained amphiphilic molecules with a Lennard–Jones potential have been adopted to investigate the structure of a membrane, where the molecular structure can be partly considered [4]. Regarding vesicles, Monte Carlo simulations using a lattice model [5], Brownian dynamics simulations [6], and analysis with the time-dependent Ginzburg–Landau (GL) equation [7] have been adopted.

In the present study, we perform a mesoscopic simulation to study the self-assembly of amphiphilic molecules into vesicles using a dissipative particle dynamics (DPD) program [8].

26.2 DPD Model for Amphiphiles

In the present work, we focus on three different amphiphiles built from simple spheres as illustrated in Fig. 26.1. The A_1B_3 model consists of one hydrophilic head group A and three hydrophobic tail groups B. The A_2B_5 model is another

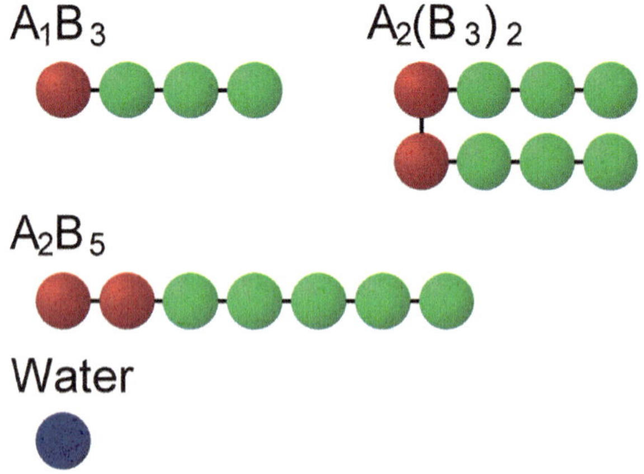

Fig. 26.1 DPD particle model for amphiphilic molecules and water

single-chain model but has two head groups and five tail groups and is somewhat larger than the A_1B_3 model. The head/tail ratio of A_2B_5 (2/7 = 0.286) is larger than that of A_1B_3 (1/4 = 0.250). $A_2(B_3)_2$ is a two-tailed model whose head/tail ratio is the same as that of A_1B_3. Each particle represents several atoms in an amphiphile molecule. The size of a particle is roughly 0.5 nm. A water particle is modeled as having the same size as a component of the amphiphile models. In this case, the water particle contains several water molecules.

26.3 Simulation Conditions

We perform extensive simulations to study the process of spontaneous vesicle formation from two initial conditions, (a) a randomly dispersed system and (b) a bilayer structure (Fig. 26.2). The bilayer structure of (b) corresponds to an initial condition of a filmlike bilayer in an experiment on the synthesis of ribosome [9]. All particles are drawn in Fig. 26.2a, whereas water particles are not drawn in Fig. 26.2b. For simulation from the bilayer structure, the cell size should be enlarged to prevent formation across the periodic boundary, which may occur at a high concentration of the amphiphile (higher than 15 vol.%). For the simulation from initial condition (a), the system dimensions, nondimensionalized by particle size, are $24 \times 24 \times 24$, the concentration of the amphiphile is 9.7 vol.%, and the numbers of particles are given in Table 26.1. The total number of particles is the same under all conditions. For the simulation from initial condition (b), the system dimensions are $30 \times 30 \times 30$ and the number of particles is given in Table 26.2.

Interaction parameters used in the simulation are summarized in Table 26.3. Using a time step of 0.05, the time evolution of the DPD particles is calculated

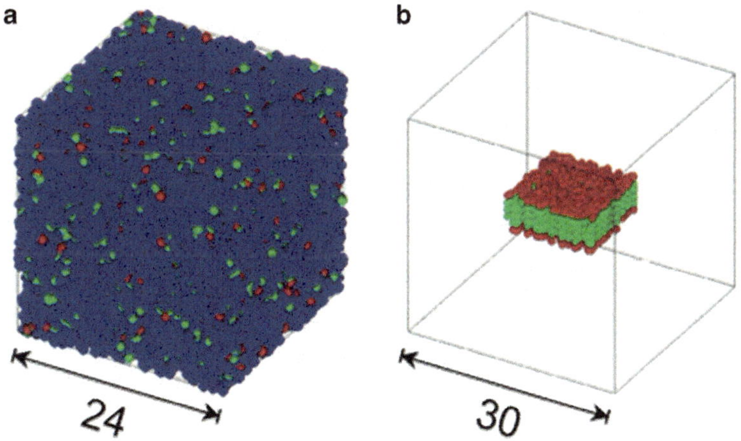

Fig. 26.2 Two initial conditions: (**a**) a randomly dispersed system and (**b**) a bilayer structure of the amphiphile A_1B_3

Table 26.1 Number of particles for initial condition (a)

Model	Molecules	Particles of amphiphiles	Concentration (vol%)	Water particles	Total particles
A_1B_3	1008	4032	9.7	37,440	41,472
$A_2(B_3)_2$	504	4032	9.7	37,440	41,472
A_2B_5	576	4032	9.7	37,440	41,472

Table 26.2 Number of particles for initial condition (b)

Model	Molecules	Particles of amphiphiles	Concentration (vol%)	Water particles	Total particles
A_1B_3	1,008	4032	5.0	76,968	81,000
$A_2(B_3)_2$	504	4032	5.0	76,968	81,000
A_2B_5	576	4032	5.0	76,968	81,000

Table 26.3 Interaction parameter a_{ij}

	Head (red)	Tail (green)	Water (blue)
Head (red)	25	40	25
Tail (green)	40	25	40
Water (blue)	25	40	25

until $t = 20{,}000$, or completion of the vesicle formation in the case of (a), and until $t = 3000$ or vesicle formation in case of (b). The calculation is performed using an SGI Origin 200 workstation. The processing time for $t = 1000$ is 22 h for the $24 \times 24 \times 24$ system and 33 h for the $30 \times 30 \times 30$ system.

26.4 Results and Discussion

Simulated snapshots of the process of aggregation and vesicle formation are shown in Fig. 26.3 for the case of A_1B_3. A slice of the system in the final frame is depicted. In the early stage, spherical micelles merge into large ones, and finally one large vesicle is observed in the cell. In the case of $A_2(B_3)_2$, a similar evolution of the structure is found. In these processes, a large oblate micelle (bilayer membrane) is observed before vesicle formation. The vesicle is aspherical and contains a low number of water particles because of the formation process from a randomly dispersed system.

In contrast, for the large amphiphile A_2B_5, vesicle formation is not observed, but two micelles form after a very long period at $t = 20{,}000$. These two micelles are stable, and they do not fuse together even in a contact situation. It seems that a much larger number of amphiphiles or a longer period is required to form the vesicle of A_2B_5.

Figure 26.4 shows the dynamics of aggregation and the vesicle formation process from the randomly dispersed systems, showing the time evolution of the maximum cluster size. The aggregation process with two-tailed amphiphiles $A_2(B_3)_2$ is faster because the radius of gyration is larger than that for the single-tailed model and

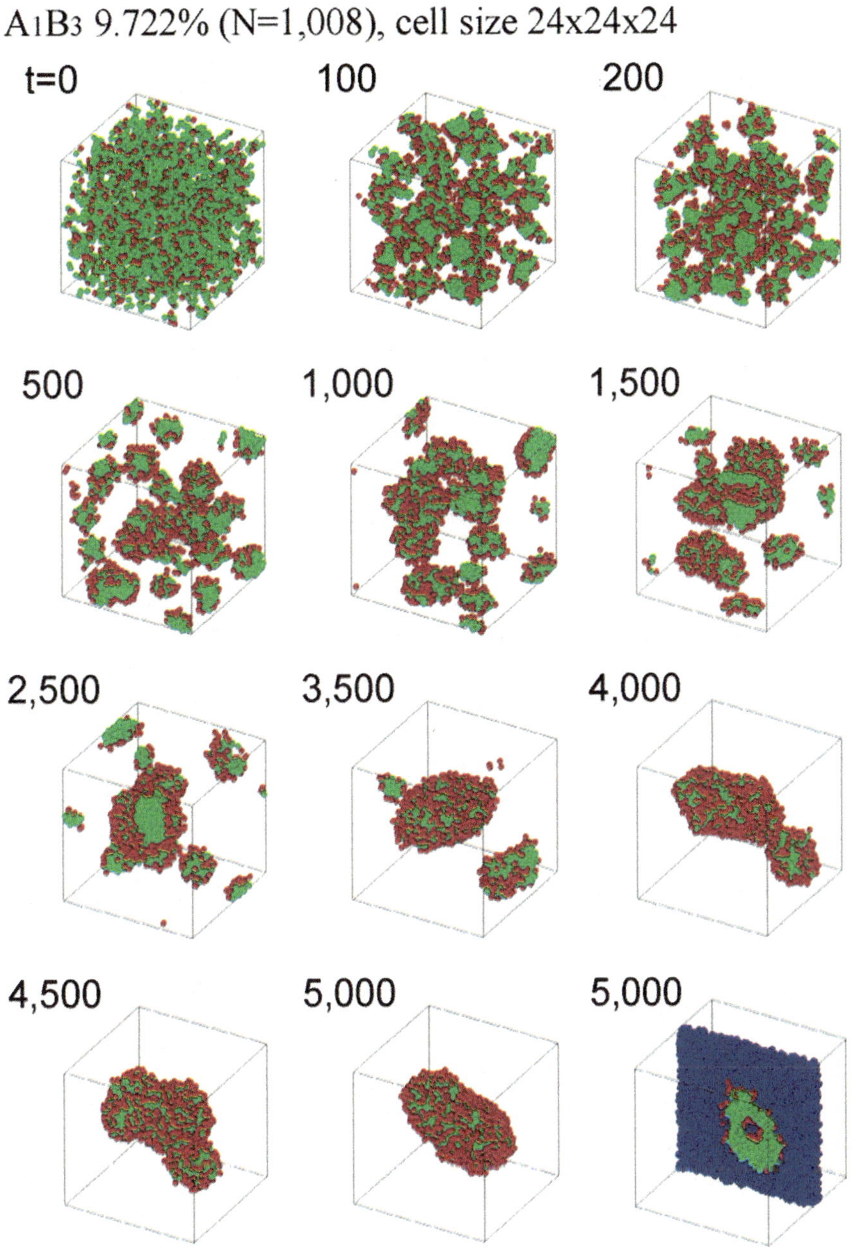

Fig. 26.3 Time-dependent morphology of the A_1B_3 model

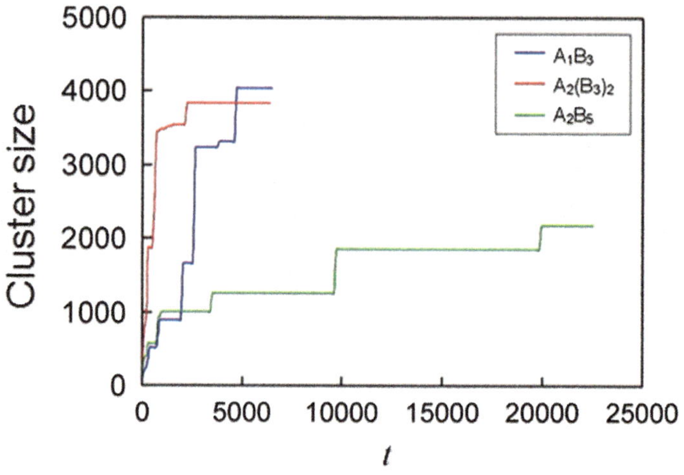

Fig. 26.4 Time dependence of the size of the largest cluster of amphiphiles

there is frequent contact. In contrast, the relatively large single-tailed model A_2B_5 has slow dynamics of aggregation, even though the growth rate of aggregation at an early stage is as high as that of the other two systems. This is because the head/tail ratio is higher and the micelle is more stable owing to a high concentration of hydrophilic particles on the surface. Therefore, the probability of fusion when two micelles come into contact is lower than that for other types of systems.

Next, snapshots of the vesicle formation process from the bilayer structure of system (b) are shown in Fig. 26.5 for A_1B_3. Vesicle formation is clearly observed, and it is similarly observed in the case of $A_2(B_3)_2$. The initial bilayer structure is artificially constructed, and hydrophobic tail particles at the side of the membrane come into contact with water and are unstable. Therefore, the shape of the bilayer becomes disklike at an early stage, and the bilayer then encapsulates water and forms a vesicle. In the figure, a slice including water particles is depicted at the final stage.

In the case of the A_2B_5 model, however, an oblate micelle remains at $t = 3000$. An additional simulation is performed under the condition that the number of amphiphiles is 1.5 times the number in the original system. As a result, vesicle formation similar to that for the other two systems is observed. Spherical vesicles are observed throughout system (b) in contrast to the case for system (a).

Next, the conservative energy regarding the amphiphiles during the time evolution is estimated according to

$$V^{C} = \sum_{\text{amphiphile}} \left\{ \frac{1}{2} a_{ij} \left(1 - r_{ij}\right)^{2} \right\} / N_{\text{amphiphile}}, \tag{26.1}$$

and we discuss why amphiphiles change into vesicles. The summation is for all amphiphiles, and $N_{\text{amphiphile}}$ is the total number of particles of amphiphiles. The

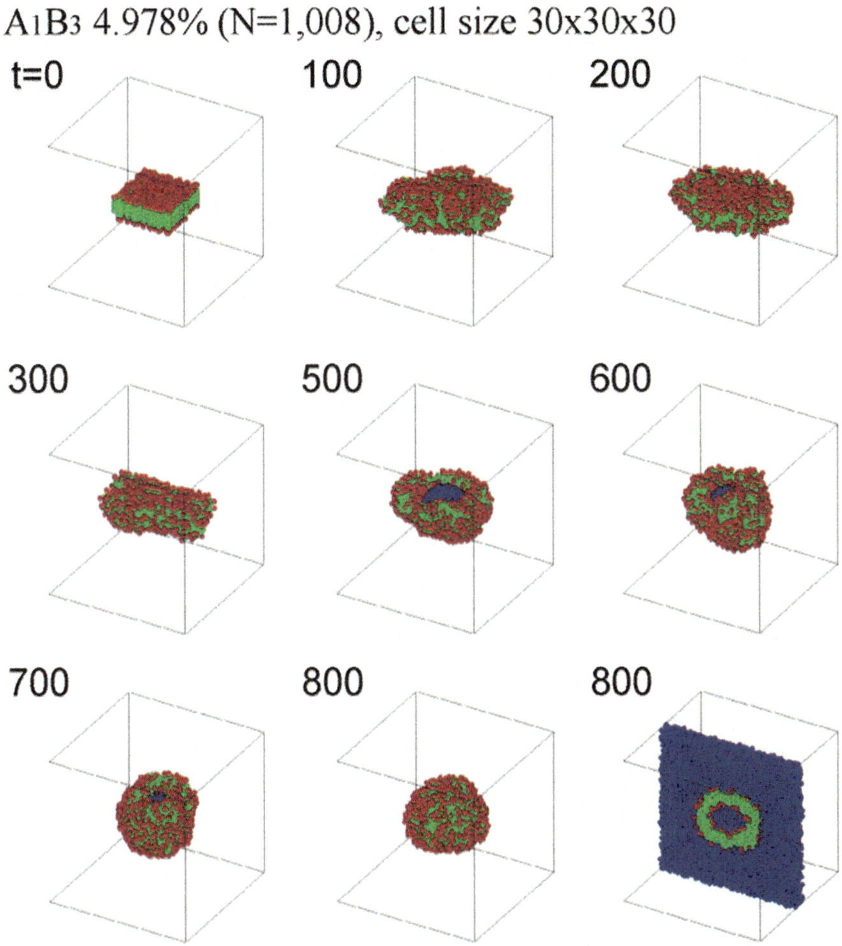

Fig. 26.5 Snapshots of the time evolution of the A_1B_3 model

result is shown in Fig. 26.6 for A_1B_3. The energy is averaged for each period of 30 units. During the process of vesicle formation, the conservative energy diminishes by 0.02–0.03 kT. It is confirmed that the vesicles form via an intermediate state of a bilayer membrane to minimize the hydrophobic interaction energy between the amphiphile and water.

We performed a DPD simulation for the vesicle formation of amphiphiles and reproduced the procedure in which a bilayer encapsulates water and forms a vesicle similar to the procedure of the synthesis of ribosome observed experimentally. It was also clarified that a large amphiphile has slow dynamics and needs a large number of molecules to form a vesicle.

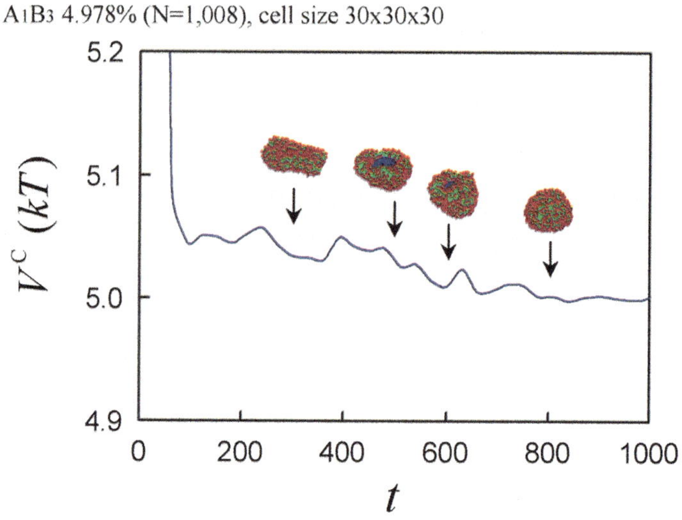

Fig. 26.6 Transient change in the interaction energy of the A_1B_3 model

26.5 Conclusions

DPD was applied to study the vesicle formation of amphiphiles. From both a randomly dispersed system and a bilayer structure of amphiphiles in water particles, spontaneous vesicle formation via an oblate micelle state was recognized. Water particles were encapsulated to form a vesicle. During vesicle formation, it was clarified that the hydrophobic interaction energy is minimized. It was also recognized that the process of aggregation is faster for the two-tailed amphiphiles than in the case of single-tailed models. This study confirmed that DPD simulation is an effective technique for analyzing the dynamics and vesicle formation process of amphiphiles.

26.6 OCTA Example Run

A follow-up calculation was performed using OCTA (COGNAC). The main points regarding the creation of the initial structure and calculation conditions are explained below:

1. Creation of the initial structure and calculation conditions (sample file: "vesicle.udf")
   The vesicle formation from a bilayer structure of the amphiphile can be calculated by applying the zooming method described in Chap. 25. Here, as a simple case study, vesicle formation is simulated from the randomly dispersed

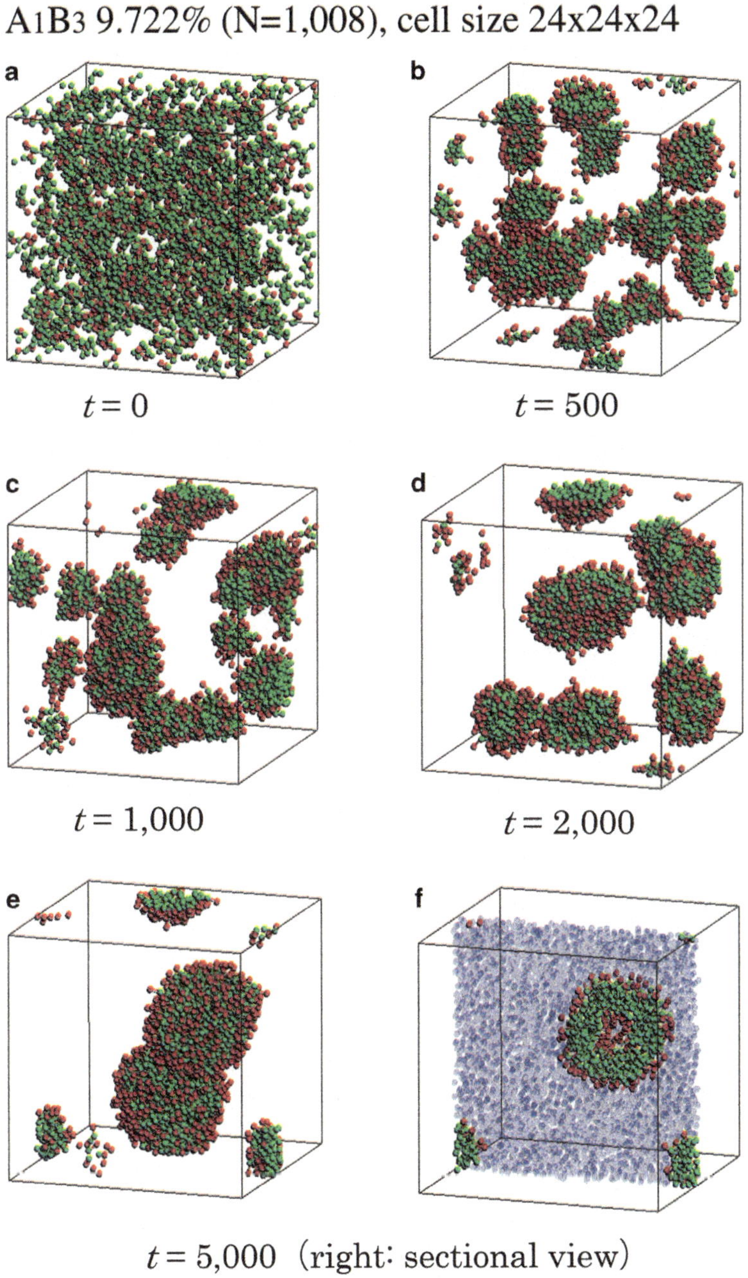

Fig. 26.7 Formation of vesicles of the A_1B_3 model using COGNAC

structure as the initial state. The amphiphile molecule is A_1B_3 and the calculation conditions are described above. According to the erratum information of the original paper, the system temperature is set as $T = 0.43$.

2. Calculation results (sample file: "vesicle_out.bdf")
   Figure 26.7 shows calculation results. Although the thermal fluctuations affect the structure at each time step, the vesicle structure is similar to that shown in Fig. 26.3.

26.7 Download Instruction

Sample files are available from the OCTA website. See the instruction on http://octa.jp/download_bookdata.html. Password for the sample (zip file) of this chapter is "WPL2wBd3".

References

1. Y. Murakami, *Basic and Advanced of Supramolecular Chemistry* (NTS, Tokyo, 1996)
2. S.L. Kimzey, Biomedical results from Skylab, Chap. 28. (NASA, 2015), http://lsda.jsc.nasa.gov/books/skylab/Ch28.htm. Accessed on 10 Nov 2015
3. E. Farge, P.F. Devaux, Biophys. J. **61**, 347 (1992)
4. R. Goetz, R. Lipowsky, J. Chem. Phys. **108**, 7397 (1998)
5. A.T. Bernardes, Langmuir **12**, 5763 (1996)
6. H. Noguchi, M. Takasu, Phys. Rev. E **64**, 041913 (2001)
7. T. Taniguchi, Phys. Rev. Lett. **76**, 4444 (1996)
8. S. Yamamoto, Y. Maruyama, S. Hyodo, J. Chem. Phys. **116**, 5842 (2002). 117, 2990
9. S. Sugai (ed.), *Biophysics Engineering* (IPC, Tokyo, 1992)

Chapter 27
Electrolyte Membranes

Satoru Yamamoto and Taku Ozawa

27.1 Introduction

In the mesoscale structure of hydrated Nafion, which is used as the electrolyte membrane in polymer electrolyte fuel cells, it is presumed from the monomer molecular structure and the results of small-angle X-ray scattering experiments that water clusters surrounded by sulfonic acid groups are linked in a spongelike arrangement. Furthermore, the material's proton conductivity is understood to be expressed as proton movement through the water channel. However, although several models that describe the detailed structure have been proposed, they are still being debated; with respect also to the mechanism of proton conductivity, although hopping conductivity via the sulfonic acid groups is the most likely explanation, the details are still not yet understood [1, 2]. Here, the mesoscale structure of the hydrated Nafion membrane is calculated using dissipative particle dynamics (DPD) and compared with the results of experiments reported to date [3]. In addition, anisotropy in the swelling behavior of some hydrocarbon-based electrolyte membranes has been reported recently, and the possibility of an anisotropic mesostructure has been identified [4, 5]. A characteristic of these hydrocarbon-based electrolyte membranes is that their main chain skeletons are rigid; hence, their molecular chains easily become oriented, such that formation of an anisotropic structure is expected. We therefore investigated the mechanism of expression of anisotropy in hydrocarbon-based electrolyte membranes by incorporating molecular rigidity into the coarse-grained model.

S. Yamamoto (✉)
Dassault Systemes Biovia K.K., Tokyo, Japan
e-mail: satoru.yamamoto@3ds.com

T. Ozawa
JSOL Corporation, Tokyo, Japan

Japan Association for Chemical Innovation, *Computer Simulation of Polymeric Materials*, DOI 10.1007/978-981-10-0815-3_27

27.2 Method

Nafion, whose chemical structure is shown below, is a fluorinated polyethylene main chain with a fluorinated vinyl ether side chain that has a sulfonic acid group attached to its end.

$$-[(CF_2\text{-}CF_2)_n\text{-}CF\text{-}CF_2]_m\text{-}$$
$$O\text{-}CF_2\text{-}CF\text{-}O\text{-}CF_2\text{-}CF_2\text{-}SO_3H$$
$$CF_3$$

Here, n and m generally fall within the ranges of 5–14 and 200–1000, respectively, and vary according to the length of the side chains and spacing at which side chains are bonded to the main chain [1]. The sulfonic acid group at the end of the side chain easily dissociates in the presence of water to form SO_3^-. Nafion is thermally, chemically, and mechanically stable. Using the molecular structure of the Nafion monomer as a reference, DPD are introduced to construct a coarse-grained molecular model as shown in Fig. 27.1. In DPD, a single group of atoms is expressed as a sphere, and in this case, sphere A largely corresponds to $\text{-}CF_2CF_2CF_2CF_2\text{-}$, sphere B to $\text{-}OCF_2C(CF_3)FO\text{-}$, and sphere C to $\text{-}CF_2CF_2SO_3H$. The size of each sphere is approximately 0.5 nm. Nafion's actual degree of polymerization is approximately 230, but owing to limitations on the model size used for calculation, a model with a degree of polymerization of 5 is used. If the length of the main chain is too short, there is a concern that the polymer properties of the main chain may not be expressed. However, in the phase diagram of the block copolymer obtained by mean field theory, the conditions under which phase separation occurs are such that, when the number of phase segments N reaches a certain size, a further increase in N does not affect the phase structure [6]. Hence, if a stable structure can be obtained with the model used, it may be assumed that the relevant polymer properties have been adequately considered. In fact, because the same structure formation was confirmed for the dimer model used in preliminary calculations (refer to the results below), it

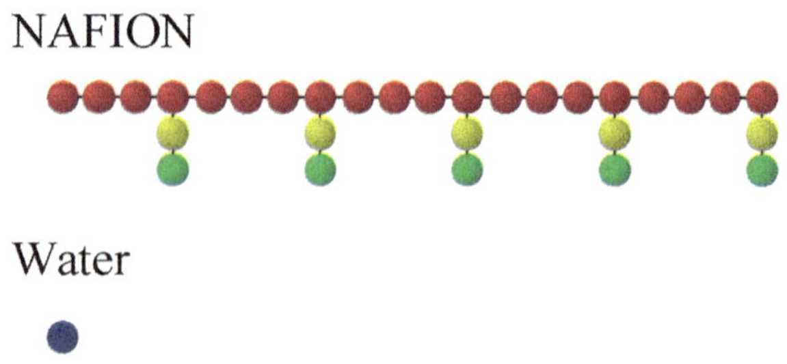

Fig. 27.1 DPD models of Nafion and water

may be concluded that the 5-mer model shown in the diagram is fully adequate. For water, the sphere W is used, as it has the same size as the spheres constituting the Nafion model.

When performing the DPD calculation, the coefficients for the interaction between individual spheres must be decided; these are determined as shown below. The DPD conservative force a_{ij} is related to the parameter χ_{ij} for the interaction between chemical species as follows (in the case of a particle density of 3) [7, 8]:

$$a_{ij} = a_{ii} + 3.27\chi_{ij} \tag{27.1}$$

Furthermore, χ_{ij} is related to the solubility parameter δ obtained from the cohesive energy for each chemical species:

$$\chi_{ij} = \frac{V_{\mathrm{seg}}}{RT}\left(\delta_i - \delta_j\right)^2 \tag{27.2}$$

$$\delta = \left(\frac{E_{\mathrm{coh}}}{V}\right)^{0.5} \tag{27.3}$$

Here, V_{seg} is the volume of each segment making up the polymer. Cohesive energy can be calculated by subtracting the energy of the chemical species without the cell from the energy when the chemical species is within the cell with periodic boundary conditions [9].

Table 27.1 gives the results of calculating the solubility parameter from the cohesive energy of the three types of spheres constituting Nafion. Materials Studio (Dassault Systemes Biovia) is used for this calculation. Table 27.2 gives the interaction coefficients, determined using the results in Table 27.1. Strongly hydrophobic A has poor affinity with strongly hydrophilic C; a strong separation of A and C is thus expected. In addition, because the sulfonic acid group dissociates and dissolves in water, water and C are considered to be similar particles; this consideration is used to determine the interaction coefficient with respect to water.

Taking a sphere having a diameter of 1, a basic cell of 40 × 40 × 40 is used in the calculations. As the initial condition, Nafion and water are dispersed randomly, and

Table 27.1 Solubility parameters of spheres constituting the Nafion DPD model

	Corresponding segment	δ (J/cm^3)$^{0.5}$
A	$-CF_2CF_2CF_2CF_2-$	12.49
B	$-OCF_2C(CF_3)FO-$	19.80
C	$-CF_2CF_2SO_3H$	33.36

Table 27.2 Interaction parameters a_{ij}

	a_{ij}
-B	27.0
A-C	41.0
B-C	31.8

the time development of the system is calculated. Three moisture contents of 10, 20, and 30 vol.% (corresponding to 5.26, 11.11, and 17.65 wt.%, respectively) are used. The particle density inside the cell is taken to be 3, and the total number of particles is 192,000.

In this calculation, time of 1 unit corresponds to approximately 10 ps; for each condition, the calculation is continued up to a time of 1000 units (approximately 10 ns). An SGI Origin 200 workstation is used to perform the calculation, and the computation for each condition takes approximately 80 h.

27.3 Results and Discussion

The results of calculation of the mesostructure for moisture content of 20 vol.% are shown in Fig. 27.2a. In the figure, water is depicted as spheres, but the A, B, and C spheres that constitute Nafion are represented only by lines indicating the bonds connecting them. Small water clusters form in the initial stage; these gradually become larger as time progresses, reaching a steady state at a time of approximately 200 units, at which point the water clusters have formed a dispersed structure. The formation of water clusters surrounded by sulfonic acid groups can be observed. From the results of detailed observation of the internal structure, it is seen that areas with water have a spongelike structure containing connected channels.

In a manner similar to the case of 20 vol.% moisture content, results are also calculated under conditions of 10 and 30 vol.%; doing so reveals that the water cluster size increases as the moisture content increases. It is already clear from past small-angle X-ray scattering experiments that changing the moisture content affects the size and spacing of water clusters, and it has been reported that changing the moisture content from 5 to 20 wt.% increases the cluster spacing from 3 to 5 nm

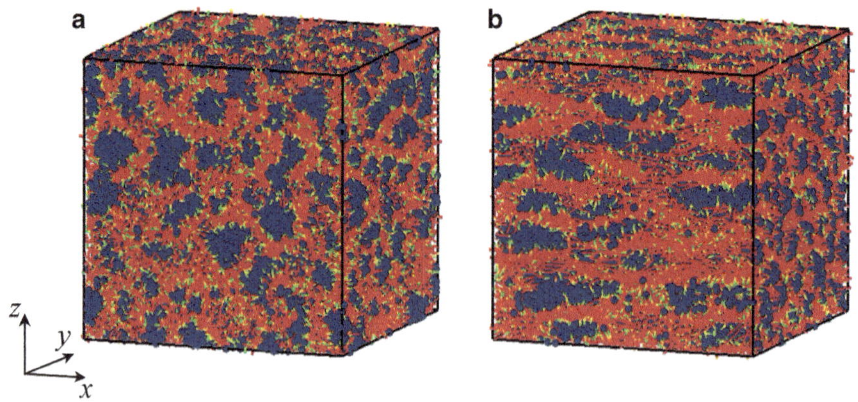

Fig. 27.2 Mesostructure of an electrolyte membrane: (**a**) without rigidity, (**b**) with rigidity (moisture content 20 vol.%)

[2]. Thus, the radial distribution $g(r)$ of water is calculated for each model, and the dependency of the water cluster size and spacing on moisture content is investigated.

For water clusters, the cluster size (diameter) is defined such that the radius is given by the point at which $g(r)$ becomes smaller than 1 near the first peak. The results of investigating its relationship with moisture content indicate that, when the moisture content increases from 10 vol.% (5.3 wt.%) to 30 vol.% (17.6 wt.%), the cluster size increases from 5.9 to 9.5. Considering the sphere size in the DPD model, this corresponds approximately to an increase from 3.0 to 4.7 nm.

In addition, the results of the investigation of the relationship between cluster spacing and moisture content indicate that, depending on the moisture content, the cluster spacing changes from 7.0 to 11.0, corresponding approximately to a change from 3.5 to 5.5 nm. These calculation results agree with experimental reports on cluster spacing and moisture content dependency.

Next, the mesostructure is calculated for the case in which the rigidity is equivalent to that of the main chain poly(p-phenylene) skeleton in a hydrocarbon-based electrolyte membrane. Here, a comparison is carried out presuming rigidity is conferred using the same chemical structure as that of Nafion. Rigidity is represented by constraining the equilibrium value of the main chain A-A-A bond angle and side chain A-B-C bond angle to 180° and the equilibrium value of the side chain bond angle with respect to the main chain (A-A-B bond angle) to 90°. The results of analyzing the obtained mesostructure indicate that rigidity alone is insufficient to cause anisotropy to appear. Hence, it is possible that, in the actual membrane production process, shearing that occurs owing to the flow of solvent may trigger the appearance of anisotropy in the mesostructure; therefore, after imposing a shear flow in the x direction and calculating the mesostructure, the shear is removed and the calculation is performed until equilibrium is reached. The results are shown in Fig. 27.2b. In case (a), in which molecular rigidity is not considered, the mesostructure is isotropic even though a shear is imposed. However, in case (b), which considers molecular rigidity, the polymer and water regions are extended in the x direction, as clearly confirmed by the examination of the water particle radial distribution function. The reason is assumed to be that, if the molecules are rigid, the molecular chains have a strong tendency to become oriented in the shear direction, which forms a mesostructure extended in the x direction. A mesostructure with this kind of anisotropy is expected to be strong with respect to deformation in the in-plane direction such that, in wet/dry cycling, the dimensional changes in the direction of thickness will be larger than those in the in-plane direction. This result supports the swelling anisotropy [4, 5] that occurs in actual membranes.

27.4 Conclusion

The structure of the polymer electrolyte membrane Nafion was calculated using DPD, which is a mesoscale calculation method. A DPD model based on the molecular structure of the Nafion monomer was created, with the coefficients of interaction between the DPD particles determined by calculating the cohesive energy of the segments corresponding to each particle type. The results of calculating the structure of Nafion starting from random initial conditions showed the appearance of a spongelike structure similar to what has been postulated from experimental results. The water cluster size and distribution spacing were approximately equal to those obtained from experiments, and the results of experiments involving changes in structure depending on moisture content were also reproduced. These findings indicate that a model that possibly represents the actual structure of Nafion has been successfully demonstrated through calculation. In addition, results obtained with the current model suggest that the anisotropy observed in hydrocarbon-based electrolyte membranes may be explained by considering molecular rigidity.

27.5 OCTA Example Run

A follow-up calculation is performed using OCTA (COGNAC). The main points regarding the creation of the initial structure and calculation conditions are explained below:

1. Creation of the initial structure and calculation conditions
   COGNAC's "Random" function is used to create the initial structure (coordinates). The structure of the Nafion molecule is as shown in Fig. 27.1, and the calculation conditions are as described above. Here, the rigidity of the molecules is not considered, but molecular rigidity can easily be introduced in COGNAC by setting the angle potential.
   As a sample, there is a Python script that sets Nafion and water molecules in the input file "setNafion.py". At the beginning of the script, the structure of one molecule is described, and the numbers of different molecule types are set to create a simulation system with an arbitrary volume fraction. This script should be run on "nafion.bdf".
2. Calculation results
   Figure 27.3 shows the calculation results. Figure 27.3a shows a snapshot in which a water cluster structure similar to that shown in Fig. 27.2 has formed. Figure 27.3b shows an isosurface (50 %) of the volume fraction of water, calculated from the position of water particles. By performing a visualization in this way, the complicated cluster structure can be evaluated.

To obtain the volume fraction distribution, a COGNAC function ("Density_Output", $32 \times 32 \times 32$) is used. Instantaneous volume fraction distributions can also be obtained using the action function, but taking a time average smoothens the shape of the interface, allowing a volume fraction distribution that reflects the structures that can occur within the specified time to be obtained.

The water particle's radial distribution function is evaluated, which allows estimation of the average cluster size and distance between clusters. COGNAC's action function is used to obtain the radial distribution function. As shown in Fig. 27.4,

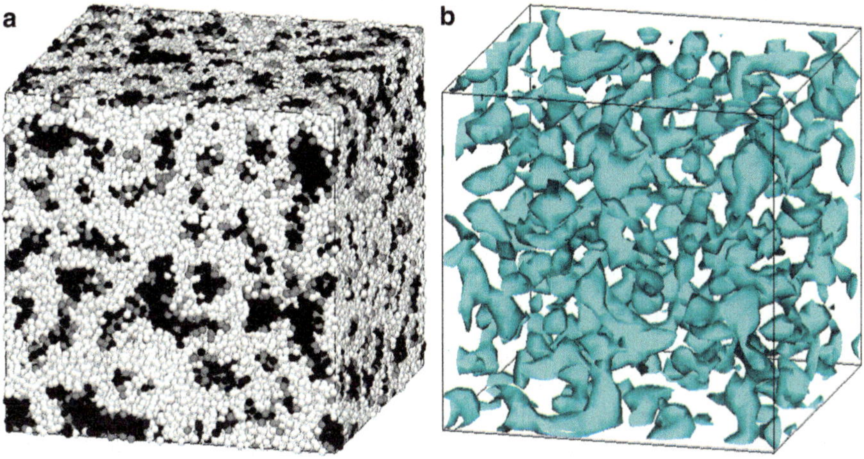

Fig. 27.3 Electrolyte membrane mesostructure calculation using COGNAC. (**a**) Depiction of particles (*black*, water; *gray*, hydrophilic group; *white*, hydrophobic group). (**b**) Depiction of the concentration distribution (water area isosurface)

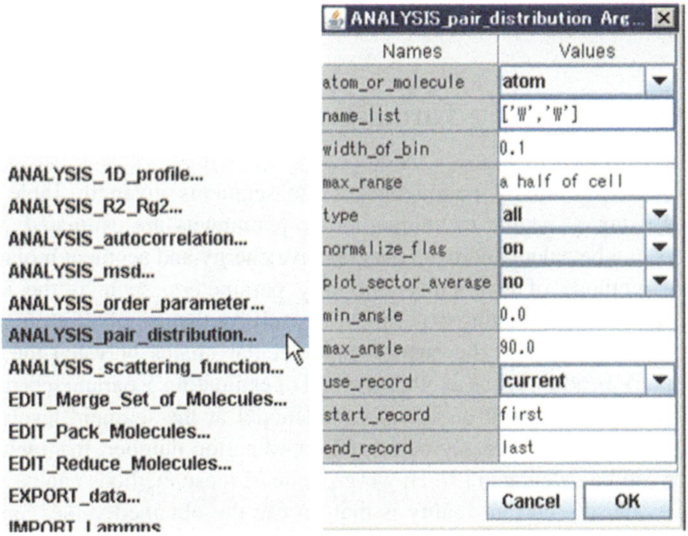

Fig. 27.4 Action function used to obtain the radial distribution function

the dialog (Fig. 27.4, right) is displayed when **ANALYSIS_pair_distribution**... is selected. Here, the radial distribution function between water particles is obtained by setting the name_list field to "['W','W']"; the result is expressed as a graph like that shown in Fig. 27.5.

In Fig. 27.5, the values along the horizontal axis where the curve crosses a value of 1 on the vertical axis show the water cluster size (at approximately $r = 3$) and

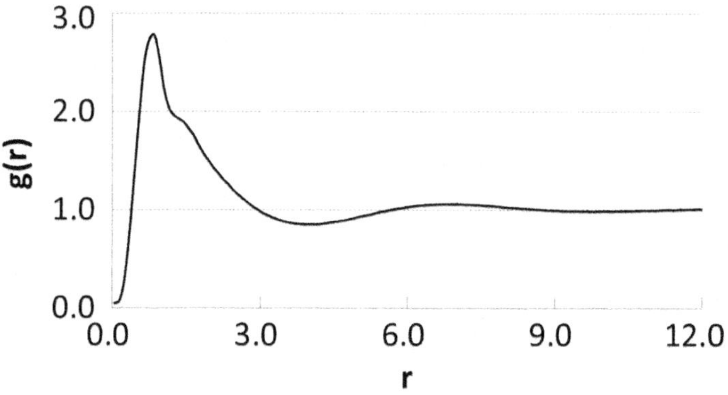

Fig. 27.5 Water particle radial distribution function (where the horizontal axis represents the dimensionless length)

distance between clusters (at approximately $r = 6$). Furthermore, when this graph no longer changes with time, it can generally be concluded that the system is approaching a state of equilibrium.

27.6 Calculation of χ Parameters

In this section, solubility parameters of the segments given in Table 27.1 are calculated using a specific technique, and χ parameters are estimated. Solubility parameters can be calculated from the cohesive energy and segment molar volume. There are methods of estimating solubility parameters, such as the use of an empirical database containing experimental results or the use of all-atom molecular dynamics (comparison of the energy of molecular chains between the bulk and isolated state). In addition, as another method of estimating χ parameters, the Monte Carlo method can be used on the all-atom model at the segment level to obtain the contact energy between segments and coordination number, from which the χ parameter can be calculated [9]. However, none of these methods can be said to be conclusive, and the current reality is that, to use the obtained values, appropriate modifications must be made.

However, as seen from the examples in other chapters, χ parameters are often set arbitrarily to appropriate values to reproduce phenomena of interest. Care is necessary with respect to cases in which small changes to values have large effects on the results; however, where such sensitivity is low, and where qualitatively similar trends can be obtained, it is customary to adopt such an approach.

27.7 Download Instruction

Sample files are available from the OCTA website. See the instruction on http://octa.jp/download_bookdata.html. Password for the sample (zip file) of this chapter is “s2ewFTHR”.

References

1. H.G. Haubold, T. Vad, H. Jungbluth, P. Hiller, Electrochim. Acta **46**, 1559 (2001)
2. T.D. Gierke, G.E. Munn, F.C. Wilson, J. Polym. Sci. **19**, 1687 (1981)
3. S. Yamamoto, S. Hyodo, Polym. J. **35**, 519 (2003)
4. K. Kaneko, Y. Takeoka, M. Rikukawa, K. Sanui, Polymer. Preprints Jpn. **54**, 4515 (2005)
5. S.J. Mumby, P. Sher, B.E. Eichinger, Polymer **34**, 2540 (1993)
6. R.D. Groot, P.B. Warren, J. Chem. Phys. **107**, 4423 (1997)
7. R.D. Groot, T.J. Madden, J. Chem. Phys. **108**, 8713 (1998)
8. B.E. Eichinger, D.R. Rigby, M.H. Muir, Comp. Polym. Sci. **5**, 147 (1995)
9. C.F. Fan, B.D. Olafson, M. Blanco, S.L. Hsu, Macromolecules **25**, 3667 (1992)

Chapter 28
Orientation Birefringence

Makoto Wakabayashi

28.1 Introduction

Polycarbonate resin (PC) is a resin with a main chain that contains carbonate bonds. Polycarbonate, for which bisphenol A (BisA) is the raw material, has excellent properties such as transparency, impact resistance, heat resistance, and dimensional accuracy, and it is for these characteristics that it has been widely used as an optical resin for digital versatile disks (i.e., DVDs) and other media technologies. However, it suffers from a high degree of optical anisotropy, and birefringence, which can easily occur during molding, is a well-known problem. The speed of light in a material is reduced by the interaction with an induced electric field generated by the vibration of electrons in that material. The ratio (c/v) of the speed of light in a vacuum (c) to the speed in the material (v) gives the refractive index (n). The easier it is for the electrons in the material to move (i.e., the higher the polarizability), the larger the refractive index. If there is anisotropy in the molecular structure, then the polarizability differs depending on the direction of polarization. Differences in the speed of light in the material will thus occur, yielding anisotropy of the refractive index. This is birefringence.

Denbigh reported a method of obtaining birefringence from polarizability [1] in which the anisotropy of the polarizability (P) is expressed by the sum of the bond polarizabilities of the individual chemical bond structures in the molecule.

Electronic supplementary material The online version of this chapter (doi: 10.1007/978-981-10-0815-3_28) contains supplementary material, which is available to authorized users.

M. Wakabayashi (✉)
Idemitsu Kosan Co., Ltd., Tokyo, Japan
e-mail: mnet1150@cc.rim.or.jp

Japan Association for Chemical Innovation, *Computer Simulation of Polymeric Materials*, DOI 10.1007/978-981-10-0815-3_28

Aikawa used a method that employs the vector sum of bond polarizabilities and a semiempirical molecular orbital calculation to obtain the intrinsic birefringence of the polymer [2]. Miyazaki et al. reported the molecular dynamics simulation of the orientation birefringence of polystyrene [3]. The current case study describes a technique for predicting the orientation birefringence of PC through molecular dynamics simulation using OCTA [4].

28.2 Method of Calculating Orientation Birefringence

Orientation birefringence is computed by calculating the polymer's uniaxial elongation using COGNAC, OCTA's molecular dynamics engine.

First, using molecular dynamics, a polymer in an amorphous state is generated, and that polymer's uniaxial elongation is calculated. From the elongation calculation results, the sum of bond polarizabilities for the molecules in all the bonds is calculated; after converting this to a refractive index, the orientation birefringence is obtained.

To calculate the birefringence, the polarizability vector **P** for the entire polymer system is obtained according to

$$\mathbf{P} = \sum_i \left(b_{1i}\cos^2\theta_i + b_{2i}\sin^2\theta_i\right), \tag{28.1}$$

where θ_i is the angle formed between bond i and the orientation axis, taking the orientation axis as being along the incident light path; b_{1i} is the bond polarizability along the direction parallel to bond i; and b_{2i} is the bond polarizability along the direction perpendicular to bond i. The values in Table 28.1 are used for b_{1i} and b_{2i}.

The components of the refractive index vector **n** are given by the Lorentz–Lorenz equation:

$$\frac{\mathbf{n}^2 - 1}{\mathbf{n}^2 + 2} = \frac{4}{3}\pi\frac{\mathbf{P}}{V} \tag{28.2}$$

Table 28.1 Bond polarizability

Bond	b1 [$\times10^{-25}$ cm^3]	b2 [$\times10^{-25}$ cm^3]
C_{aliph}–C_{aliph} [1]	18.8	0.2
C_{aliph}–C_{arom} [1]	18.8	0.2
C_{arom}–C_{arom} [1]	22.5	4.8
C–O [5]	14.6	1.7
C=O [1]	19.9	7.5
C–H [1]	7.9	5.8

and birefringence Δn is calculated as

$$\Delta n = n_z - \frac{n_x + n_y}{2} \tag{28.3}$$

28.3 Input Data

J-OCTA (JSOL Corporation, Tokyo, Japan) is used to generate the input user definable format (UDF) file. The molecular structure is shown in Fig. 28.1. BisA-PC is taken to be a 25-mer. Figure 28.2 illustrates the calculation procedure:

Step 1: Generation of the molecular structure and setting of the force field parameters

The molecular structure is generated using J-OCTA's modeler function, and the force field parameters are set. Although OCTA includes SILK, an input data generation support tool, it is necessary to prepare a script, and a commercial modeler is thus used here. The Dreiding force field [6], included in J-OCTA, is used for the force field parameters:

Step 2: Positioning of molecules and generation of an amorphous structure

The obtained eight molecules are positioned randomly inside unit cells having sides of length 100 Å such that adjacent molecules do not overlap; a compression calculation at 1,000 MPa is then performed:

Δt: 1 fs

Total calculation time: 100,000 (0.1 ns)

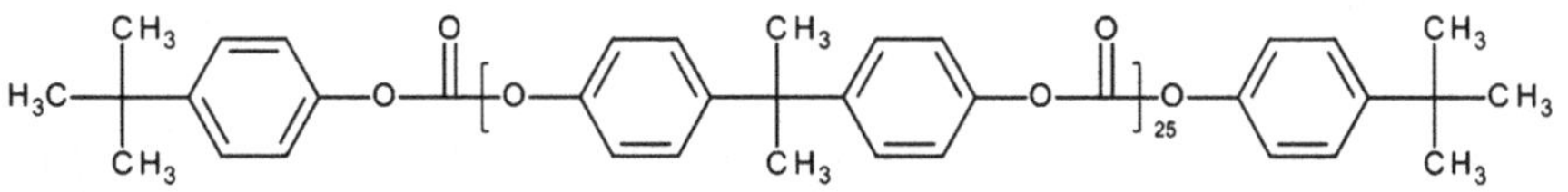

Fig. 28.1 Molecular structure of polycarbonate resin (BisA-PC)

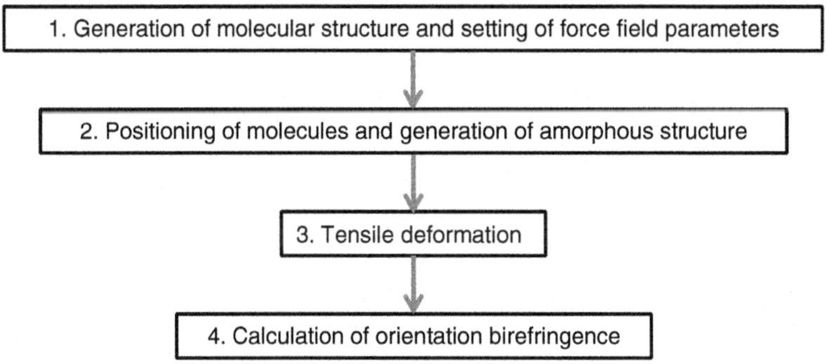

Fig. 28.2 Calculation procedure

Temperature: 428 K
Pressure: 1E9 Pa (1,000 MPa)
Algorithm: NPT_Andersen_Nose_Hoover

Next, the pressure is set to 1 atm (0.1 MPa), and an equilibration calculation is performed:

Δt: 1 fs
Total calculation time: 100,000 (0.1 ns)
Temperature: 428 K
Algorithm: NPT_Andersen_Nose_Hoover
Pressure: 1E5 Pa (0.1 MPa)

An initial structure generated in this way for use in tensile deformation calculations is attached as the input UDF file "BisA25PC_in.udf". Figure 28.3 depicts how this input UDF file is displayed when opened from GOURMET.

Step 3: Tensile deformation

Tensile deformation is calculated under the following conditions. The drawing speed is set to 1,000 m/s so that, for the time scale of the molecular dynamics calculation, the stress optical constant is close to measurements. In addition, Poisson's ratio is set to a representative value of 0.4:

(1) SI should be selected for [**unit**] [**Select Unit Set**. . .]
(2) *delta_T*: 0.1 fs
Total_Steps: 160,000 (0.016 ns)
(3) *Temperature*: 428 K
(4) Select *Method*: Cell_Deformation
(5) Select *Method*: Simple_Elongation
(6) *Elongation_Rate*: 1,000 m/s
(7) *Poisson's ratio*: 0.4
(8) Select *Dynamics_Algorithm*: NVT_Nose_Hoover

The initial structure can be confirmed in Viewer by opening the attached input UDF file "BisA25PC_in.udf" from GOURMET, right-clicking the input UDF file name in (9) the Name field, clicking *VIEWER_show* . . . (located second from the bottom), and then clicking OK.

28.4 Output Data

The calculation results are attached as the output UDF file "BisA25PC_out.udf". When this UDF file is opened from GOURMET, it is displayed as shown in Fig. 28.4. The elongation structure can be confirmed using the same operations as for the input UDF file. Figure 28.5 shows a captured image of Record #151 (12 ps) for BisA-PC. Note that the PC's benzene rings tend to be oriented parallel to the direction of elongation.

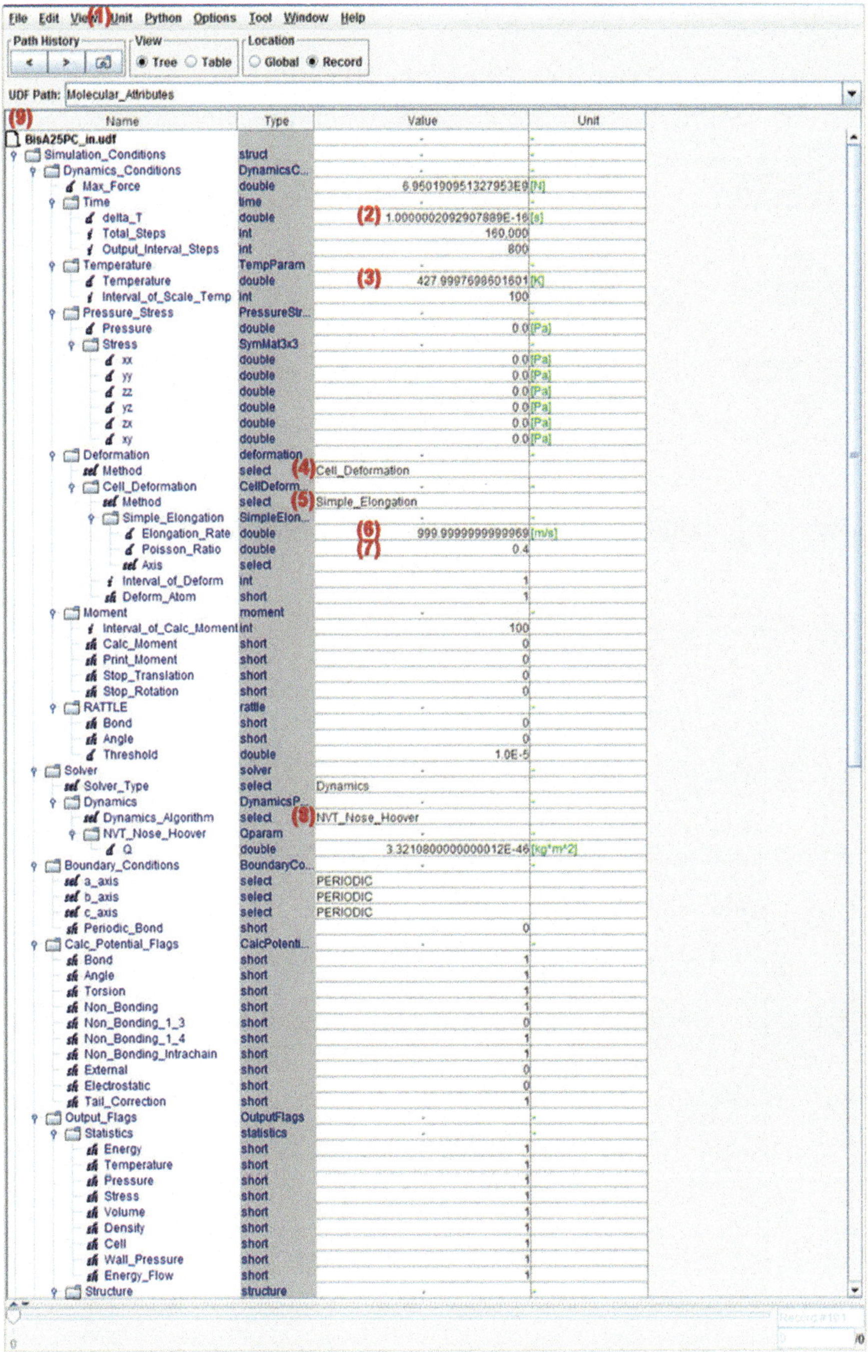

Fig. 28.3 Input UDF file “BisA25PC_in.udf”

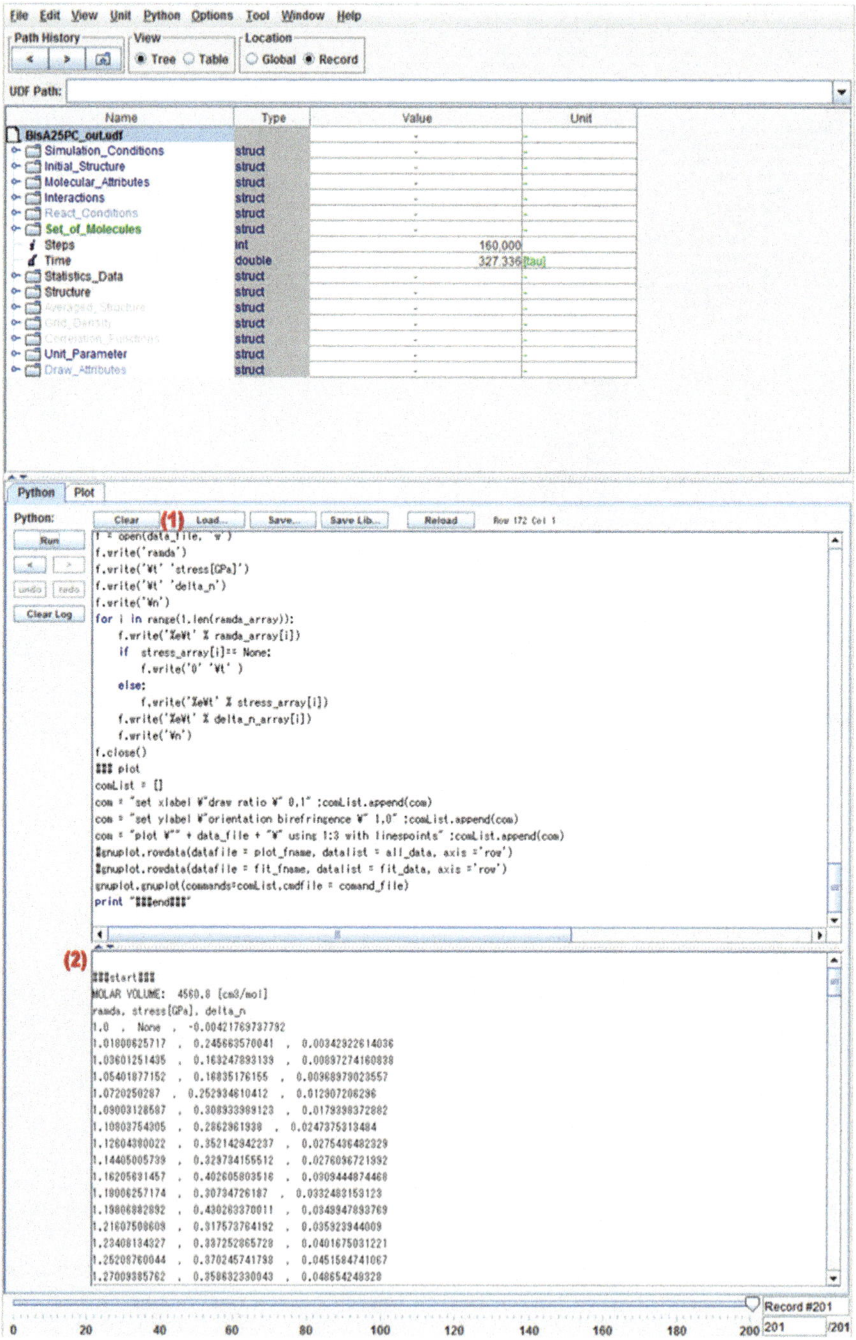

Fig. 28.4 Output UDF file “BisA25PC_out.udf”

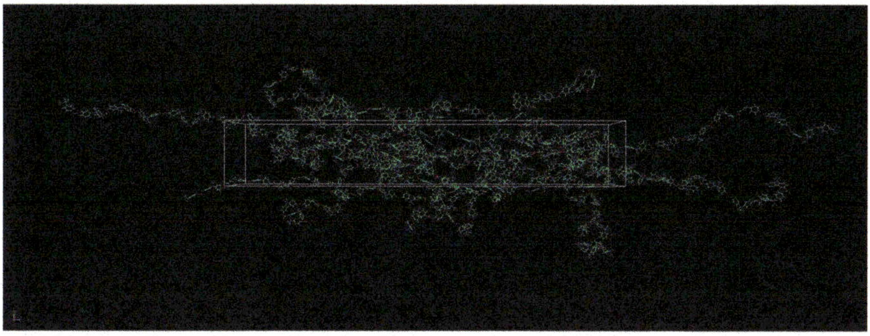

Fig. 28.5 Captured image of BisA-PC at 12 ps

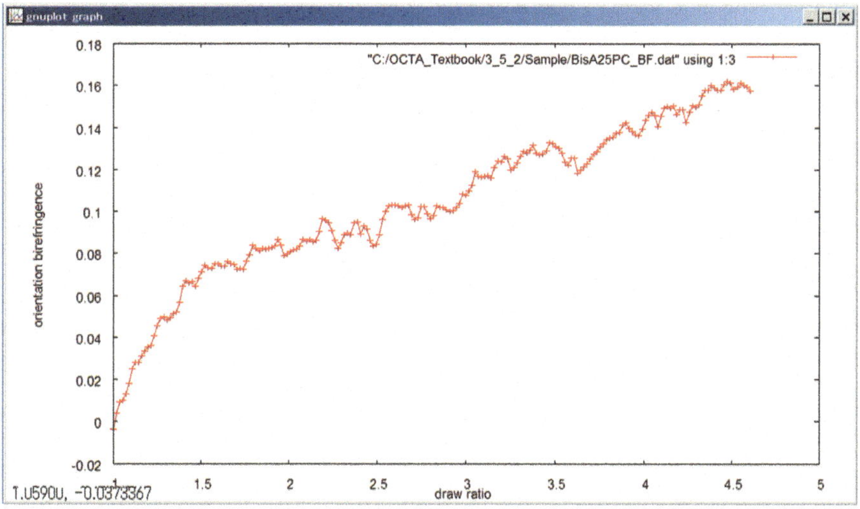

Fig. 28.6 Relationship between the draw ratio (λ) and birefringence (Δn) for BisA-PC (gnuplot)

Step 4: Calculation of orientation birefringence

With the output UDF file "BisA25PC_out.udf" shown in Fig. 28.4 open, we click (1) **Load**... in the Python Scripting Window to load the Python script for calculating orientation birefringence "calc_birefringence_A25.py"; we then run the script. Afterward, the draw ratio, stress, and birefringence are output in (2) the Python Log Window, and a gnuplot of the draw ratio and birefringence is displayed. The gnuplot for BisA-PC is shown in Fig. 28.6. Note that birefringence increases in accordance with the draw ratio. Output results are stored in "BisA25PC_BF.dat".

28.5 Results and Discussion

In general, when strain is applied to rubber, stress and birefringence have a proportional relationship, which is known as the stress–optic law. This law does not hold in glass transition and glass regions; instead, it has been reported that a modified stress–optic law applies in such cases [7].

Figures 28.7 and 28.8 illustrate the relationship between draw ratio and stress and between stress and birefringence for BisA-PC. Given that the yield point for BisA-PC is observed at a draw ratio of approximately 1.4, the stress optical constant is calculated using the data up to the yield point of 1.4, giving a result of 110 BR ($\times 10^{-12}$ Pa^{-1}). Under the configured conditions (temperature of 155 °C and drawing speed of 1,000 m/s), it can be assumed that, if the time temperature superposition principle applies, the current calculation reflects the glass region state. Although the time scale of the molecular dynamics calculation makes it difficult to quantitatively evaluate measurement data, the value of the stress optical constant obtained by scaling the draw rate is shown to be of the same order as actual measured values.

28.6 Download Instruction

Sample files are available from the OCTA website. See the instruction on http://octa.jp/download_bookdata.html. Password for the sample(zip file) of this chapter is “oVPy2kC3”.

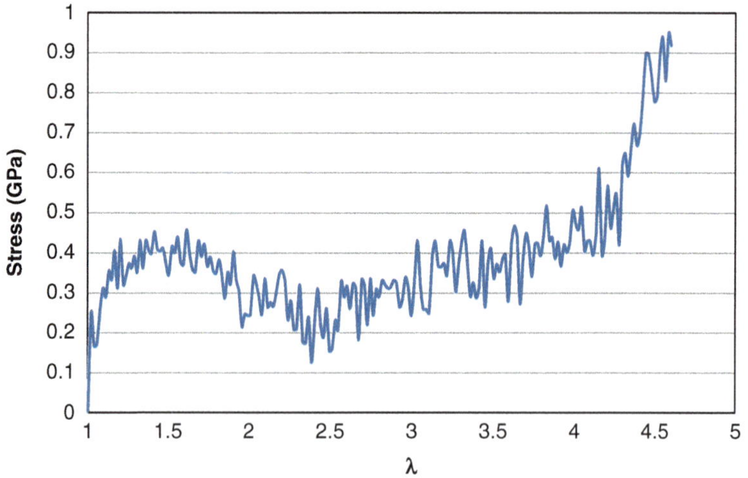

Fig. 28.7 Relationship between the draw ratio (λ) and stress for BisA-PC

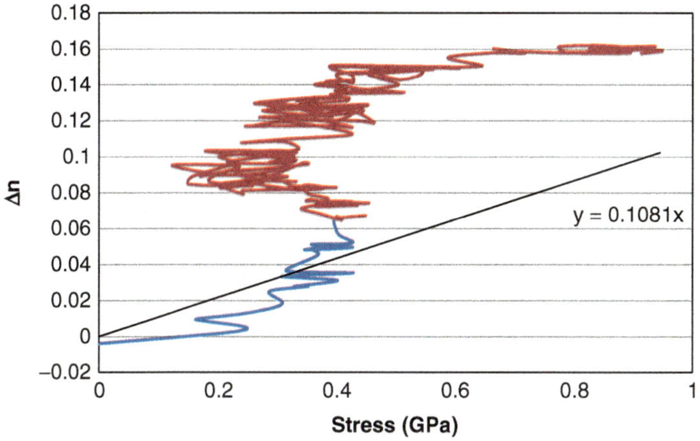

Fig. 28.8 Relationship between stress and birefringence (Δn) for BisA-PC

References

1. K.G. Denbigh, Trans. Faraday Soc. **36**, 936 (1940)
2. Y. Aikawa, Kobunshi Ronbunshu (Jpn. J. Polym. Sci. Technol.) **51**, 237 (1994)
3. Y. Miyazaki, Y. Togawa, Y. Masubuchi, In *Polymer Processing Society 22nd Annual Meeting*. (Yamagata, 2006)
4. C. Iwashimizu, M. Okubo, T. Ozawa, H. Kimizuka, In *14th JSPP Autumnal Meeting*. (2006)
5. C.W. Bunn, *Chemical Crystallography*, 2nd edn. (Oxford University Press, London, 1961)
6. S.L. Mayo, B.D. Olafson, W.A. Goddard III, J. Phys. Chem. **94**, 8897 (1990)
7. T. Inoue, K. Osaki, H. Morishita, H. Tamura, S. Sakamoto, J. Soc. Mater. Sci. Jpn. **52**, 314 (2003)

Chapter 29
Lithography

Hiroshi Morita

29.1 Introduction

The line widths of lithography patterns have become increasingly smaller, reaching values less than 20 nm [1]. As the pattern width decreases, the pattern size becomes closer to the gyration radius of the resist polymer. Thus it is more difficult to develop the resist polymer [2]. Line edge roughness (LER) is a problem related to the resist polymer that is difficult to solve. The problem involves the roughness at the side wall of a pattern being transferred to the circuit pattern, resulting in inhomogeneous electric resistance of a circuit. LER must therefore be reduced. LER is derived from the interfacial structure after the resist polymer dissolves, and one of the essences of LER is the control of the dissolution of the resist polymer without the problem of photoirradiation. To understand and control the dissolution process of resist polymer, as a virtual experiment, a coarse-grained simulation is conducted for the lithography process.

Chapter 20 described an application study of the evaporation process conducting coarse-grained simulation and demonstrated that the coarse-grained model is applicable to the simulation of solvent evaporation. Furthermore, as previously mentioned, the pattern width has recently become smaller than 20 nm, which is within the applicable range of coarse-grained simulation. This chapter introduces the application of dissipative particle dynamics (DPD) simulation to the lithography process.

Traditional lithography is done in the following way: coating of resist film, baking, photoexposure, development, rinsing, and etching. Although the processes of exposure, development, and rinsing are repeated several times in recently developed double and triple patterning methods, the basic procedure of lithography

---

H. Morita (✉)
National Institute of Advanced Industrial Science and Technology, Ibaraki, Japan
e-mail: h.morita@aist.go.jp

Japan Association for Chemical Innovation, *Computer Simulation of Polymeric Materials*, DOI 10.1007/978-981-10-0815-3_29

is not so different from the described method. Because it is difficult for us to simulate all the procedures of lithography, in this simulation, the process related to LER, that is, the development process, is simulated here [3–7].

Once the coarse-grained simulation is performed, the dynamics of each chain, which cannot be observed in experiments, can be obtained. In this chapter, the development process is simulated using the DPD method on COGNAC. The development process is the dissolution process of resist polymer. The chapter focuses on the model simulation of the dissolution of polymers. Here the simulations described in [3] are realized. Owing to the size limitation of User Definable Format (UDF) files, the system size, which is directly related to the total number of particles, must be smaller than that in [3]. Note that the effect of using a smaller system may affect the results, and these results must be treated carefully. If readers want to study along this way, it is recommended to conduct much bigger system of simulation.

29.2 Calculation Model

In this section, we begin by making the initial structure of the thin film, including the film and liquid phases. As described in Chaps. 19 and 20, an initial structure is prepared employing the node density biased Monte Carlo (NDBMC) method. The density profile is obtained by a one-dimensional self-consistent field (SCF) simulation using SUSHI, where the profile is the phase-separated structure of the polymer and liquid phases. In the SCF simulation, the χ parameter must be set, and a larger χ parameter is suitable for this simulation because it avoids the mixing of solvents into the polymer phase. Conditions in the SCF simulation, such as the boundary conditions, are almost the same as those for similar cases in Chaps. 19 and 20. The phase-separated structure in which the polymer film and liquid phases are separated to both left and right sides is appropriate for the NDBMC method. The total volume of the polymer is directly related to the thickness of the thin film used in the development process simulation and the total volume of polymer must be controlled for the purpose of lithography simulation. The obtained one-dimensional density profiles of polymer and solvent are used in the NDBMC simulation of COGNAC. Although the two-phase structure having the film and liquid phases is obtained by NDBMC simulation, solvents are sometimes mixed in the film (polymer) phase. This is because the density value of the liquid in the polymer phase is not zero in the SCF simulation and this result is transferred through NDBMC simulation. The liquid (solvent) particles must be moved to the liquid phase using a Python script, and following the relaxation simulation, the initial structure for the development simulation is obtained.

The relaxed structure is modified using a Python script to set the dissolved polymers in the polymer thin film. This treatment has the same meaning as virtual photoexposure, although the diffusion of acid derived from the photoacid generator determines the dissolved pattern in a real experiment. The polymers in the vertically lined part of the film, which is shown later in Fig. 29.1, are changed to dissolved

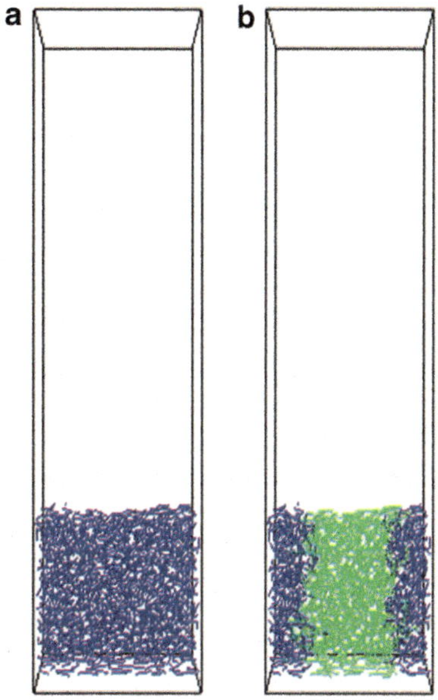

Fig. 29.1 Snapshots of the initial structures of the lithography process simulation. Illustrations (**a**) and (**b**) show the structures of a thin film and the initial structure for development simulation obtained using the Python script "LithoPattern.py"

polymers by changing the interaction parameter a_{ij}. In the dissolved region, a_{ij} for each particle of polymer is changed to a smaller value, while it is changed to a larger value for each particle of insoluble polymers. In the dissolution simulation, insoluble polymers are fixed in position. To fix the particles, a function of COGNAC, *Constraint_Condition*, is used. If these polymers are not fixed, the pattern melts. Melting of the pattern means that the polymer is getting a wetting onto the substrate owing to the addition of the attractive interaction from the substrate. There are other techniques of avoiding the melting of patterns, and another technique is used in [1].

29.3 Tips in Creating Input Data

The specific procedure of lithography simulation is described. First, a one-dimensional SCF simulation along the axis of the height direction is performed. In the calculated system, the polymer and solvent are included: one boundary, that is, the model for the substrate, is treated using the *WALL* condition, and another boundary, that is, the boundary at the liquid phase, is treated using the *NEUMANN* condition. The polymers are distributed at the side of the *WALL* boundary and form as a film. A larger χ parameter for the polymer and solvent is set to make the thin interfacial structure between polymer and solvent. It is well known that a high χ simulation to obtain the phase-separated structure makes the convergence of the SCF equations difficult, and much care is needed for this simulation. Here, one polymer is made from ten particles and the χ parameter is set as 4.2. The input and output files for SUSHI simulation are "Film1D-DPD_uin.udf" and "Film1D-DPD_uot.udf", respectively.

Obtained density profiles of polymer and solvent are used for the NDBMC simulation. Details of the parameter settings can be referred to in the UDF file "Film-DPD-01_in.udf." The boundary conditions for the vertical (height) and horizontal directions are set as *REFLECTIVE1* and *PERIODIC*, respectively. After the NDBMC simulation, some solvent particles are found inside the polymer film. These solvent particles are moved to the liquid phase using the Python script "B_pos_conv.py." In the DPD simulation, even if the particles overlap, the potential energy does not diverge, and there is little problem in this treatment. After applying the Python script, the relaxation simulation is conducted for a simulation time of 20,000 τ. It needs the care for this relaxation simulation. That is the setting condition of the interaction parameter between polymer and solvent. Here $a_{SP} = 60.0$ is set, which corresponds to a strong segregation condition. The value of a_{SP} affects the density of both the solvent and polymer phases. In the DPD simulation, the total density of the simulation system is 3.0. However, the density of each phase is shifted from 3.0 owing to the balance of the chemical potential of each phase, and the chemical potential of each phase is decided by a_{ij} parameters. If the density of the polymer phase increases, the diffusion of solvent becomes difficult, and, as a result, the dissolution of the resist polymer takes much longer time. This indicates that the result of the dissolution process simulation is affected by the initial structure. It is thus considered that a value of a_{SP} that is too large is not favorable for the relaxation simulation. The output file of this relaxation simulation is "Film-DPD-03relax_out.udf."

Using the relaxed structure, the dissolution process is simulated. This simulation is conducted after providing the soluble and insoluble polymers in the thin film. As described before, the soluble and insoluble polymers are set by position, and this is done using the Python script "LithoPattern.py". Figure 29.1 shows the structures before and after applying "LithoPattern.py". The soluble polymers are indicated by green particles, and their positions in the x-direction range from one-fourth to three-fourths of the system size. Other polymer particles shown in blue are insoluble particles and it remains as a pattern after the dissolution process simulation. The soluble and insoluble particles are distinguished by the a_{SP} parameter. a_{SP} for a soluble particle is set smaller than that for an insoluble particle. To avoid the melting down of the residual pattern, the positions of insoluble particles are fixed by the *Constraint_Condition*, which is already inputted by the Python script "LithoPattern.py."

Here four kinds of input files, "Film-DPD-04Litho1_in.udf" to "Film-DPD-04Litho4_in.udf," are prepared. Three hundred chains of 10 particles and 9000 solvent particles are included and the system size is $10 \times 10 \times 40$.

29.4 Analysis of Output Data

The simulation results are checked by viewing a snapshot for each time step, and the final dissolved structure is checked by viewing a snapshot and the density profile of the polymer and solvent. In particular, the depth profile in the height direction of the film is important to the analysis of the dissolution of polymer. The depth profile for each density is analyzed using the action program **ANALYSIS_1D_profile**. The use of this script has already been explained in Chap. 20.

29.5 Analysis of Results: Structural Analysis of Solvent Evaporation

Four simulations are performed. The simulation conditions of the a_{ij} parameter are listed in Table 29.1. A and B refer to insoluble and soluble species of polymers, respectively. C refers to solvent particles.

To avoid mixing between A (insoluble polymer) and C (solvent), a_{AC} is set as 60.0. Meanwhile, a_{BC} for mixing between B (soluble polymer) and C (solvent) is changed among the UDF files from 10.0 to 25.0. If a_{ij} is larger, the interaction between species i and j becomes more repulsive. Therefore, the condition of a smaller a_{BC} is confirmed as the more soluble situation in simulation of the dissolution process. Figure 29.2 shows snapshots of the dissolution simulation.

All four simulations realize the dissolution of the middle part. Furthermore, at time $= 50\ \tau$, from Fig. 29.2a–d, the number of chains dissolved into a solvent phase decreases and the dissolution becomes slower. In the case of the second figure from the left in Fig. 29.2a, the soluble polymers diffuse into the solvent phase more quickly. In this plot, the soluble polymers are still in the swollen phase only at the surface. These results indicate that the dissolution rate can be controlled by the interaction parameter a_{BC}.

The dissolution of polymer chains can be described as having two steps. The solvents first penetrate the polymer phase. Then the swollen polymers diffuse into solvent phase. In the penetration step, the chemical potential difference between

Table 29.1 Parameter values of a_{ij}

i	j	Film-DPD-04Litho1_in.udf	Film-DPD-04Litho2_in.udf	Film-DPD-04Litho2_in.udf	Film-DPD-04Litho4_in.udf
A	A	25.0	25.0	25.0	25.0
A	B	25.0	25.0	25.0	25.0
A	C	60.0	60.0	60.0	60.0
B	B	25.0	25.0	25.0	25.0
B	C	10.0	15.0	20.0	25.0
C	C	25.0	25.0	25.0	25.0

polymer and solvent phases is the driving force and is controlled by the a_{ij} parameter. This DPD model reproduces the dissolution of polymer as the potential difference model. Meanwhile, though the diffusion of polymer into the solvent phase is strongly affected by the flow in the solvent phase, flow in the solvent phase

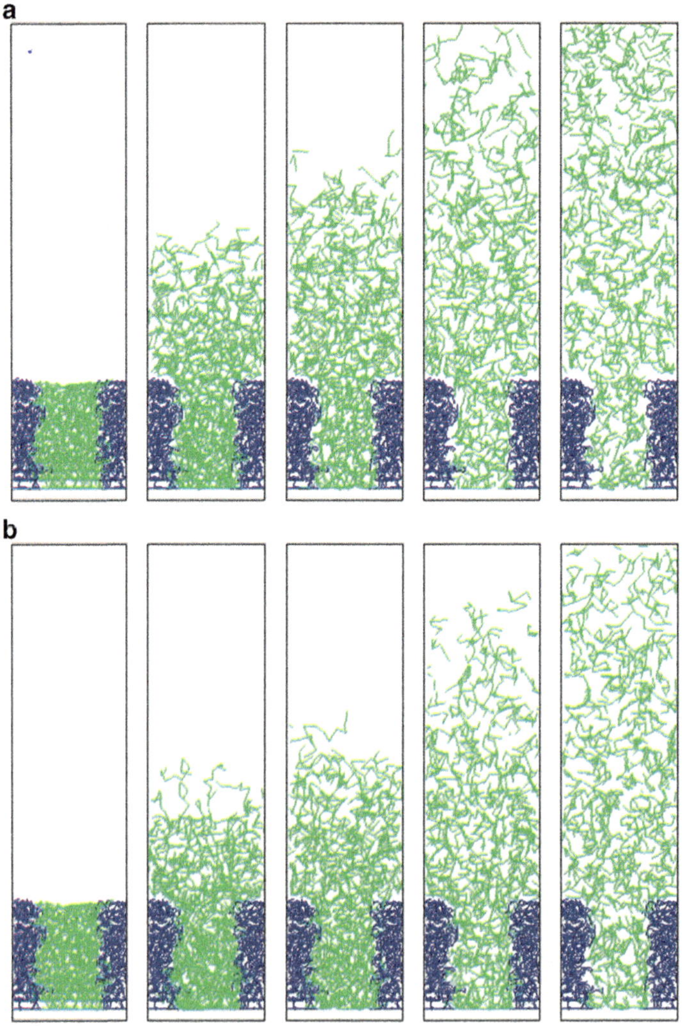

Fig. 29.2 Snapshots of the results of lithography process simulation. From left to right figures, time proceeds. (**a–d**) show the results of "Film-DPD-04Litho1_out.udf" at time = 0, 50, 100, 300, and 600τ; "Film-DPD-04Litho2_out.udf" at time = 0, 50, 100, 300, and 600τ; "Film-DPD-04Litho3_out.udf" at time=0, 50, 100, 300, 600, and 1000τ; and "Film-DPD-04Litho4_out.udf" at time=0, 50, 100, 300, 600, and 1000, 2000τ, respectively

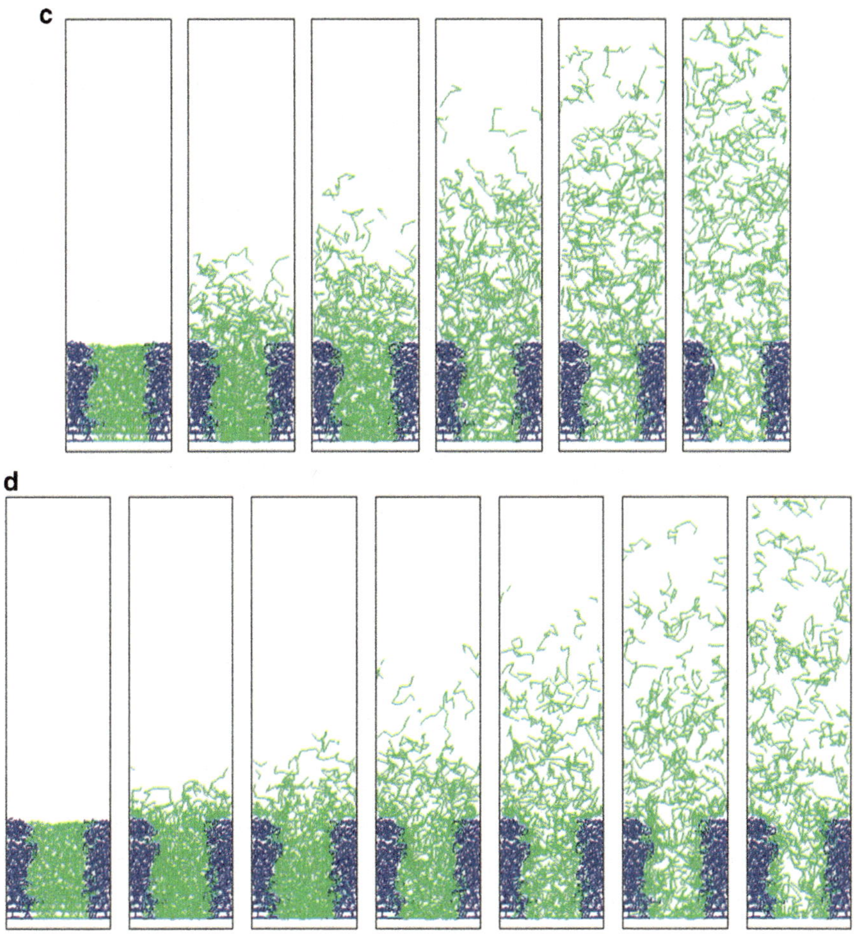

Fig. 29.2 (continued)

is not realized in such a small system. Therefore, this model underestimates the diffusion of polymer. This is just one kind of model, and the further modification of model will be needed.

29.6 Download Instruction

Sample files are available from the OCTA website. See the instruction on http://octa.jp/download_bookdata.html. Password for the sample (zip file) of this chapter is "2z5ko3Hj".

References

1. The International Technology Roadmap for Semiconductors, http://www.itrs.net/ITRS%201999-2014%20Mtgs,%20Presentations%20&%20Links/2012ITRS/Home2012.htm. Accessed 21 Dec 2015.
2. H. Ito, *Resist Materials* (Kyoritsu Syuppan, Tokyo, 2005) (in Japanese)
3. H. Morita, M. Doi, Proc. SPIE **7273**, 727337 (2009)
4. H. Morita, M. Doi, Proc. SPIE **7639**, 763932 (2010)
5. H. Morita, M. Doi, J. Micro-Nanolitho Mem. **9**, 041213 (2010)
6. H. Morita, Proc. SPIE **7972**, 79720W (2011)
7. H. Morita, I. Okabe, S. Agarwal, V.K. Singh, Proc. SPIE **8325**, 83250J (2012)

Index

A
Action mechanism, 21–22
Adhesive properties, 315–325
Adsorbed structure, 337–345
Adsorption, 327–335
Advection–diffusion equation, 162, 163
Affine deformation, 276
Aggregated structure, 274, 275
Algorithm of Magatti, 173
Amphiphilic molecules, 359
Autocorrelation functions, 41–42, 172, 173

B
Bead-spring model, 8–9, 171, 172
Bjerrum length, 165
Blending law, 189
Blend system, 284, 289
Boltzmann distribution, 164
Boltzmann's relation, 117
Boundary conditions, 34–35, 165
Branch point, 352
Bridge chains, 249
Bulk modulus, 191, 193, 196

C
Carbon nanotube (CNT), 212
CCR. *See* Convective constraint release (CCR)
CGMD. *See* Coarse-grained molecular dynamic (CGMD)
Charge distribution, 162
Chemical potential, 163, 164, 301
Chemical reactions, 37
CLF. *See* Contour length fluctuation (CLF)
Cluster size, 362, 373
CNT. *See* Carbon nanotube (CNT)
Coarse-grained molecular dynamic (CGMD), 201
COGNAC, 109
Colloid, 149
Colloidal particle, 149
Comb block chain, 338
Commensurate, 309
Compatibilizer, 201
Compressive deformation, 193, 196
Concentration distributions, 332, 333
Conservative energy, 364
Conservative forces, 349
Constitutive equation, 160
Constraint force, 157
Constraint release (CR), 102
Continuum model, 129
Contour length fluctuation (CLF), 102
Convective constraint release (CCR), 102
Core–shell structure, 190, 196
CR. *See* Constraint release (CR)
Critical nucleus, 181
Crossing dynamics, 351–352
Cross-link, 229–248
Crystal growth, 186–187
Crystalline lamellae, 6
Crystalline polymers, 179–188
Crystallization, 305–313
Cygwin, 56

Japan Association for Chemical Innovation, *Computer Simulation of Polymeric Materials*, DOI 10.1007/978-981-10-0815-3

D

Data section, 18
DBMC. *See* Density-biased Monte Carlo (DBMC)
Debye function, 72
Debye–Hückel approximation, 165
Debye length, 166
Definition section, 18
Deformation, 239–241
Density-biased Monte Carlo (DBMC), 38, 250
Density functional theory (DFT), 67–99
Depth density profile, 301
Derjaguin approximation, 341
Development process, 390
DFT. *See* Density functional theory (DFT)
Diffusion constant, 350
Direct numerical simulation, 162
Dissipative particle dynamics (DPD), 102, 211, 297, 349–350, 389
Dissolution process, 389
Dreiding force field, 181
Droplet, 130
Dynamic modulus, 117
Dynamics Manager, 129

E

Editor window, 18
Elastic material, 132
Electric double layer, 162
Electrolyte, 150
Electrolyte solution, 131
Electrophoresis, 162–166
Electrostatic fields, 327
Elongation, 239, 263, 275–280
End cross-link, 231
End segment distribution, 285, 289
End-to-end distance, 40
Entangled dynamics, 101
Entanglement, 101
Entanglement molecular weight, 172
Entanglement point, 348
Epitaxy, 311
Equations of motion, 32–33
Ethylene-propylene rubber (EPR), 189
Excess free energy, 341

F

FDM. *See* Finite difference method (FDM)
FEM. *See* Finite element method (FEM)
Fiber formation, 186
Filler-filled rubber, 269–281
Filler model, 270, 271
Finite difference method (FDM), 129
Finite element method (FEM), 129
Flory–Huggins–de Gennes model, 75–76
Flory–Huggins free energy model, 69
Flory–Huggins parameter, 69
Flow, 184
Flow region, 171
Fluctuation–dissipation theorem, 158
Force curve, 341, 343
Fracture toughness, 208
Free-standing film, 291

G

G(t), 173
Gaussian chain model, 71–72
Gel, 132–133
Generating an initial structure, 173
Ginzburg–Landau theory, 74–75
Glass region, 171
Glass transition temperature *(Tg)*, 291, 294–296
GOURMET, 17, 161
GourmetTerm, 56
Graphene, 212, 221
Green–Kubo formula, 117, 172

H

Hydrodynamics, 162
Hydrodynamics effect, 81–82

I

Ideal chain model, 9
Impact strength, 189
Incommensurate, 309
Incompressible, 157
Insoluble polymer, 393
Interaction sites, 31
Interface, 150, 318–319, 323–324
Interfacial fracture, 201
Isotropic elastic material, 193

K

Kyoto Advanced Particle Simulator for Electro-hydrodynamics (KAPSEL), 149–166

L

Lamellar crystals, 180
Lamella thickening, 187

LAMMPS. *See* Large-scale atomic/molecular massively parallel simulator (LAMMPS)
Langevin, 157
Large deformation, 184
Large-scale atomic/molecular massively parallel simulator (LAMMPS), 39
Lennard-Jones potential, 159
LER. *See* Line edge roughness (LER)
Light transmittance, 134
Linear elastic material, 193, 197
Linear relaxation modulus, 116, 117
Linear viscoelasticity, 112–123
Linear viscosity growth curve, 117
Line edge roughness (LER), 389
Lithography process, 389–395
LJ atomic wall, 309
LJ flat wall, 307, 308

M
Macrophase-separated structures, 6
Macrophase separation, 91–92
Materials Studio, 371
Maximum relaxation time, 171, 173, 174
Maxwell-type relaxation, 347
MD. *See* Molecular dynamics (MD)
Mean square displacement (MSD), 41, 294
Mesostructure, 373
Microphase-separated structure, 6, 249
Microphase separation, 93–95
Micro-reactors, 131
Micro-TAS, 131
Moisture contents, 372
Molecular dynamics (MD), 221
Molecular mechanisms of crystallization, 180
MSD. *See* Mean square displacement (MSD)
Multi-phase flow, 130–131

N
Nafion, 369
N-alkanes, 305–313
Nanocomposite, 211, 221
NAPLES, 101–126
Navier–Stokes equation, 162, 164
NDBMC. *See* Node density biased Monte Carlo (NDBMC)
Newton–Euler equation, 162
Newtonian, 164
Node density biased Monte Carlo (NDBMC), 202, 292, 298, 390
Nonadecane, 306
Nonlinear viscoelasticity, 123–126
Normal separation, 221
NPT ensemble, 273

O
OCTA, 15–27
Oligomers, 315–325
Order parameter, 42
Orientation birefringence, 379–387

P
Packing length, 109
Pair distribution functions, 42–44
Parallel model, 189
PASTA, 101–126
PCN model. *See* Primitive chain network (PCN) model
PE. *See* Polyethylene (PE)
Phase diagram, 83, 97–98
Plateau modulus, 108
Plot tab, 20
Poisson–Boltzmann equation, 164
Poisson equation, 329
Poisson_Ratio, 277
Polycarbonate, 379
Polyelectrolytes, 327–335
Polyethylene (PE), 190
Polymer blend, 189–210
Polymer crystallization, 179–188
Polymer dispersants, 337
Polymer processing, 184
Polypropylene (PP), 189
Potential functions, 29–32
Potential_map.udf, 46, 173
PP. *See* Polypropylene (PP)
Primitive chain network (PCN) model, 106
Python, 153
Python tab, 20

R
Radial distribution function, 373, 374
Radius of gyration, 40
Random cross-link, 230–231
Random phase approximation (RPA), 71, 72
Reaction, 235–238
Record, 25–27
Relaxation spectrum, 171, 174
Reptation, 102
Reynolds-number, 150
Rheology, 149
Rigby–Roe force field, 182, 183
Rigid-body model, 7–8

Rouse chain, 101
Rouse relaxation, 174
RPA. *See* Random phase approximation (RPA)
Rubber-dispersed material, 189
Rubber region, 171, 174

S
SCFT. *See* Self-consistent field theory (SCFT)
Sea-island structure, 189
Sedimentation, 157
Segregation, 315–325
Select mechanism, 23–25
Self-consistent field theory (SCFT), 76–81, 201
Self-propelled particle, 159
Series model, 189
Shear deformation, 193, 196
Shear modulus, 191, 193, 196
Slip-link model, 9–10, 103–107
Smoothed profile method (SPM), 150
Smooth profile function, 157
Solid substrate, 305
Solubility parameter, 371
Soluble polymer, 393
Solvent evaporation, 297
Spherulites, 134
SPM. *See* Smoothed profile method (SPM)
Spontaneous vesicle formation, 366
Star-branched polymers, 105
Stress optical constant, 386
Stress relaxation, 172–173
Stress-strain curve, 263–266, 276, 277
Stress tensor, 107–108
Surface force, 337–345
Surface structure, 283–289
Surface *Tg*, 291
SUSHI simulation, 315–325

T
Table View, 20
Takayanagi model, 189
Tapered growth front, 187
Temperature unit, 296
Temporal network model, 102
Tensile modulus, 193
Ternary blend, 189
Tg. *See* Glass transition temperature *(Tg)*
Thermal conductivity, 217
Thermal fluctuation, 157
Thermoplastic elastomer, 249–267
Thin films, 305–313
Thread-like micelles, 347–358
Total strain free energy, 193
Transient energy, 352
Transition region, 171
Tree View, 20
Tube model, 102

U
UDF. *See* User definable format (UDF)
UDF path, 19
Undecane, 306
Uniaxial elongation, 239, 263
Unit conversion, 22–23
United atom model, 6–7
User definable format (UDF), 16

V
Vesicle formation, 359–368
Viewer window, 23
Viscoelastic behavior, 347
Viscoelastic materials, 171

W
WLF equation, 111

X
X-ray scattering, 369

Y
Young's modulus, 191

Z
Zooming, 38

Printed by Printforce, the Netherlands